VICTORIAN RELATIVITY

VICTORIAN RELATIVITY

Radical Thought and Scientific Discovery

Christopher Herbert

The University of Chicago Press
Chicago and London

CHRISTOPHER HERBERT is professor of English at Northwestern University. He is author of *Trollope and Comic Pleasure* (1987) and *Culture and Anomie: Ethnographic Imagination in the Nineteenth Century* (1991), also published by the University of Chicago Press.

The University of Chicago Press, Chicago 60637
The University of Chicago Press, Ltd., London

Printed in the United States of America

10 09 08 07 06 05 04 03 02 01   1 2 3 4 5
ISBN: 0-226-32732-9 (cloth)
ISBN: 0-226-32733-7 (paper)

Library of Congress Cataloging-in-Publication Data

Herbert, Christopher.
Victorian relativity : radical thought and scientific discovery / Christopher Herbert.
p. cm.
Includes bibliographical references and index.
ISBN: 0-226-32732-9 (alk. paper)—ISBN: 0-226-32733-7 (pbk : alk. paper)
1. Relativity—History—19th century. 2. Knowledge, Theory of—History—19th century. I. Title
BD221 .H47 2001
115—dc21
00-012177

Portions of chapter 4 appeared in "Science and Narcissism," *Modernism/Modernity* 3 (1996): 129–35; © The Johns Hopkins University Press, reprinted by permission. Portions of chapter 5 appeared in "Frazer, Einstein, and Free Play," in *Prehistories of the Future: The Primitivist Project and the Culture of Modernism*, ed. Elazar Barkan and Ronald Bush (Stanford: Stanford University Press, 1995); © 1995 by the Board of Trustees of the Leland Stanford Junior University, reprinted by permission.

♾ The paper used in this publication meets the minimum requirements of the American National Standard for Information Sciences—Permanence of Paper for Printed Library Materials, ANSI Z39.48-1992.

For George

friend, interlocutor, cicerone of the bird world

Contents

Acknowledgments

I often despaired of being able to bring this book to completion, but only because of the difficulty of the subject relative to my own capacities—never because I lacked excellent colleagues and friends to provide advice, encouragement, and moral support; and never because Northwestern University, where I have been privileged to teach for my entire career, failed to provide needed facilities for the pursuit of long-term research. Jules Law and Michal Ginsburg kindly read sections of the manuscript and provided their usual shrewd but always benign criticism. A person whose immediate work environment includes such gifted and unselfish colleagues can only be called very fortunate. My valued old friend David Simpson, as though he had no other demands on his time, insisted on reading the entire manuscript from afar and then sent many pages of detailed, searching comments that were of the greatest help to me in revising; I hereby apologize to him for not being able to respond to all of these comments as they deserved. An interlocutor in a class by himself has been George Levine, spotter of the wily masked duck, upholder of the ideal of objectivity, and eminent fellow laborer in the archives of Victorian scientific literature. I only hope that the completion of this long project will not result in too much diminution of the lively, almost daily interchange of scholarly e-mail on Karl Pearson and allied topics that I have so much enjoyed for several years past.

I wish to extend my gratitude also to the two original readers of this manuscript for the University of Chicago Press. I will not thank by name the one whose identity I know, for fear of implicating him in the shortcomings of the book on which he reported with extraordinary conscientiousness and generosity. The frequency with which his name appears in my footnotes is testimony to my indebtedness to his work.

I am lucky to have had a long association with the University of Chicago Press and with its admirable editor, Alan Thomas. Any scholarly author would be proud to appear in the distinguished company represented by the Chicago list, and would feel fortunate to be published by such an outstandingly efficient press. I am well aware that the high standard of professionalism that prevails at the University of Chicago Press is not to be taken for granted.

My appreciation is due to Dean Eric J. Sundquist of the Weinberg College of Arts and Sciences of Northwestern University for giving me the leave of absence in 1998–99 that allowed me to finish this book. In common with all other members of Northwestern's Department of English, I owe a special debt of gratitude to Kathy Daniels, our nonpareil department assistant, for her untiring efforts in keeping daily office life on an even keel.

For helping to keep me relatively if not absolutely sane in the course of this writing project, I must thank the members of that venerable institution, the book group: Curt and Linda, George and Ellen, Richard and Susan.

Sophie, Stéphane, and Bernadette did not so much provide assistance in this latest writing enterprise of mine as furnish the context of meaning that makes it all seem worthwhile.

PREFACE

Relativity and Ideology

In a book titled *Culture and Anomie*, I sought to trace and to theorize the development of the ethnographic concept of "culture" in the nineteenth century. I realized in the course of that project that it belonged properly, for all its irreducible specificity, to the more inclusive project of tracing the development of what might be called the relativistic imagination in the same period. It is safe to say that the very existence of Victorian relativism will come as news to some who study nineteenth-century literature, and that a scripting of the period in which this line of thought is assigned a leading role may possess at least the potential interest of novelty. At all events, I address this broader topic in the present work, taking as a principal point of reference theoretical physics rather than, as in *Culture and Anomie*, anthropology and the other social sciences. The literature of physics turns out to offer in the period under consideration a vivid and especially precocious enactment of that radical, self-dramatizing break with tradition called "modernism." Only in psychoanalysis—the scientific status of which is much more problematic—do we find a comparable phenomenon, though of course other scientific fields from biology to economics underwent revolutionary change in this period.

My project takes its thematic cues and its polemical *raison d'être* from postmodern literary theory and, in particular, from the critique of science defined variously by the work of such scholars as Ludwik Fleck, T. S. Kuhn, Paul Feyerabend, Bruno Latour, and Donna Haraway. In *Victorian Relativity*, hoping to acquit some of the intellectual debt I owe to this tradition, I restore to it a part of its distinguished ancestry. The restoration does not come without cost. Though the point is not belabored in what follows, I suggest that it is a mistake—not necessarily an innocuous one—to suppose that the intellectual condition defined by polyvalency, indeterminacy, constructivism, *différance*, and the ideological critique of knowledge represents a radical initiative of our own day, prefigured by a few heroic pioneers like Saussure and Nietzsche.[1] This foreshortened version of the history of contemporary ideas has enfeebled these ideas by depriving them of their anchorage in a rich and impressive, but long occluded, supporting literature. It has also compromised them by lending itself to the tendency of progressive intellectual enterprise to foster within itself cults of authority-worship—precisely the habit of mind that the nineteenth-century relativity movement sought to dispel.[2] H. L. Mansel, a Victorian clergyman who figures in my own version, commented on this phenomenon. "The so-called freethinker is as often as any other man the slave of some self-chosen master," he observes, "and many who scorn the imputation of believing anything merely because it is found in the Bible, would find it hard to give any better reason for their own unbelief than the *ipse dixit* of some infidel philosopher" (Mansel, *Limits of Religious Thought*, 49). Much current humanities scholarship would be hard put to shield itself from this deadly gibe. By restoring to view certain trends of radical nineteenth-century speculation, we can, I hope, lessen our susceptibility to the pseudofreethought indicted by Mansel; in particular, we can expose to critical scrutiny the amnesiac tendency to appropriate central elements of Victorian intellectual history and set them up polemically under the proprietary sign of that semi-misnomer, "postmodernism."[3] Such scrutiny need not call into question the contemporary viability of the themes of relativistic analysis that come down to us from Victorian and post-Victorian precursors. On the contrary: it seeks to free them from the liabilities imposed on them by a self-deluding intellectual history, and in this way to help give them a renewed lease on life.

The story told, in a fragmentary form and from a largely British perspective, in this book reflects the principle that intellectual history with a claim to scientific rigor must abjure originary myths as far as possible, in favor of models of broadly collective movements of thought emerging with incremental slowness: in favor of an evolutionary rather than a creationist or a catastrophist model. "We know of nothing which has had any *bona fide* beginning," said Samuel Butler, the Victorian novelist and evolutionary theorist (*Collected Essays*, 1:137), defining a cardinal rule of the history of ideas. (It is important to note that evolutionism in the domain of ideas is by no means incompatible with the occurrence of "paradigm shift" or "epistemic breaks.") Accordingly, this book casts its net of reference fairly widely, giving significant attention to a core cast of writers numbering around thirty, many of whom (for example, Mansel, Sir William Hamilton, Alexander Bain, J. B. Stallo, Karl Pearson, W. K. Clifford, F. C. S. Schiller, and others) do not usually figure in the pantheon of canonical Victorian authors and far less in that of the forerunners of contemporary theory. I compensate for the danger of diffuseness my choice of materials poses by focusing a series of chapters following a general introduction on one or two principal authors apiece, though with reference to surrounding networks of others. The first of two chapters on the doctrine of the relativity of knowledge thus centers on the key figure of Herbert Spencer, and the second on Feuerbach and Stallo; a chapter on the relativistic critique of logic centers on Newman; a chapter on relativity and scientific anthropocentrism on Karl Pearson. These thematically organized chapters present Victorian relativity as a system of closely interlinked motifs that could be and often were portrayed as logically entailed by one another, but that appear from the perspective of intellectual history to cling together according to a richly creative and unpredictable cultural dynamic never wholly subject to logical necessity. In a final chapter, I venture (at risk of *lèse majesté)* a comparative reading of various works of Frazer and Einstein in the light of this concatenation of nineteenth-century themes. The book concludes with a brief afterword on the problem of history-writing and on the illustrative figure of George Grote, radical Victorian politician and historian of Greece.

This book narrates an extended case history of the ideological construction of knowledge; it both takes for granted as its fundamental postulate and seeks to demonstrate by the production of evidence the inseparable continuity of scientific discovery with a multiplicity of fields of discourse. Unlike the works of many of the scholars who have blazed these paths before me, it employs the concept of "ideology" in a nontendentious way: it makes no presumption of any authentic states of affairs as they might be known apart from the supposedly delusive inflections of the ideological, nor does it identify ideology a priori with the maintenance of systems of vested social interests.[4] It simply presumes that cognition and theory-spinning are bound to be expressive of densely patterned systems of imagination, that these systems are to an overwhelming degree culturally dictated, and that they have therefore an overwhelmingly ethical character—that discoveries of how things are organized can hardly fail to be informed by cultural principles of how they ought to be. Such principles assuredly do not enter into the formation of knowledge in the various disciplines in any deterministic or predictable way, however: the path from ethical imagination, particularly at the level of ideological formations, to scientific objectivity "is labyrinthine, not straightforward," to quote Paul de Man somewhat out of context (*Romanticism*, 26). In all interesting cases, these paths are bound to elude definitive scholarly mapping. Yet the effort to explore them can be enlightening even so.

I argue that the development of the relativity principle in physics and in other fields of inquiry in the nineteenth century was inseparably intertwined with themes of militant emancipationism and anti-authoritarianism. Relativity was in this sense not so much discovered as a scientific property of nature as it was implanted there by a deeply moralized scientific imagination. In pursuing this argument, one might easily fall into the clichés of a style of scholarship anatomized and joked about by the physicist Alan D. Sokal in his now famous 1996 hoax perpetrated in the pages of the cultural-studies journal *Social Text*. Quoting Niels Bohr to the effect that in twentieth-century physics "a complete elucidation of one and the same object may require diverse points of view which defy a unique description," Sokal's satirical persona, mimicking what the author takes to be a hopelessly fatuous formula of contemporary cultural studies, declares that quantum physics "provides a powerful refutation of the authoritarianism and elitism inherent in traditional science" (Sokal, "Transgressing the Boundaries," 229). One need not subscribe to any such logic to insist on not only the validity but the necessity of scholarship that considers the sciences, among their other properties, as cultural institutions. To recognize their function as very prominent agencies for imagining and propagating values—despite their professed code of value-neutrality—is to see them as bound up in ramified two-way relations with (all) other institutions, and therefore as *immersed in discourse* and in all the figurative operations that constitute the discursive universe. In the nineteenth century, establishmentarian science was indeed perceived by critics to be imbued with a dangerously authoritarian creed that preserved itself by the promotion of a mystificatory cult of "absolute truth," and that was intimately allied in these respects with its supposed adversary, dogmatic religion. Therefore, according to these critics, it carried within itself a logic of complicity in violent abuses of power. The movement for a reformed science based on relativistic principles consequently signified at each step not just a call for a perfected rationality free of "metaphysical" obfuscation but also an act of cultural reconstruction. This argument, sometimes made distinctly by Victorian writers and sometimes only by implication, formed the basis—so I seek to show, at the risk of incurring Sokal's laughter—of the relativity movement that transformed a broad spectrum of nineteenth-century sciences, produced a bonanza of new knowledge, and ultimately caused or enabled Bohr to promulgate as scientific doctrine the fantastically anticommonsensical principle he states in 1934: that the most we may ever know of reality is a set of differing points of view "which defy a unique description."

According to this version of the story, much of the most valuable intellectual legacy of the period that I study is the outgrowth of a movement of radical moral activism. I do not argue, however, that this association is a necessary or an intrinsic one that can be taken for granted—that in adopting the relativity principle as a law of analysis one professes radical politics (or any politics) willy-nilly. Indeed, I stress evidence revealing this principle to be a potentially polymorphous one, subject to perverse-seeming inversions (H. L. Mansel's own being among the more perverse, as we shall see). Relativism is itself wholly subject to relativity, in other words. Yet the burden of my study is to show how deeply a particular cultural history has invested the relativity theme with political and didactic imperatives that may be discounted only with a significant loss of understanding. In the process of this demonstration, *Victorian Relativity* questions the assumption, sometimes made uncritically or dogmatically by activist scholars of our own day, that scientific discourse in any period is bound to express and to reinforce dominant ideologies and established structures. Victorian scientific literature, I argue, serves as the conduit for a challenge to reigning values so extreme, so vertiginous as to risk seeming indistinguishable from nihilism and anarchism—a challenge that retains the power to startle us from ideological somnambulism even today.

INTRODUCTION

The Conspiracy against Truth

> Relativism has become the great bugaboo. It's almost like Communism. It has the same damning association, whether you're a party member or not.
>
> —M. Norton Wise, on allegedly being blackballed at the Institute for Advanced Study on the grounds of his perceived relativistic tendencies[1]

In a 1992 essay titled "The Image of Objectivity," Lorraine Daston and Peter Galison examine late-nineteenth-century scientific atlases as indices of the prevailing mentality, or "creed," of science at the time. They document a pervasive and, they say, wholly unprecedented "fear of interpretation" among late-Victorian scientists and a corresponding insistence on a rigidly puritanical code of objectivity as the prerequisite of achieving "truth to nature" in scientific representations. Scientists were required by this code to display at all times the relentless self-discipline necessary to keep their work free of any trace of contamination by personal interpretation. Natural science thus became the field of a fervent moral didacticism. "Self-discipline or self-control was of course the cardinal Victorian virtue," Daston and Galison remind us, with reference to Samuel Smiles ("Image," 121, 103, 84, 118).

This account of "the moralization of objectivity" ("Image," 81) in the late nineteenth and early twentieth centuries is a valid and important one, but it raises further questions and finally seems incomplete in significant respects. For one thing, the apogee of the Victorian cult of puritanical self-control was reached in the 1850s, after which the cultural climate, particularly in Britain, grew markedly more lenient and permissive, as testified by such phenomena as the aesthetic movement. The trend of late-century scientific discourse toward a theory of intensified self-repression, portrayed in Daston's and Galison's essay as a natural emanation of the moral sensibility of the period, thus seems at first glance reactionary and puzzling in its historical context. One would have expected a trend in the other direction. The anomaly makes more sense once one recognizes the ideological instability that had come to characterize late-Victorian scientific discourse. The concept of scientific objectivity had become increasingly contested, increasingly volatile, over the course of the second half of the century, and by the 1890s was not at all as monolithic as Daston's and Galison's narrative might lead one to assume. In the 1890 first edition of *The Golden Bough* [hereafter *GB*1], for instance, James Frazer, who prided himself on his devotion to the ideal of scrupulous scientific method, remarked somewhat offhandedly as he expounded the orthodox theory of the discovery of truth by the rigorous testing of hypotheses that "after all, what we call truth is only the hypothesis which is found to work best" (*GB*1, 212)—hardly a doctrine that accords with a strict, moralized concept of scientific objectivity or with an obsessive fear of interpretation. Samuel Butler had ventured a similar thesis several years earlier, declaring that the notion of "truth" in science should be replaced by that of conceptual "convenience" and that scientific work was inherently a process of risky speculation akin to gambling (*Collected Essays* 1:158–60, 139–40). The concept of rigorous scientific "truth to nature" and the obvious charge of anxiety that propels it do indeed, as Daston and Galison say, flow from Victorian moral culture, but they can only be fully understood as responses to a new movement, intimated by Frazer and Butler, that for some time past had called objectivity sharply into question and seemed to pose a frightening danger not just to the practice of science but to the whole national system of values. Not to highlight this movement is to condemn oneself to a static and two-dimensional view of the Victorian intellectual scene and to a faulty view of our own relation to it—the former inevitably stereotypical, the latter inevitably self-aggrandizing.

The name of the movement was "relativity."

Only Relations

J. S. Mill, defining in 1865 a powerful new school of thought, declared that the "doctrine of the Relativity of our knowledge" was such that it "impresses a character on the whole mode of philosophical thinking of whoever receives it" (*Sir William Hamilton,* 25). Writing in the same year, Walter Pater proclaimed the "cultivation of the 'relative' spirit in place of the 'absolute'" to be the defining characteristic of "modern thought"; Coleridge, "with his passion for the absolute, for something fixed where all is moving," is for Pater the embodiment of a spirit hopelessly ill-attuned to the modern (*Appreciations,* 66, 104). The year before, Matthew Arnold had announced the advent of "the 'relative' spirit" in his own way, portraying it as a sign of the contemporary slide toward anarchy. In "The Literary Influence of Academies," he decries the tendency of the day "to spread the baneful notion that there is no such thing as a high correct standard in intellectual matters" and thus to foster the theory "that every one may as well take his own way" (267). In subsequent decades, many underwent the radical transforming influence to which Mill, Pater, and Arnold testified; from about the middle of the nineteenth century through the first third of the twentieth, the increasing salience of the idea of relativity was a defining feature—one is tempted to say, with Pater, *the* defining feature—of European intellectual history. To apprehend reality in a distinctively modern mode came in this period to mean, in effect, apprehending it relativistically. "The doctrine of the Relativity of Knowledge . . . has been accepted by so many schools of thought for their basis," observes Frederic Harrison late in the century, "as almost to have passed into the sphere of subjects which are little liable to question" (*Philosophy of Common Sense,* 21). One field of inquiry and of discourse after another (political economy, biology, sociology and anthropology, philosophy, physics, mathematics, linguistics) had been decisively altered by this doctrine by the turn of the century—and now, in the age of postmodernity, various species of relativity have enjoyed a dramatic recrudescence once again. The absence, so far, of a comprehensive history of this phenomenon, especially one that takes full account of its nineteenth-century genesis, is a notable phenomenon in itself, deserving not only remediation but an attempt at an explanation.[2] These are among my aims in this book.

If there is such a thing as an essential concept of relativity, it has seemed to some commentators to defy cogent definition (and has been subject to calculated misdefinition by many others, as we shall see). The mathematician Sir Arthur Eddington, writing in 1939 with special reference to the "revolutionary movement" that had transformed twentieth-century physics, made the same point that Mill implied at the opening of the modern relativity era almost seventy-five years before: that contemporary relativity theory was best considered not as something so limited as the outgrowth of a specific doctrine, but as a unified, all-transforming mode of thinking and of sensibility. "For my own part I have regarded relativity as a new outlook whose consequences must gradually unfold themselves, rather than as a particular axiom or hypothesis to be translated once for all into definite formulation," Eddington declared. He nonetheless then ventured a working definition of the new outlook. "Perhaps the nearest approach to a formulation of the [relativity] principle is the statement that we observe only *relations* between physical entities" (*Philosophy of Physical Science*, 31; like all italicized words and phrases in quoted material throughout this book, the italicized word in this quotation is italicized in the quoted source also.). Things do not possess intrinsic properties of their own (at least none that we can ever observe), in other words, only comparative properties. In a world where the relativity principle holds sway, we must therefore say that the very existence of a thing depends on the coexistence of other things—indeed, of all other things.

Eddington's formula, the implications of which thus "unfold" so rapidly, expresses in the vocabulary of physics the doctrine of the original expounder of relativity theory, Protagoras the Sophist. The essential principle of this philosopher, Plato tells us in the *Theaetetus*, is "that nothing is one thing just by itself"—from which follows the notorious dictum that "a man is the measure of all things" (*Theaetetus*, 17, 16). The Protagorean principle is invoked with increasing frequency and distinctness, and given ever greater extension of meaning, by Victorian and post-Victorian intellectuals. The principle of relativity, says Mill, who calls it unreservedly "a fundamental truth," is "that all consciousness is of difference; that two objects are the smallest number required to constitute consciousness" (*Sir William Hamilton*, 65, 14). Earlier he had promulgated this rule as basic to the modern science of economics. "Value is a relative term," he says.

> The value of a commodity is not a name for an inherent and substantive quality of the thing itself, but means the quantity of other things which can be obtained in exchange for it. The value of one thing must always be understood relatively to some other thing, or to things in general. (*Principles of Political Economy*, 1:549)

Logical entities are relative, subject to the rule of Difference, in the same way that physical and economic entities are, insisted John Henry Newman in 1870. In all processes of classification and abstraction, he declares, "we regard things, not as they are in themselves, but mainly as they stand in relation to each other" (*Grammar*, 44). "The aim of science is not things themselves, as the dogmatists in their simplicity imagine," stated the mathematician Henri Poincaré in 1902, "but the relations between things; outside those relations there is no reality knowable" (*Science and Hypothesis*, xxiv). When Ferdinand de Saussure in the years after 1907 founded modern linguistics on the principle that "in language there are only differences *without positive terms*" (*Course in General Linguistics*, 120), he was thus importing into his own scholarly field a premise that by then had a long history, and had become, indeed, as these few quotations may already suggest, an orthodoxy among progressive intellectuals. The same is true of Einstein's new science of space and time, in which, as Ernst Cassirer explained, "the 'here' gains its meaning only with reference to a 'there,' the 'now' only with reference to an earlier or later contrasted with it. . . . They are and remain *systems of relations*" (*Substance and Function*, 172). Similar phraseology—strategically unmoored from its history of development in the benighted nineteenth century—now has the force of dogma in the various domains of postmodern theory, where the law of *différance* presides. Are illustrative instances needed? The point of departure of contemporary gender theory, for instance, is the denial that such a term as "woman" possesses a natural referent and the claim that it must be construed instead as "a relational term in a system of difference" (Tickner, "Men's Work," 45). Physical objects, logical categories, space and time, human genders: all have been progressively subsumed within the relativistic order of thinking. In what follows, I use "relativity" in the expansive sense suitable to this ramified development to refer to a broad field of intellectual phenomena organized around the principle that "nothing is one thing just by itself."

As a topic of intellectual and cultural history, relativity thus defined presents a set of riddles. The first of these is a riddle of magnitudes. How can one make intelligible the seeming disparity between, on the one hand, the narrowly technical philosopheme, the axiom of Protagoras, on which relativistic thinking is founded and, on the other, the vast elaboration it has received in modern discourse? (Included very prominently in this latter category is the literature of antirelativity, which if anything exceeds in scale—as it does also in polemical militancy—that which it strives to suppress.) The key to understanding this problem forms the primary thesis of the present study. I argue that the relativity principle, from the time of its formulation in fifth-century Athens to that of its reemergence in the second half of the nineteenth century, and then to its postmodernist manifestations, has belonged very ambiguously to two incompatible worlds of discourse. On the one hand, it has served as an operator in various specialized fields of scholarship, scientific work, and philosophical dialectics far removed from the spheres of political affairs and practical life. Its expositors, eager to protect it from contamination by intellectually unpurified elements (and also eager to protect themselves from various forms of incrimination), sometimes have been at pains to insist on its disinterested remoteness from worldly concerns. Yet the relativity idea has from the first been so deeply embedded in politics, and in such rhetorically violent politics, as to make one wonder whether its professed unworldliness and innocuousness can ever have been more than a pose, whether it can ultimately be viewed as anything other than a vehicle for ideological enterprises—even when it expresses itself in the technical vocabulary of, say, physics or structural linguistics or marginalist economics or pragmatic philosophy. Certainly its adversaries have persistently represented it as infested with this kind of bad faith, and as the instrument of subversive interests. One need not accept any vulgar version of such a conspiracy theory to see that the scholastic aspect of relativity is belied by its perpetual involvement in violence, particularly political violence, and that relativity could even be seen as a method of concerted philosophical intervention in the problem of violence in human life. There is in fact substantial evidence to support this thesis; presenting this evidence is one of the main objects of this book.

Even allowing for the possibility that relativity doctrine has always served undeclared political and philosophical agendas, the enigma of its influence remains, for it comes saddled with notorious disabilities as an analytical method. As Socrates demonstrates in the *Theaetetus*, and as philosophers have reiterated ever since whenever the need arose, relativity is logically incoherent and self-refuting. (If propositions can only be relatively true, how can one legitimately make such an assertion to begin with?) By the same token, it is easier to state the principle that all consciousness is of difference than to understand what this principle can mean in practice: for how can one speak intelligibly of differences unless one can speak of determinate entities among which relations of difference can occur? "From this contradiction how shall we extricate ourselves? How solve the paradox, of two things, each depending on the other?" asks Mill, in particular reference to the conundrums of a relativistic political economy (Mill, *Principles*, 1:533). The principle of Protagoras, open to grave objections as it thus is on the plane of theory, resists being put effectively in practice as a method of concrete inquiry. Despite wide endorsement of Saussure's dogma that languages are purely differential systems, for example, "when one actually analyzes a language it becomes extremely difficult to avoid speaking as if there were positive terms. It is difficult to analyze a language purely as a system of relations" (Culler, *Saussure*, 92). Such analyses, if rigorously pursued, infallibly open upon the infinite regresses of *différance*, and few writers have had the nerve to attempt this challenge. Moreover, in addition to the logical predicaments they notoriously entail, relativistic doctrines have long radiated a phosphorescence of dangerous moral depravity, as we shall see. Respectable people fear them, or affect to. And even for those willing to run the risk of association with "the great bugaboo" of relativism, finding in it the basis for a trenchant critique of this or that set of "absolute values," relativity makes at best an awkward polemical tool, likely as it is not only to expose us and our causes to philosophical disrepute but to reveal our own covert absolutisms and to destabilize our own certitudes in the process. So the puzzle of its irresistible appeal to many modern thinkers is worth contemplating.

Champions of the Absolute

In trying to unpuzzle this problem, I have considered nineteenth- and early-twentieth-century relativity literature not as a purely dialectical construct (as though the soundness of its philosophical reasoning were the point at issue), but as the elaboration of a radical scheme of values anchored in the exigencies of a specific historical world. My angle of analysis is the one identified subsequently with Ludwik Fleck, Gaston Bachelard, Michel Foucault, T. S. Kuhn, and others, but that, following Spencer, was laid out by Thorstein Veblen in essays of 1906 and 1908, "The Place of Science in Modern Civilization" and "The Evolution of the Scientific Point of View." Veblen argues that the history of scientific knowledge takes the form of a sequence of discontinuous theory-worlds, each largely incommensurate with those that come before and after. "Any marked cultural era will have its own characteristic attitude and animus toward matters of knowledge," says Veblen, "and will seek answers to . . . questions only in terms that are consonant with the habits of thought current at the time" ("Evolution," 38). The knowledge system of any epoch is presided over, he theorizes, by a distinctive metaphorical system that figuratively reproduces the pattern of existing society. "The habits of thought that rule in the working-out of a system of knowledge are such as are fostered by the more impressive affairs of life, by the institutional structure under which the community lives." Even today, says Veblen, "[the scientist's] canons of validity are made for him by the cultural situation; they are habits of thought imposed on him by the scheme of life current in the community in which he lives" ("Place of Science," 10, 17).[3] Such a theory exemplifies the principle of "the relativity of knowledge" in an extreme sociocultural inflection, for what drops out of it without a trace is that supposedly hegemonic norm of turn-of-the-century theory of science, "objectivity"—or rather, the contemporary ideal of objectivity is seen to be merely a passing cultural fashion. What counts as truth is historically variable, Veblen says; therefore knowledge must always be indeterminate and temporary. "Every goal of research is necessarily a point of departure; every term is transitional" ("Evolution," 54, 33).

What is absent from this exposition is what Marx insisted upon in the concept of "ideology," namely, the way the cultural system acts reciprocally back upon society by establishing a fabric of common sense within which the social status quo takes on the appearance of an immutable natural order. To admit such a principle is to imagine that altering the structure of the intellectual world in radical ways could have social effects. This is what the activists of the nineteenth-century relativity movement did imagine. The literature of this movement everywhere takes as its key premise the conviction that the historical aftermath of the Enlightenment and of the fall of the *ancien régime* and of the divine right of autocracy has been, bewilderingly, the growth of menacing new totalitarian structures. These structures are alleged to appear not only or even primarily in the form of authoritarian systems of government, but in the mental forms such systems both concretize and exploit: what Blake called "the mind-forg'd manacles" of modern Europe, what Mill called "the engines of moral repression" (*On Liberty*, 13), what W. S. Jevons called "the noxious influence of authority" (*Theory*, 273), what Nietzsche called the slave mentality, what Gramsci called "ideology." The relativity movement, even in its most abstract and technically scientific manifestations, has been driven by the imagining of a newly emancipated order of thought amid a context of growing and (its distinctive characteristic) ever more insidious repression, and it has always been inseparable from "moral relativism." So at least I argue, contrary to prevailing views.

Running through Victorian relativity literature, imparting a note of polemical urgency, was a paranoid-seeming fear of a great reversion to something like the feudal era as Veblen described it, a phase of "predaceous life" governed brutally by agencies of "prescriptive authority" (Veblen, "Place of Science," 10, 12). One striking motif of this body of writing (highlighted in chapter 5) is thus a recurring vision of a collapse of European civilization into a reign of violence and terror. The attempt by relativity theorists to effect a radical reform of the absolutist philosophy of science must be seen in part as a response to such fears, though it is hard to describe their scholarly enterprise in these terms without making it seem hopelessly quixotic. Quixotic or not, they saw themselves as struggling to emancipate their society from a great structure of ideologically saturated common sense of which the ultimate tendency was to sustain despotic regimes.

It is true that in many of the texts to which I shall refer, the emancipatory theme is not explicitly marked, and even that the sorts of value-laden connections I try to establish among different modes of literary expression have sometimes been pointedly disavowed by the authors I study. Einstein (though a radical anti-authoritarian like Feuerbach, Mill, Spencer, Grote, J. B. Stallo, Westermarck, W. K. Clifford, Ruth Benedict, and other pioneering expositors of the relativistic thesis) firmly insisted on the partitioning of scientific inquiry from other realms of value. He once comforted the Archbishop of Canterbury, who had asked him if relativity theory posed any threat to religious faith. "None," replied Einstein. "Relativity is a purely scientific matter and has nothing to do with religion" (quoted in Frank, *Einstein*, 89–90).[4] My own study is predicated on the contrary assumption that there can hardly be such a thing as a "purely scientific [or aesthetic, or political, or religious] matter," that any isolation of the category of scientific knowledge from historical conditions and from its cultural correlates, antecedents, and consequences necessarily does violence to our understanding (and predictably will prove to be dictated by practical or ideological vested interests). Indeed, the isolation in principle of any realm of knowledge from any other is precisely what a relativistic mode of inquiry is bound to repudiate as the error of errors. Eddington, a more consistent relativity theorist than Einstein himself, thus glosses the anecdote of the archbishop with the comment that "the compartments into which human thought is divided are not . . . water-tight" after all (*Philosophy of Physical Science*, 8). "Between the logic of religious thought and that of scientific thought," he might have said with Durkheim, "there is no abyss" (Durkheim, *Elementary Forms*, 271).

What is enacted in early relativity writings is not only a rehearsal of "the struggle between Liberty and Authority" (Mill, *On Liberty*, 1), but a positing of an ideal regime of values. This ideal regime forms the inverse image of all systems of autocracy and absolutism. Its presiding values are reciprocity; interconnectedness; the privileging of diversity, dissent, and creativity; and the systematic demystification of established structures of authority. But the greatest of these is reciprocity. "All the positive definitions of religion are based on reciprocity," says Feuerbach (*Essence of Christianity*, 273). Alexander Bain in 1873 credits George Grote with having illustrated in his study of Plato "the essential *reciprocity* of virtuous conduct," a thesis Bain expressly calls "one of the many phases of the Law of Relativity" (Bain, "Intellectual Character," 119).[5] The "principle of reciprocity" (Grote, *Plato*, 4:128) is carried so far that it becomes the prerequisite for existence in relativity thinking: not to enter into two-way relations with another thing is simply not to exist. All things, in order to have identities of their own, are enmeshed in a perpetual traffic of communication with other things (which is one reason why relativity theorists from an early stage emphasized the symbolic—the differential—character of reality).[6] In special-relativity physics, beams of light flashing "signals" back and forth across the void of space form the central metaphor for the order of nature, replacing billiard balls and interlocking gears.

The law of reciprocity seems to take specific form in modern discourse first as a principle of the scientific analysis of human society; only at a later stage is it extended in distinctive ways to other fields.[7] As early as 1843, Mill proposes basing social science on the axiom that "there exists a natural correlation" or a "*consensus*" among "the various parts of the social body" (*Logic*, 2:509)—a thesis in which modern scientific analysis is made to coincide at least on the level of poetics with the imagining of a kind of utopian moral regime in which consensus, not authority and compulsion, creates social order. What Mill suggests with his language of "natural correlation" is in fact an imaginary realm presided over by Eros, the active principle of assimilation and unification, "whose purpose," as Freud says, "is to combine single human individuals, and after that families, then races, peoples and nations, in one great unity, the unity of mankind" (Freud, *Civilization and its Discontents*, 69). In the world of relativity, "nothing is one thing just by itself": as Bain put it in 1870, "all objects of knowledge . . . go in couples" (*Logic*, 1:255). In its erotic character, its antagonism toward every form of dogmatic authority, its elevation of ironic discrepancy to the level of an essential principle of understanding (in relativity physics, for example), its almost obsessive love of paradox, and its impulse to thumb its nose scandalously at respectability (as embodied in the rule of logic, for example), the world of relativity literature is a comic world through and through. It is nonetheless a deeply serious world that springs from the sense that the sinister principles of totalitarianism, absolutism, and concentrated coercive power are in the ascendant, and that the supreme mission of modern thinkers in every domain is to combat such principles by building models of thought (which is to say, models of human relations) radically emancipated from their influence.

From the point of view of its adversaries, however, the comic realm of relativity takes on the aspect of melodrama or gothic romance. In this version, the virginal image of "Truth," the object of a cult of adoration, is subjected to criminal assault by the forces of relativism. As in Spenser's allegory (I refer to Edmund Spenser, celebrant of the absolutist Elizabethan state and of the merciless campaign of terror waged by his patron, Lord Grey, upon the Irish; not to Herbert Spencer the arch-relativist, denouncer of the regression toward militarism in late-nineteenth-century industrial society and analyst of the replacement of "the principle of voluntary co-operation" by "the extension of centralized administration and of compulsory regulation"—Spencer, *Principles of Sociology*, 1:562, 570)—as in *The Faerie Queene*, that is, the distinguishing characteristic of Truth for antirelativists is that she is Una, one, exclusive, while falsity is expressly identified with the malign principle of multiplicity. Duessa the evil sorceress bears the very name of the Protagorean theory that every truth is in a sense necessarily duplicitous, that "nothing is one thing just by itself" and that "two objects are the smallest number required to constitute consciousness." The loathsome dragon Errour is first and foremost a source of horrible *proliferation:*

> Of her there bred
> A thousand young ones, which she dayly fed,
> Sucking upon her poisnous dugs, eachone
> Of sundrie shapes, yet all ill favored.
>
> (*Faerie Queene*, I.i.15)

In the "Areopagitica" Milton rehearses much the same mythology in a different set of images (and to point a very different political moral). Truth came into the world—gendered female—with Christ, but subsequently "[there] arose a wicked race of deceivers, who . . . took the virgin Truth, hewed her lovely form into a thousand pieces, and scattered them to the four winds," from which fragmentary condition it is impossible to redeem her until the Second Coming ("Areopagitica," 30). It is the very sign of the fallen state of mankind that Truth is no longer to be found in her authentic single and unified state, but in a multitude of fragments only.

Neither Spenser nor Milton had any notion of modern relativity as a possible rival of the Christian regime of Truth, but they each exemplify in different ways Michel Foucault's argument that modern European sensibility is fixated on the dread of uncontrollable polysemy, of "the cancerous and dangerous proliferation of significations," and on the determination to contain this frightful danger at all costs (Foucault, "What Is an Author?" 118). This dominating ideological principle was mercilessly diagnosed in 1841 by Ludwig Feuerbach, who identified it particularly with religious faith. "Faith," he says in a text that plays a significant role in the propagation of Victorian relativity,

> discriminates thus: This is true, that is false. And it claims truth to itself alone. . . . Faith is in its nature exclusive. One thing alone is truth, one alone is God, one alone has the monopoly of being the Son of God; all else is nothing, error, delusion. Jehovah alone is the true God; all other gods are vain idols. (*Essence of Christianity*, 248)

In *The Faerie Queene*, preserving the monopolistic exclusiveness of Truth is figured as the unleashing of a carnivalesque orgy of sadistic violence, lovingly fantasized by the poet (I.xi.20–55), upon the purveyors of false doctrines (allegorical representations of Roman Catholics in particular); all Spenser's emphasis on Una's mild and loving character dispels as best it can the anxieties that the glorification of brutal political violence, even in defense of "Truth," might otherwise arouse in the poetry-reading public.

As an analogous instance of the patterns of antirelativity discourse, consider—an admittedly extreme point of reference—the somber figure of Philipp Lenard, the emanation of a somewhat later historical period than the one I consider hereafter.

Lenard (1862–1947), professor at the University of Heidelberg, was a distinguished German scientist who won the 1905 Nobel Prize in physics for pioneering work on a central issue for modern physics, the photoelectric effect. He squandered his scholarly honors by becoming in later years an ideologue in the service of the Third Reich, in which capacity he led a sustained Nazi campaign of propaganda attacks and terroristic conspiracies aimed at suppressing relativity theory in physics. This almost unthinkable spectacle of an illustrious scientist allying himself with the likes of the National Socialist leadership in an attempt to destroy modern physics bears striking witness to the political stakes of the modern debate over the relativity principle: after Lenard, it becomes futile (if ever it was otherwise) to maintain that relativity physics belongs to a realm of pure scientific rationality divorced from moral values and from the political. It is essential to understand that relativity theory was anathema to Nazi officialdom on more than one score. First of all, all products of duplicitous Jewish brains such as Albert Einstein's were presumed in the demented world of Hitlerism to be evil and dangerous (and were represented habitually in just the same metaphors of revolting filthiness Spenser uses for the monster Errour). In still other fundamental respects, the "new way of thinking" in science, as Bergson called it in 1922 (quoted in Pais, *Subtle is the Lord*, 163), or "Jewish physics," as the Nazis routinely called it, seemed to ideologues of the German state to be inherently subversive, a menace to the cult of Hitlerian absolutism. They knew better than to regard relativity as "a purely scientific matter," seeing it instead as a mode of awareness implacably hostile to their own.

Lenard and fellow Nazis like Nobel laureate Johannes Stark thus attacked the relativity movement in physics for its supposedly Jewish trait of uninterest in or active antipathy toward "truth." In making such a charge, Hitler's scientists meant among other things to indict contemporary relativity thinking for its pernicious abstractness, its proneness to a "predominance of theory over sensory observation" (Frank, *Einstein*, 238).[8] Wilhelm Mueller, in *Jewry and Science*, accordingly denounced Einsteinian physics for its indulgence in "spectral abstraction," a mode of reasoning in which substantial, tangible facts "are lost in unreality" (quoted in Shirer, *Third Reich*, 250). Another Nazi scientist, Ludwig Bieberback, declared similarly that relativity theory, like other Jewish productions, "conspicuously lacks understanding for the truth," unlike authentic German science, with its "careful and serious will to truth." "Jewish physics," concludes this writer, invoking as Mueller did what we shall see to be the leitmotif of antirelativity polemics then and now, "is thus a phantom" (quoted in Shirer, *Third Reich*, 251). All the mythography of book 1 of *The Faerie Queene* is here summoned up anew, all the fable of "stedfast Truth" assailed by the phantasmal delusions of the learned sorcerer Archimago, who with "his magick bookes and artes of sundrie kindes, / . . . seeks out mighty charmes, to trouble sleepy minds" (*Faerie Queene*, I.vii.1, I.i.36). Less explicit than in Spenser but still unmistakable in the antirelativity writings of Nazi scientists is the threat of the catastrophic violence in store for the authors of doctrines deemed incompatible with state ideology.[9]

The story of Philipp Lenard may scarcely seem to constitute a legitimate chapter in the intellectual history of our century, yet the campaign against "Jewish physics" represents not so much an isolated lunatic episode as an especially vivid instance (with some murderous additional baggage) of the traditional indictment of relativistic thinking in all its forms. The endeavor to suppress relativism in the name of "absolute" values and of "truth" constitutes in fact one of the abiding projects of the European intellectual tradition. The arguments and the rhetorical devices deployed by Lenard, Stark, Mueller, and Bieberback against Einsteinian science are exactly cognate with those put in circulation persistently, and especially in the last 150 years, by a broad coalition of allies in the struggle against relativism. Antirelativists wishing to dissociate themselves from such disgraceful company might be hard pressed to say exactly how their own philosophical thinking differs from Lenard's.

The war against relativism has its paradigm and prototype in Plato's philosophical attacks—and then in the direct assault of the Athenian state, foreshadowing that of the German state upon Einstein—upon Protagoras the Sophist. Socrates in the *Theaetetus* refutes the doctrine "that nothing is one thing just by itself" by advancing what has become, as we noted, the standard argument: that relativism refutes itself. In professing a doctrine so egregiously false, Protagoras becomes for Plato a willful, malicious subverter of "truth." "It would seem that Protagoras' *Truth* isn't true for anyone: not for anyone else, and not for Protagoras himself," declares Socrates. According to this wicked doctrine, which subverts religion and morality and in fact, as Plato insinuates, is only professed in shameful secrecy, "if all things . . . change, then every answer . . . is equally correct" (*Theaetetus*, 47, 63). In the *Protagoras*, Plato carries the indictment further, portraying the Sophists, the cynical destroyers of truth, as the inheritors of an ancient organized conspiracy of subversion and deceit practiced by a league of men who, "for fear of giving offence, adopted the subterfuge of disguising [their treacherous activities] as some other craft" (*Protagoras*, 9). In the archive of antirelativity literature across the ages, it is fair to say that such notions as "Sophism," "Errour," and "Jewish physics" function as equivalent and interchangeable categories, vouched for by all the same self-duplicating rhetoric of vilification and the same demonizing mythography—and, not incidentally, very often leading to the same practical consequences: suppression of the unorthodox by the political authorities.

The legend of Protagoras the criminal subverter of society and of the wickedness of the Sophists has been long-lived, for it has had a signal role to play in the maintenance of the European cultural ideology based on the singleness and absoluteness of "truth." In George Henry Lewes's mid-Victorian *History of Philosophy* (published originally in 1845–46, then in several subsequent editions in greatly developed form), for example, the pariah Protagoras is identified as the original author of the doctrine, already well known in Victorian thought, of "the Relativity of Human Knowledge." According to this principle, says Lewes, thought is identical with and limited to sensation, implying "that everything is true *relatively*" and that consequently "knowledge is inevitably fleeting and imperfect." Were the Sophists, who preached this doctrine, the immoral corrupters of youth they are typically portrayed as being, says Lewes, "Greece could only have been an earthly Pandemonium, where Belial was King" (Lewes, *History*, 122, 121). The view of Protagorean relativism in its Platonic guise, as an abhorrent perversion and a threat to civilization itself, is very much the view given by the scholar of agnosticism Robert Flint in 1887–88. The philosophy of the Sophists, he declares, exemplified the "predominant evil tendencies" of an age "in which self-interest, vanity, and ambition were the ruling motives of action" and in which "the pure love of virtue" was rarely to be found. Protagoras and the others of his school, lacking faith "in any absolute truth or goodness," "so exaggerated the relativity alike of sense and of thought as to leave no room for a reasonable trust in the certainty of any kind of knowledge." "The sophist is a man who does not care for truth," says Flint, in the formula much used by Ludwig Bieberback and others to describe Jewish physicists of a later generation (*Agnosticism*, 85, 86). Yet Protagoras was in reality no criminal and no nihilist, but a teacher of "excellent morality," declares the freethinker Lewes, displaying some of the same courage in challenging the interdict pronounced upon Protagoras by nineteenth-century defenders of "absolute truth [and] goodness" that Max Planck and Max von Laue displayed in their fidelity to Einstein's teachings in physics during the age of Philipp Lenard (Lewes, *History*, 112).[10]

The struggle against relativism in all spheres of thought has continued unabated with, if anything, increasing virulence, but always relying on the same repertoire of characteristic rhetorical structures as in Lewes's day. The biologist St. George Mivart's *On Truth* (1889) provides a representative text of this doggedly monotonous literature.[11] Mivart warns his readers that speculative thought in the late nineteenth century was in danger of nothing less than "falling into utter scepticism and absurdity." To preserve ourselves from this fate, he declares, we must cling fast to the assurance, furnished by irrefutable common sense, "that we can know absolute truth," that "truth really exists." This salvific category of "absolute truth"—a relative novelty in the philosophical lexicon of the day—is invoked in Mivart's book with what can only be called compulsive insistency. (In the course of page 67, not an atypical page, such phrases as "absolutely certain," "supremely certain," and "complete certainty" recur six times, giving one a strong sense of the displacement of reasoned argument by a sort of dogmatic incantation.) For Mivart, "absolute," in the joint sense of absolutely single and absolutely coercive, is the necessary predicate of the concept of "truth." "Truth is one, while error is manifold," he declares, imagining that he is stating a self-evident axiom or a law of nature (*On Truth*, 135, 134, 238, 240). From this perspective, any professed truth that is not "absolute" can only be a delusion, an offshoot of "Errours endlesse traine" and an instance of "the cancerous and dangerous proliferation of significations."

Both goodness and truth are characterized, Mivart says further, by their "harmony with an eternal, absolute law" that is "supreme and absolutely incumbent upon us without appeal" (*On Truth*, 248, 251). In a later work, *The Groundwork of Science* (1898), Mivart, like Plato, critiques as hopelessly self-refuting the doctrine of the relativity of knowledge, which leads us unfailingly, he declares, "into an abyss of intellectual nihilism,"[12] and he again sets up as the sole alternative to this fate a cult of a law from which there is absolutely no appeal. What the truth-loving scientist discovers in nature is not uncertainty, not change, Mivart asserts, but "permanent law and order" that is "ruled by rigid laws" (Mivart, *Groundwork*, 280–81, 215, 283). The intellectual regimen of science in such a formulation and the political regimen of, say, the National Socialist state are hardly distinguishable, at least on the level of rhetoric. In both, the scheme of permanently established despotic rule is glorified; unquestioning submission to it is held up as the paramount rule of conduct; and the concept of obedience to "rigid laws" is chronically ambiguous in its reference ("mutinies are suppressed in accordance with laws of iron which are eternally the same," declared Adolf Hitler, playing on the ambiguity, in a speech announcing the bloody Röhm purge of 1934—Hitler, *Speeches*, 321). Both systems are seen by their promoters, moreover, as though by an ideological necessity, as menaced from within by a clique of nihilists using relativistic doctrines to corrupt "truth" itself. Mivart was no protofascist, but to read his polemical prose is to grasp how readily the ideology of modern science might take on an active association with the trend of what Herbert Spencer broadly called "militant" society (in which "absolute subordination . . . is the supreme virtue"—Spencer, *Principles of Sociology*, 1:544, 548) and might slip into collaboration with outright political tyranny, as it finally did in the case of Philipp Lenard and his colleagues.

The possibility of this perilous convergence of scientific ideology, the denial of relativity, and the ethic of state tyranny stands out still more unequivocally in V. I. Lenin's *Materialism and Empirio-Criticism* (1908). In the name of scientific materialism, Lenin sweepingly assaults the so-called bourgeois and pro-clerical "reactionary" strain of contemporary trends in the philosophy of science. Focusing on the arch-relativist and "ideologist of reactionary philistinism" Ernst Mach (a primary influence upon the young Einstein) and on Russian Machians such as A. Bogdanov, Lenin in 1908 prefigures from the absolutist Left all the arguments put forward a generation later by Lenard from the fascist Right. The preeminent question in the philosophy of science, says Lenin, is "the question of *relativism* . . . which is stressed by all Machians. The Machians *one and all* insist that they are relativists" (Lenin, *Materialism*, 404, 196). Bogdanov, the nominal Marxist conspiring with "fashionable *reactionary* philosophy" (Lenin, 192), is quoted damningly by Lenin as denying "an objective truth in the absolute meaning of the word" and as professing the evil doctrine of "the relativity of truth," in the sense that truth is relative to the apprehension of different historical epochs (quoted in Lenin, *Materialism*, 185, 194). "To make relativism the basis of the theory of knowledge," asserts Lenin, "is inevitably to condemn oneself either to absolute scepticism, agnosticism and sophistry, or to subjectivism." By contrast, the claim of Marxism to possess "objective truth" is glossed by Lenin as the assertion that "by following *any other path* we shall arrive at nothing but confusion and lies" (199, 205). Truth (here, a category directly equivalent to a certain political ideology) is Una, sole, exclusive, unique; to deviate from "objective truth in the absolute meaning of the word" is to fall not only into intellectual error but into conscious perfidy.

In *Materialism and Empirio-Criticism*, the equivocal character of the discourse surrounding the relativity movement is thus strikingly manifest: the professorish debate about points of highly abstruse philosophical theory, particularly in connection with the interpretation of modern science, proves again to be entangled inextricably in issues of politics—to be political at the core. Not to mince words, the assault on Machian relativity in this text of Lenin's, as in those of Philipp Lenard, is once again directly, not metaphorically or implicitly, a function of the establishment of a system of state tyranny. Once he had become the absolute ruler of the Soviet Union, Lenin was able to translate the rhetorical violence of *Materialism and Empirio-Criticism* directly into the unbridled physical violence that it intimates. Launching in 1921–22 a wave of political terror to be carried out "in the most decisive and merciless manner," he issued a directive whose continuity with his strident critique of "reactionary" Machian relativity is plain to see. "The greater the number of representatives of the reactionary clergy and reactionary bourgeoisie we succeed in executing," he declared, shifting from the academic to the performative register of discourse, "the better" (Pipes, *The Unknown Lenin*, 153, 154). In the 1930s, his anathema pronounced upon relativity physics came to fruition in the savage assault launched by the Soviet authorities against the politically suspect community of physics—a purge in which perhaps 10 percent of all Soviet physicists were killed (Josephson, *Totalitarian Science*, 54). Were some disaffected person to propose the theory that the ideology of "objective truth in the absolute meaning of the word" serves typically as a vehicle for at least potential and threatened violence and has ultimately *no other function*, Lenin would furnish a striking case in point.

The American philosopher Paul Carus—himself denounced by Lenin as a notorious pro-clerical Machian reactionary (Lenin, *Materialism*, 282)—opens his 1913 *Principle of Relativity in the Light of the Philosophy of Science* with a sophisticated endorsement of modern relativity thinking as expounded by Mach and others: "relativity is the principle of all real and actual being," he declares (shrewdly highlighting Herbert Spencer as one of the most influential early expositors of relativity theory—a key notation for modern intellectual history). But Carus, whose devotion to Machianism was less ardent than Lenin believed, then abruptly shifts into a sweeping attack on contemporary "relativity physicists" who, he declares, "deny all objectivity" and treacherously assail the traditional ideal of scientific inquiry. They "deny the legitimacy of the ideal of objectivity, or as they call it, the concept of the real." He truculently (and justifiably) defends his conflation of modern relativity physics with the philosophical school of pragmatism and what he calls "other antiscientific tendencies." With the advent of relativity theory in physics, he declares, switching in the usual way into the language of gothic or Spenserian terror, "the entire realm of science was almost panic stricken for scientists seemed to have lost the *terra firma* under their feet; they felt as if they were sinking into a bottomless abyss and were left without a standing place in the whirl of a universal flux."

Nor is this terrifying state of affairs merely the result of intellectual error, for the relativity principle in physics is—of course—the work of a diabolical conspiracy of nihilists. "The relativists are quite serious and are aware of the gravity of the consequences of their subversive work," says Carus. For all their powerful deployment of scientific methodology, "their aim is after all a denial of the old ideal of science, of the objectivity of truth, and of clearness of thought." Carus thus rehearses the dreary formula of Plato's and Robert Flint's argument against Protagoras, Spenser's against Archimago and the Roman Catholic Church, Philipp Lenard's against Einstein, and Lenin's against Mach: relativity physicists are, at bottom, scorners of "truth" using a clever simulacrum of sophisticated scientific reason for perverted purposes (Carus, *Principle of Relativity*, 7, 4, 42, 46, 85, 77, 85).

My point is to highlight the extent to which the longstanding attack on relativity has depended for its considerable power of persuasion not so much on the framing of logical argument as on the deployment of an array of self-duplicating figurative motifs and the repetition of a certain mythic narrative. The rhetorical sign of all this literature is the way its main themes manifest themselves, even amid contexts of supposedly highly rationalistic discourse (and discourse expressly in praise of rationality), in the eruption of a syndrome of imagery that seems to spring directly from fantasy materials of popular culture. One characteristic image is that of the persecution or dismemberment of an idealized heroine called, as in *The Faerie Queene* or the "Areopagitica," "Truth." The persecutors, the doctrines of relativism, are imagined persistently in the form of "a wicked race of deceivers" and of malignant ghosts and specters. A closely related motif stands out vividly in Carus and Mivart and a string of other writers before and since: nightmare imagery of the disintegration of solid reality and of the terrifying fall into "an abyss" of flux and indeterminacy.

Antirelativity literature has focused more recently on preserving natural science, imagined as the stronghold of true, objective knowledge and as protecting truth claims in almost any other domain from the peril of "intellectual nihilism." (If scientists in their laboratories have achieved absolute understanding of the working of nature, various writers assert or imply, then we can plausibly claim some cultural values—notably those produced by science-intensive societies—to be absolutely better than others.) The case of relativity physics presents an obvious and crucial challenge to this project. No reputable philosophical writer could today follow Lenard's or Carus's tack of portraying Einsteinian relativity as antiscientific. Rather, latter-day expositors of relativity theory in physics join in the enterprise of preserving "objective truth in the absolute meaning of the word" by vying with each other to deny any commonality between relativity physics and any version of "relativism" (the antiscientific and subversive animus of which is widely taken for granted). In this account, relativity physics is in fact implacably hostile to all relativism, and ought properly to bear the name Einstein is alleged to have originally given it: "the theory of invariance." The intent of Einsteinian theory, we are advised by many commentators, is to attain rigidly fixed objective values for physical phenomema—values distinct from the variables different observers in their differing "coordinate systems" will necessarily assign. To imagine that the theory gives credence to notions that "truth is relative," or that it has anything in common with such notions, is to commit a grotesque intellectual error. "The intellectual strategy of relativistic physics is quite contrary to relativism" (Toulmin, *Human Understanding*, 1:90). "The entire point of Einstein's relativity . . . was to construct absolute relations, not demolish them"; special-relativity theory properly understood is hence a monument to belief in "absolute truth" (Friedman and Donley, *Einstein*, 17, 65). "Einstein was not a relativist . . . he believed . . . in a single, absolute truth" (Josephson, *Totalitarian Science*, 44). One who has surveyed the uninspiring history of modern antirelativity polemics is likely to confront with a sinking heart all this peremptory instruction that to admire and be a believer in science is to worship perforce in the church of "absolute truth." In this same spirit, Hans Reichenbach scornfully denies any possible link between relativity physics and contemporaneous forms of moral relativism, and insists, as though replying on behalf of relativity physics to the standard slogans of antirelativists from Plato to Philipp Lenard, that "relativity does not mean an abandonment of truth" (Reichenbach, "Philsophical Significance," 289, 296). Arnold Sommerfeld declares categorically that Einstein's 1905 special-relativity paper "has, of course, absolutely nothing whatsoever to do with ethical relativism" (Sommerfeld, "Einstein's Seventieth Birthday," 99)—another version of Einstein's reassurances to the archbishop. In all this polemical literature, the shift in expressive register from the scientific prose of, say, Einstein's 1905 special-relativity paper into the language of dogma and of fantastic ideological constructs like "absolute truth" (intimately allied as the latter plainly is with wholly nonscientific motives such as the revulsion from moral relativism) is too flagrant and too coercive, too insistent and Leninesque in its invocation of the dictatorial word "absolute," not to call for critical analysis. The basis of such an analysis must be a refusal to take for granted any assertions about the purity and autonomy of scientific knowledge, and still less about its "absolute" character.

Einstein himself, having completed his trajectory from youthful

Machian relativism to a reactionary deterministic realism that put him at odds ever after with post-relativity physics, declared in the terrible year 1934 (in the same book in which he actually reasserted his belief in the ether, the material form of absolute space, which he had abolished conceptually in 1905) that the four-dimensional space of special relativity "is just as rigid and absolute as Newton's space" (*World*, 91). It would be impossible to contemplate the thought that Einstein, in embracing such language, was consciously or unconsciously fashioning "Jewish physics" in terms likely to win the favor of ideologues of the newly-installed Nazi authorities. It is none the less true that with this eruption of value-laden figurative language, and especially with the glorification of the fraught term "absolute," one again quits the realm of mathematical physics and moves into that of polemics, where Einstein's portrayals of relativity theory may be no more authoritative than anyone else's, and no more immune to subtle, even unintentional, ideological effects. Subsequent writers on relativity continue, in any case, making the point he makes here, showing by their insistence in doing so the cultural stakes in these debates.

A case in point: Marie-Antoinette Tonnelat's *Histoire du principe de relativité* (1971), a work devoted single-mindedly to quashing yet again the claim that relativity theory in physics has anything in common with any form of philosophical relativism, to which it is, this author proclaims in the orthodox way, "diametrically opposed." All forms of the doctrine of "the relativity of knowledge" proceed from skepticism about the ability of science to attain "absolute truth," says Tonnelat. By contrast, the mission of relativity theory is to achieve for physics "the justification of its ideal of objectivity" and, by so doing, to permit the scientist to discover the "invariant intrinsic structure" of "objective reality"—a reality admittedly not directly perceptible, not embodied in phenomena, but one constructed by rational, mathematical procedures, and having in a sense only an intellectual existence. In this account the Einstein of 1905 is profoundly realistic, and, Ernst Cassirer and Gaston Bachelard to the contrary notwithstanding, expresses in his theory full belief in the "absolute" existence of objects, not merely in physical relations. Relativity theory thus establishes "invariant absolutes" and contradicts the fundamental principle of relativity as stated by Einstein's disciple Eddington: "that we observe only *relations* between physical entities." Tonnelat insists that so-called "proper" measurements (those taken by means of instruments located in an object's own frame of reference) constitute, in special relativity, "true (invariant) values," while measurements in another "Galilean" system of reference yield merely "apparent" values—this in apparent disregard of the supreme principle of the equivalence of all inertial frames of reference, which every other authority tells us admits of no exception (Tonnelat, *Histoire*, 17, 265, 10, 238 n., 241, 238–39, 247).[13]

Tonnelat's learned treatise thus rehearses again the traditional melodrama of the delivery of "Truth" from the evil spell of relativism, though leaving her open to the charge that she has to destroy relativity in order to rescue it—from what may be only imaginary enemies after all. As for the category of the "absolute," it is enough for now to observe that it is invoked with the same compulsive insistence by this author as it was in the nineteenth century by theoretically unsophisticated writers of a pietistic bent like St. George Mivart. Any term pronounced by Einstein in elucidation of relativity theory is entitled to a full presumption of respect, but at the risk of overstatement we must repeat that "absolute" in this context is more easily seen as the carrier of ideological and crypto-theological intent than as the definition of any proper scientific category.[14] For one thing, this term, the site of so much cultural urgency, is afflicted with notorious and by now incurable ambiguity. Various commentators on the literature of Einsteinian physics have pointed out that it is employed there in so wide a range of meanings as almost to render it useless for philosophical purposes.[15] In their critique of the "absolute," these writers pursue a nineteenth-century project of deconstructing what Mill referred to drily as an "obscure . . . word of several meanings" and more bluntly as "a heap of contradictions" (*Sir William Hamilton*, 49, 63–64). Beginning in the early nineteenth century, one of these meanings identified the word with the idea of political authoritarianism.[16] This semantic parasite has been inseparable from "absolute" ever since, so much so that the word can now scarcely be used in good faith in uncomplicated reference to a code of scientific reason. Yet, difficult as its sense may be to grasp—in fact, its best definition may be, according to various authorities in the Kantian tradition,[17] "that which is ungraspable"—the "absolute" remains a supreme fetish of scientific and philosophical literature alike. Its Victorian critics would claim that whenever "absolute truth" is summoned in debate, we are in the presence not of rational language but of the mystified form of the ancient gods Authority and Coercion.

Meanwhile, the broad assault continues on the supposedly widespread evil empire of relativism (which to all appearances has in fact failed utterly in extending its influence beyond a narrow sector of academic culture). It would be hard to exaggerate either the stridency of this campaign or the extent of its reliance upon what may seem to be intellectually uncouth lines of polemic. So perverse and diseased is relativistic thinking, so patently unjustifiable by civilized norms of rationality, so dangerous to society, that it can only be explained by criminal malice: so we continue to be told, by writers who rely as much as ever on the characteristic imagery of the uncanny, of "spectral abstraction" and evil phantoms, so favored by Nazi polemicists.[18]

In *Beyond Objectivism and Relativism*, R. J. Bernstein declares his intention "to confront the specter of relativism" that haunts contemporary philosophical disputes, lest we continue to be "ineluctably led to relativism, skepticism, historicism, and nihilism." "The fashionable varieties of relativism that are spreading everywhere frequently lead to cynicism," he declares (without citing examples). Indeed, "the specter that hovers in the background" of contemporary philosophy "is not just radical epistemological skepticism but the dread of madness and chaos where nothing is fixed" (Bernstein, *Beyond Objectivism*,x, 2–3, 4, 18).[19] Philosophy is to be the instrument for exorcising the ideas upon which this panicky dread has fixed, and which carry at their every iteration subliminal imagery of civil and political as well as metaphysical disorders. Ernest Gellner, praising the "miraculous success" of modern science and of its presumption of what he terms, in resonant Spenserian style, "One True Vision," alerts us to the same danger in the same rhetorically extravagant terms. "A spectre haunts human thought," he announces: "relativism." He defines this terrible menace in just the same Platonic formula rehearsed in Mivart, in Matthew Arnold, in Lenin, in Carus, in Philipp Lenard and his associates: relativism, says Gellner, is the nihilistic doctrine that since "there is no unique truth," then by logical necessity all ideas whatever, good and bad alike, are equally valid; evidence and reason count for nothing; and truth itself vanishes. "If truth has many faces, then not one of them deserves trust or respect," declares Gellner (*Relativism and the Social Sciences*, 90, 93, 84, 83). This doctrine, with its cult of "truth" and its explicit language of moral judgment, is so plainly ideological and so plainly a construct of mythographic formulae that it requires not even the most perfunctory logical demonstration—nor does this writer offer one. Nor does James W. Tuttleton offer one in declaring that in Gerald Graff's "entirely relativist" willingness to sanction pluralities of conflicting views in the field of literary interpretation, "truth becomes a casualty to the view that all claims are equally meritorious" (Tuttleton, review of Graff, 1344). The "cancerous and dangerous proliferation of significations" is taken for granted as the antithesis of legitimate philosophy.

The literary scholar Colin Falck assaults Saussurean linguistics and the postmodern and deconstructive interpretive trends that derive from it on the same grounds. Being purely relational, "the structuralist or post-structuralist tradition of linguistic . . . meaning in effect *abolishes reality*," Falck declares. He calls for a new mode of literary interpretation that "will enable us to restore the [concept] of truth," which is to say, the principle that only one reading of a text can be the correct one, and that, to borrow Lenin's phraseology, "by following *any other path* we shall arrive at nothing but confusion and lies." Falck denounces Derridean analysis in much the same terms Lenard and his associates use in their attacks on "Jewish physics": deconstruction is said to represent, just as Einstein's physics was, "a disembodied process or system" designed to produce a "truth-free or reality-free condition"; it results in the dangerous and fallacious principle of literary study "that any critic's reading may be as interesting as any other's." As Wilhelm Mueller denounced relativity physics for its "godless" abstracting away of "the living . . . world of living essence" into a state of "spectral abstraction" (quoted in Shirer, *Third Reich*, 250), so Falck makes an explicit plea for renewed religious values to combat the "disembodied and contextless process" involved in "the Saussurean emphasis on the relational aspects of sign-systems," an emphasis that perversely denies the capacity of literature to reveal things "in their essential forms and their essential rhythms" (Falck, *Myth, Truth and Literature*, xii, 19, 28, 33, 30, 11, 27, 33). Even in so innocuous-seeming a venue as linguistics, the relativistic mode of thought is proclaimed from the essentialist-absolutist point of view to be a menace to truth and to religion. For Colin Falck, Saussure and Derrida are not scientific inquirers of good intent whose findings are subject to verification or disproof, but two modern forms of the evil conjurer Archimago, as Bogdanov was for Lenin or as Einstein was for Lenard. They are the enemies of God.

The depiction of relativism as intellectual and moral nihilism, and its condemnation as the cause of the supposed spiritual deterioration of modern society, has also spread downward in the scholarly and literary scale to become a staple of politically-inflected academic writing and of journalistic party propaganda. That is to say, it has become a significant public institution with real-life consequences. Crude stuff it may be from a scholarly point of view, yet it echoes quite faithfully the themes of antirelativity writing as we have briefly traced them from Plato onward. Allan Bloom's 1987 best seller *The Closing of the American Mind*, which blames a supposed epidemic of nihilistic "relativism" (perhaps provoked by too-intense study of the dangerous works of Edward Westermarck and Franz Boas) for a sweeping intellectual and spiritual collapse among today's college students, would be a case in point.[20] So, too, would be the columnist John Leo, who approvingly quotes James Wolcott on "the blithe disregard of truth" in current intellectual culture and then rehearses anew the stereotyped charge. "In the post-modern . . . worldview, there is no objectivity or truth. Everything is relative. Nothing is better or truer than anything else. . . . All stories are valid and all cultures are equal." Moving up several notches in the hierarchy of literature, one could even cite a 1995 episode of Bill Watterson's comic strip *Calvin and Hobbes*, where the six-year-old hero—here representing the morally incapacitated and selfish American youth evoked by Bloom and other pessimistic analysts of the current scene—picks up the refrain, denying that moral values have any meaning for the philosophically sophisticated. "As we all know," he declares, "values are relative. Every system of belief is equally valid and we need to tolerate diversity. Virtue isn't 'better' than vice. It's just different."

In what sense is relativity a specter, as its opponents so often declare?[21]

For one thing, it has always led an ambiguous existence at best in antirelativity literature, where the characteristics attributed to it—always without specific reference—seem typically to have scant basis in reality. Nowhere does any "relativist," to my knowledge, assert that all views are equally valid or (Nietzsche perhaps excepted)[22] that there is no such thing as truth. In habitually calling it a specter, however, its would-be exorcists mean other things. They mean, first, that in theorizing a world in which only relations can be known, relativity causes solid objects (physical phenomena, texts, meanings) to seem to evaporate into incomprehensible abstractions. Also they mean that as a philosophical position it is so frail and insubstantial that it is scarcely a real entity with definite outlines. Relativism, as we saw, notoriously contradicts itself; it dissolves logically the moment it is formulated, and in this sense it is, in Sir Karl Popper's phrase, merely a "verbal spook" (*Open Society*, 393). The argument from logical incoherence, directed at Protagoras by Plato, is often repeated in current polemics. Harvey Siegel's *Relativism Refuted* leans almost entirely on it, branding relativism as fundamentally irrationalistic, thus as being "the radical and potentially destructive doctrine that it is perceived by its critics to be" (Siegel, *Relativism*, 10). W. V. O. Quine similarly uses this ploy.[23] How can Protagoras's ideas continue to enjoy their manifest vitality? Put to death time after time, relativism displays an uncanny ability to rise from the grave and stalk about. "Disembodied" as it is, weapons of logical refutation seem unable to wound it.

It is spectral too, and primarily, in its ability to inspire aversion and panic. "Relativism is a repugnant doctrine—reprehensible, repulsive and easily refuted," says James Robert Brown, highlighting the odd way in which the seeming logical feebleness of relativity theory appears to heighten its scary repulsiveness, as the feebleness of the zombies in George Romero's 1968 film *The Night of the Living Dead* heightens theirs (Brown, *Smoke and Mirrors*, 29). Nor is this an especially far-fetched analogy, given the terms in which antirelativity writing frequently presents itself. In one of the most familiar corollaries of the general indictment, moral relativism—allegedly the doctrine that no values are better than any others, and that every form of human conduct is therefore indiscriminately condoned—is identified as the origin of contemporary moral paralysis. "The world abounds with moral relativists who often as a result of their relativist beliefs display the most callous indifference toward the well-being of others" (Brown, *Smoke and Mirrors*, 70)—a population of dehumanized moral zombies, in effect. Paul R. Gross and Norman Levitt follow Popper not only in blaming relativism for "modern irrationalism" (Popper, *Myth of the Framework*, 33) but in going so far as to equate postmodern relativistic thinking with the moral nihilism of fascism itself (Gross and Levitt, *Higher Superstition*, 73).[24]

There is no need to argue for very long about the symptoms of personality degeneration so often said to accompany "relativism," since evidence is of course nonexistent to support claims that those of this "repugnant" philosophical orientation are marked to an unusual degree by cynicism, callousness, or "apathy about the state of their souls" (Bloom, 35). However, it seems worthwhile to examine the textual history of the relativity movement as I propose to do in the following chapters, if only as a corrective to the almost hysterical modern fixation on relativity as the enemy of essential cultural values. Those espousing relativistic theories in the nineteenth century and after never tire of protesting against the equation of their views with any species of nihilism. Far from upholding "the doctrine that all views are equally good" (Nozick, *Philosophical Explanations*, 21) or from being justly chargeable with "the reproach of being deficient in love of virtue or in hatred of vice" (Grote, *Plato*, 4:105), they proclaim devotion—often an uncompromisingly radical devotion—to a range of orthodox values: enlightenment, implacable critical rigor, tolerance, freedom. It is impossible to justify from historical evidence the claim that relativism has advocated any lessening of moral consciousness or that relativism is a form of cynicism or anomie. The contrary is very obviously the case. Herbert Spencer begins *First Principles* (1862), one of the most remarkable early manifestos of all-encompassing philosophical relativism, by declaring his aversion to "that error of entire and contemptuous negation" to which skeptical modern speculation is prone (*First Principles*, 8), and presents his philosophy specifically as the remedy for this error. J. S. Mill may have embraced extreme philosophical relativism in *Sir William Hamilton* (1865), yet he was not disabled by this act from protesting bitterly, in energetic public opposition to such vociferous jingoists as Tennyson, Kingsley, and Carlyle, against the sanguinary tyranny of Governor Eyre in the Jamaica rebellion of the same year (Packe, *Life of Mill*, 469). "Could it be brought home to people that there is no absolute standard in morality," says Edward

Westermarck in *Ethical Relativity* (1932), "they would perhaps be on the one hand more tolerant, and on the other hand more critical in their judgments" (59). This is not the language of nihilism.

The antithesis of relativity, as relativists in the militant tradition of Protagoras and Herbert Spencer variously argue, is not a rigorous concern for truth but, rather, "that absolutism which insists there is but one Truth and that truth one already revealed and possessed by some group or party" (Dewey, *Freedom and Culture*, 96). The discoveries modern relativity has produced in many scientific fields, and its interventions in issues of practical morality (Mill's, Spencer's, and Karl Pearson's pleas for full political citizenship for women, Westermarck's precocious plea for an end to the persecution of homosexuals, Ruth Benedict's pleas for recognition of the legitimacy of "primitive" cultures), all flow from this intuitive revulsion from "absolutism"—a revulsion on one level from its intellectual coarseness and ultimately, most profoundly, from its implicit and often explicit code of punitive and purifying violence. Relativity is a philosophical position menaced by potential incoherence on both logical and practical planes, but it is incalculably remote from nihilism. With few exceptions, the vilifying attacks on it fail so completely to meet minimal standards of legitimate debate as to seem like object lessons of just that tendency to brutal violence against which relativistic philosophy persistently warns. The Nazi and Soviet campaigns against Einstein and modern physics are reminders that if the oft-alleged social evils proceeding from "relativism" are for the most part risible fantasies, the continuity between the cause of intellectual absolutism and that of violent political tyranny is easy to document.

Such considerations point to issues that can scarcely be adjudicated by any historical study of texts. The long and relentless campaign against relativism may be prone to intellectually disreputable methods, yet it may be an all-too-facile exercise to disprove the charge of nihilism brought against relativism and to condemn the efforts of its great coalition of adversaries to suppress it. After all, relativism may be open to this charge not because it denies moral values or truth or contends that "all ideas are equal," none of which it ever has done, but because it is a scandal to what seems to be a profound, almost insuperable human instinct: the instinct to attack—and to hate and find "repugnant"—those whose ideas differ from our own, and whose claims of legitimacy for those ideas are felt even in the most innocuous circumstances to pose something like a mortal threat to us. Some such imperative, based conceivably on the evolutionary need for tribal solidarity, may underlie the incontrovertible law that "few things are more liable to arouse people's indignation than opinions that differ from their own" (Westermarck, *Ethical Relativity*, 161). To argue as relativists have done that different, even logically incompatible truths may and do properly coexist may be an affront to this instinct, to which totalitarian systems give naked expression but which in reality may just be the human patrimony—the passionate longing to inflict violence upon those who hold different theories, worship other gods, have different table manners. Certainly the ferocity of the general antipathy toward relativistic thinking suggests that it springs from deeper and more recalcitrant sources than merely intellectual disagreement.

The sense that relativistic philosophy may run at some level counter to the grain of human nature may help to explain the spectacle of writers widely described as relativists or crypto-relativists rushing to denounce their own theories and to disavow publicly "the great bugaboo": the philosophers W. V. O. Quine, T. S. Kuhn, Richard Rorty, Paul Feyerabend, and Bruno Latour all furnish instances of this proceeding. Relativity, for all its alleged successes, has largely become the philosophy that dare not speak its name. "It is important not to fall into the trap of relativism," announces the relativist Feyerabend, for example ("Nature as a Work of Art," 8).[25] The anthropologist Clifford Geertz, having declared that "relativism . . . serves these days largely as a specter to scare us away from certain ways of thinking," and having acknowledged that cultural relativism represents essentially the constitutive principle of his own discipline, dares come no closer to an endorsement of this principle than an exquisitely hedged "anti-anti-relativism" (Geertz, "Distinguished Lecture," 263). Sir Isaiah Berlin constructs in *The Crooked Timber of Humanity* the usual scarecrow parody of "relativism" (here, a sort of radically deterministic "framework relativism" presupposing total incomprehension of other systems of thought than one's own and based on the principle "that there are no objective values") and denounces it, making clear his freedom from this false philosophy; then, under the name "pluralism," he puts forth a program of thought that Popper and most of the other antirelativists quoted above would call a fairly extreme form of relativism. He takes as his target the Platonic ideal of truth, as widely promulgated by eighteenth-century rationalism: notably, the principle "that, as in the sciences, all genuine questions must have one true answer and one only, all the rest being necessarily errors." This ideology based on "faith in universal, objective truth in matters of conduct," he argues, is the recipe for totalitarian tyranny. In its place, he advocates a view of life "as affording a plurality of values, equally genuine, equally ultimate, above all equally objective; incapable, therefore, of being ordered in a timeless hierarchy, or judged in terms of some one absolute standard" (Berlin, *Crooked Timber*, 81, 5, 237, 79). The Philipp Lenards,

Allan Blooms, and John Leos of the world, not to mention the Platos, would find Berlin's disavowal of relativism far from convincing.

Repressions of History

The instance of Berlin brings into focus what may be the central element, after all, of the "spectral" quality of relativism. According to Freud, the panicky sensation of the uncanny is a telltale sign of the recurrence of something in one's history that has been repressed (Freud, "The Uncanny," 241). This formula applies in a striking way to relativity theory, a strain of modern thinking the history of which, as a function of its banishment from the community of legitimate philosophical discourse, has largely been repressed. This process has involved not just scholarly obliviousness to important authors and texts, but an active dismantling of the historical record, particularly in the form of the severing of relativistic science from its cultural environment and from the general intellectual history of which it forms a part.

For Pater, it seemed apparent as early as 1865 that the two movements of moral and scientific relativity were inseparably embroiled with one another. "The moral world is ever in contact with the physical," he observed, "and the relative spirit has invaded moral philosophy from the ground of the inductive sciences. There it has started a new analysis of the relations of . . . good and evil" (*Appreciations*, 67). This model of a simple one-way transmission may seem doubtful in the light of a more sophisticated analysis of socio-intellectual phenomena such as Veblen's or Fleck's, but no attempt to construct a more holistic one would be acceptable to many scholars for whom the separation of science and culture, particularly in the area with which we are concerned, is a point of dogma. Enforcing this separation is part of Arnold Sommerfeld's aim in asserting that special relativity in physics "has, of course, absolutely nothing whatsoever to do with ethical relativism" (Sommerfeld, "Einstein's Seventieth Birthday," 99), and of Denis Dutton's aim in pronouncing that no responsible scholar could seriously entertain "the tired old canard that Relativity Theory can be meaningfully related to cultural relativism" (Dutton, "Knowledge Replacement," 213). The two schools of thought are not just logically discontinuous, these scholars imply, but historically discontinuous as well. The point is made explicit in Paul R. Gross and Norman Levitt's declaration that "the idea that something in the ambient culture" generated special-relativity theory is "wildly implausible" (*Higher Superstition*, 103). The thoughtful antirelativist Stephen Toulmin dismisses any such idea no less categorically. "By a historical coincidence, men finally admitted the thoroughgoing relativity of moral and intellectual standards, at the very time when physicists were developing their own so-called Theory of Relativity," he reports (*Human Understanding*, 1:89). So blunt a disenfranchising of historical analysis in the face of what has every appearance of being a significant historical phenomenon—the simultaneous emergence of relativism in different fields of study—signals the ideological stakes that are felt to be involved in suppressing or at least dismembering the history of the modern relativity movement. The urgency of this project is such as to justify the invoking by a distinguished historian of that category antithetical to all historical understanding, "coincidence."

This chapter thus returns to its point of departure, the absence of a comprehensive history of relativism in modern European thought. To my knowledge, there never has been an attempt to reconstruct the discursive environment of the later nineteenth and early twentieth centuries in such a way as to give the rise of "the new outlook" the salient role and thus to illuminate the intellectual world out of which the discoveries of relativity physics, among other things, arose. The repression from memory of the late-nineteenth-century relativity movement must be understood as a strategic action in which diverse parties have conspired for their own reasons. For relativistic postmodern theorists, historical amnesia lends a character of twentieth-century radicalism to ideas handed down in fact from a consortium of Victorian intellectuals—Sir William Hamilton, Alexander Bain, Feuerbach, Mill, Newman, Spencer, J. B. Stallo, Poincaré, and many others including Darwin—in defiance of the founding myth of modernism as a sweeping rejection of Victorian values.[26] For spokesmen for the orthodox ideology of science, on the other hand, denying historical continuities is crucial to the project of annihilating relativism in all its dangerous manifestations. Gross and Levitt's *Higher Superstition* makes this denial a crux of its polemic against relativistic critics of science, treating the latter-day relativism movement purely as an aberrant outgrowth of post-Vietnam academic politics in America, with *no history to speak of*. The absence of history, of a lineage of reputable authorities, is subtly equated in this work with the alleged perversion of intellectual values by the relativity movement and with its alleged disparagement of scientific inquiry. Harvey Siegel's hostile account of relativism leaps in the same implicitly tendentious way from Plato to T. S. Kuhn, with nothing in between.

This fragmentary picture must be modified in essential ways if we are ever to get beyond the notion of relativity theory in physics as springing miraculously whole, and of course immaculately free of moral or ideological motivations, from Einstein's brain in 1905.

Insistence on the "absolute" and "objective" character of relativity physics has particularly required, as we have noticed, that Einstein's process of discovery be isolated as far as possible from the varied currents of relativistic speculation that pervaded nineteenth-century thought. Einstein is thus portrayed as developing special relativity out of his analysis of a set of mathematical problems of physics, in an intellectual space where cultural influences do not penetrate. According to this model, the new theory arises from his reflections on Maxwell's equations of the electromagnetic field, Fizeau's calculations of the speed of light in a medium, the Michelson-Morley experiment, Mach's critiques of Newton, Lorentz's and Poincaré's work in electrodynamics. Ascribing even this degree of natural genesis to special relativity is hard for modern scholarship to countenance, and it has taken further steps to remove special relativity from the explanatory aegis of intellectual history. It is thus argued that the role played by Machism in the emergence of special relativity is in fact "negligible" (Zahar, *Einstein's Revolution*, 137) and that very likely Einstein was unfamiliar with Michelson-Morley[27] and also with key relativity publications of Lorentz and Poincaré that seem to prefigure his theories in striking fashion. The historian of science Gerald Holton has done invaluable work in situating Einstein's theory in the context of fin-de-siècle physics, but he never raises the possibility that special relativity might be continuous with broader and older trends of speculation that might connect physics with the histories of other fields or of other cultural sectors altogether. Indeed, except for the possible influence of an obscure physicist named August Föppl, Holton finds scant evidence for Einstein's indebtedness to any scientists of his day. He thus portrays the 1905 special-relativity paper as essentially "a Minerva-like creation with no direct preparatory antecedent"; "in a real sense," he goes on to assert, in a proposition that would be fatal to much intellectual history if historians observed it, "genius does not have predecessors" (Holton, "Influences," 69, 75).

As a first step toward freeing ourselves from so dubious a principle, we can observe that several of the key postulates of special relativity had been distinctly formulated a half century before, and that they emerged not from strictly scientific research or from pure mathematics, but from a broad context of Victorian relativistic thinking—one strain of which was in fact, as we shall see in chapter 3, a concerted critique of orthodox scientific rationality. To cite just one example: in *First Principles* (1862), Spencer derives from his fundamental relativity postulate—"thinking being relationing, no thought can ever express more than relations"—the conclusion that the theory of the luminiferous ether, the crux of contemporary physical thought, is logically incoherent and that "nowhere is there that sheltering from inner and outer influences which is implied by absolute rest" (*First Principles*, 63,

44–45, 451). When in the second paragraph of his 1905 special-relativity paper Einstein puts forth his revolutionary postulate "that not only in mechanics, but in electrodynamics as well, the phenomena do not have any properties corresponding to the concept of absolute rest," and thus that "the introduction of a 'light ether' will prove superfluous, inasmuch as . . . no 'space at absolute rest' endowed with special properties will be introduced," he is drawing directly upon a tradition of avant-garde mid-Victorian philosophical discourse (a tradition transmitted to him by a rich medium of intervening works such as Mach's famous *Science of Mechanics* of 1883). Most of the complicated story of this process of transmission seems to have been left untold, out of solicitude, one guesses, for the autonomous majesty of science and of scientific genius. (Does that oft-denigrated Victorian, Herbert Spencer, a writer with no specialized scientific credentials, have a claim to the coauthorship not only of the theory of evolution but of that of special relativity as well? Are the two great theories, the twin pillars of modern scientific awareness, ultimately cognates?) Repressed, too, has been the striking tendency of a couple of generations of eminent relativity physicists not at all to proclaim allegiance to the official regime of "rigid and absolute" objectivity, but to move toward conceptions of science marked with surprisingly radical strains of philosophical relativism, as in the cases of writers like Poincaré, Eddington, Percy Bridgman, and Sir James Jeans. Relativity physics may indeed be the antithesis of relativism, as polemical commentators like Toulmin and Tonnelat insist, but relativity physicists often betray a clear disposition toward this forbidden trend of thought.

The portrayal of philosophical relativism as a nihilistic aberration akin to anarchism and free love has thus necessitated the silencing of the story this book seeks to tell—in fact, abolishing history has always been one of the chief goals of antirelativity polemics. For an idea to have a history is for it to be something other than a perversity, an aberration, or a "specter"; it is to be a normal, organic intellectual growth, one that has constituted itself step by step, gaining adherents, legitimizing itself by building upon prevailing values and bodies of knowledge. In the case of the relativity principle, to trace its Victorian course of development is to show that far from being a freak of philosophy, it must be regarded as a central element of modern and postmodern experience. It is merely propaganda to declare that relativism is incompatible with due reverence for science, with the use of stringent rational argument, with bodies of confidently-held knowledge, with "the possibility of knowing good and bad" (Bloom, *American Mind*, 40), with loyalty to the social order. So, too, any and all declarations that relativity physics has "absolutely nothing whatsoever to do" with contemporaneous moral and cultural relativisms must be seen as dogmatic attempts at prior restraint of inquiry; they are based (as statements of the absence of facts are bound to be) on no conclusive evidence. To abjure relativism and deny its connections with modern developments in a variety of domains, a filled-in intellectual history suggests, is to abjure a chief constituent of our own mental being and to leave us, to this extent, mysteries to ourselves. It is at the same time to leave us hostage to impoverished and antiquated methods of speculative inquiry.

The type of intellectual history at which I aim seeks therefore to break down the invidious segregation from one another of different fields of thought, and, at the same time, to bridge the gaps dividing Victorian, modern, and postmodern. It takes its methodological cues from relativity physics itself, the development of which meant first and foremost, as Einstein repeatedly emphasizes, the obsolescence of mechanical models of explanation based on sequences of causes and effects. Post-mechanical intellectual history will not be to every reader's taste. It has little to say about the provenance of such an entity as the relativity idea or about the "influence" of various texts and thinkers upon one another, as though there existed in the historical record a determinate, narratable causal sequence and a determinate scale of importance quantities which it is the historian's task to reconstruct. The truth in my experience is that theories of origin or causality in intellectual history rarely meet even minimal norms of scientific method, and that they rarely can claim to be anything more than erudite legends. If one takes seriously as the basis of scientific inquiry the fundamental relativistic principle of the interconnectedness of all things, then explaining any event in terms of the causal influence of one or two (or twenty) antecedents seems hopelessly self-deluding, as modern thinkers recognized at an early date. "I do not believe that the course of history can be accounted for on the principles of hydraulics," said one of the most brilliant propagandists of the Victorian relativity movement, J. B. Stallo ("State Creeds," 23). In obedience to this cautionary tenet, I leave questions of origin and influence almost entirely aside in seeking to trace the unfolding of the implications of the relativity idea across a variegated field of Victorian and post-Victorian discourse.

As any critical and relativistic—that is to say, scientific—inquiry into cultural phenomena is bound to do, my inquiry here also takes for granted the necessary primacy of imaginative interpretation in this kind of research. It presumes that texts can only be read as functions of other texts; it abolishes presumptive hierarchies of literary-philosophical worthiness and takes any point of departure within a given cultural field as in principle just as valid as any other; and it freely admits, in fact insists upon, the need for interpretive "anachronisms" (the reading of later texts into earlier) as the alternative to the hoax of mechanical explanation.[28] For the benefit of any reader who will be reassured by the admission, I will concede the methodological dangers of proceeding as I do in this book, constructing intellectual history out of a congeries of remote-seeming texts on the strength of what may seem like freely interpretive reading, unbuttressed by claims of demonstrable direct "influence." I plead in defense of such a method that the greater danger may be that of leaving unrecognized the manifest coherence of a large configuration of writings, of allowing the growth of relativity imagination to seem merely a random or haphazard thing, and of collaborating at least passively in this way in the ideological cleansing of the historical record.

To approach the history of relativity in such terms is unavoidably to embrace a polemical project, for it declines to condemn out of hand a way of thinking so often described by learned and less learned authorities (all of whom bear witness that the intensely moralized Victorian scientific mentality described by Daston and Galison continues to thrive) as pathological, irrationalistic, immoral. Yet this book does seek to help rid current debate of the specter of relativism—not, however, by driving a stake of logic into its heart (it is invulnerable to this treatment, anyway), not by chasing it away with curses (which also has proved inefficacious), but simply by lifting the veil of repression and amnesia, by acknowledging it as a normal and reputable thing, with a history.

CHAPTER ONE

Difference, Unity, Proliferation

Sir Joshua Reynolds: You cannot do better than have recourse to nature herself, who is always at hand. Blake: Nonsense—Every Eye Sees differently As the Eye—Such the Object

—*Poetry and Prose of William Blake*

The child born in 1900 would, then, be born into a new world which would not be a unity but a multiple.

—*The Education of Henry Adams*

No doubt unintentionally, Einstein misled the archbishop in assuring him that relativity in physics was purely a scientific matter unrelated to religion. In fact, the early literature of "the new outlook" reveals unmistakably the emergence of the modern relativity theme from the matrix of nineteenth-century religious debate. Though this theme owed much of the energy of its nineteenth-century development to its monopolization by militant freethinkers hostile to religion, it takes one of its earliest forms, very surprisingly, as nothing other than a piece of theological apparatus. Lenin's association of the scientific relativism he despised with reactionary ecclesiastical interests is to this extent not utterly fantastical.

Above all, Victorian relativity cannot be understood as a simple maxim but only as a complex, shifting, logically problematic nexus of themes—

one that has little in common in any of its versions with that philosophical chimera, "the doctrine that all views are equally good" (Nozick). It can be said to appear in the course of the nineteenth century in two aspects or phases. The first is dictated by the militant principle, "No absolutes!" The second is an attempt to define the paradoxical-seeming properties of what can be called the relativistic field. The two phases take shape in close association with each other, and in many notable texts can hardly be disentangled except as an artifice of exposition. I shall try in this chapter, in any case, to trace this broad twofold pattern of intellectual innovation, deferring to the next chapter, as much as possible, my discussion of the motive of ideological critique in the early history of relativity. Considering that issues of freedom and authority are embroiled with theoretical and scientific speculation at nearly every moment in the literature I will consider, and that writers of the Victorian relativity movement took for granted the inseparable connection between the political and intellectual regimes of their society, the division of topics I propose to follow must also be somewhat artificial and awkward, even if not rigidly adhered to. I will enforce it only up to a point.

Difference, Anglican and Utilitarian

The modern history of relativity theory extends well back into the Enlightenment, and especially into Kantian and Humean philosophy.[1] My goal is not to attempt anything like a full reconstruction of that history, and still less to identify the ultimate origins of relativity. (As I have said, I start with the assumption that abandoning the fantasy of originary moments is a prerequisite of scientific intellectual history, the other main one being the abandonment of the idea that cultural formations such as theories and doctrines possess any fixed or given ideological content which scholarship can ever take for granted.) For the purpose of tracing the destiny of relativity in the nineteenth and early twentieth centuries, however, a significant point of departure is found in the work of Sir William Hamilton (especially his once famous essay of 1829, "Philosophy of the Unconditioned") and of his follower, the clergyman and eventual Dean of St. Paul's, H. L. Mansel (especially *The Limits of Religious Thought* [1858]).[2]

For many subsequent writers, Hamilton and Mansel inaugurated modern relativity by their formulation of what we can call the law of Difference. This principle, recognizably a version of the axiom of Protagoras that "nothing is one thing just by itself," had been definitively stated by Hobbes[3] and then became an article of faith for an increasingly radicalized relativity movement in the nineteenth century. "The condition of intelligence," says

Hamilton, stating the law by means of a quotation from Victor Cousin, "*is difference*; and an act of knowledge is only possible where there exists a *plurality of terms*" (Hamilton, "Philosophy of the Unconditioned," 37). What emerges by implication, or by extrapolation, from this formula is a doctrine of what a writer to be considered later in this chapter calls "twofold relativity": the doctrine that knowledge of any thing is a function both of the relations obtaining between it and the perceiver and of those between it and other things with which it is compared. Hamilton and Mansel, following this chain of inference, argue that human knowledge is confined strictly to the relative properties of phenomena, and is therefore necessarily conditioned by the subject's cognitive faculties and circumstances and *by the act of cognition itself*. To speak of any possible knowledge of "the One, the Absolute, the Unconditioned" or of the absolute properties of things is consequently, they declare, an impossible and wholly nonsensical contradiction (Hamilton, "Philosophy," 14). "Our knowledge . . . can be nothing more than a knowledge of the relative manifestations of an existence," and "the *absolute* is conceived merely by a negation of conceivability," says Hamilton (22, 21). "Consciousness is . . . only possible in the form of a *relation*" between subject and object, echoes Mansel; but since "by the *Absolute* is meant that which exists in and by itself, having no necessary relation to any other Being," therefore "*the Absolute* . . . is a term expressing no object of thought, but only a denial of the relation by which thought is constituted" (Mansel, *Limits*, 96, 75, 97).

For Hamilton and Mansel, to pose the law of Difference as the originary principle of philosophy is unmistakably a revolutionary act; it is to call for a radical cleansing from the philosophical vocabulary of its heavy incrustation of reference to that specious category, "the absolute." Hamilton goes so far in this direction as to declare that "philosophy, if viewed as more than a science of the conditioned [i.e., the relative and contingent], is impossible" ("Philosophy," 22).[4] He credits Kant with having demonstrated the unknowability of things in themselves, but claims that philosophical absolutism—a vainglorious refusal "to limit philosophy to the observation of phenomena, and to the generalization of these phenomena into laws"—remains alive nonetheless in Kant and in later writers including Fichte, Schelling, Hegel, Cousin, and others. Their various philosophical enterprises "are just so many endeavors . . . to fix the absolute as a positive in knowledge," says Hamilton; "but the absolute, like the water in the sieves of the Danaides, has always hitherto run through as a negative into the abyss of nothing" (25–26). The demystifying principle of relativity reveals "the absolute" to be a philosophical will-o'-the-wisp, an empty signifier without any possible referent—none, at least, within the realm of human cognition. Hamilton then derives from this negation of the absolute a further principle: that objects of knowledge do not exist wholly independent of human thought. They have no intrinsic and permanent properties but are inescapably affected by the very act of cognition. "*To think* is *to condition*," says Hamilton famously (21), framing at the very early date of 1829 a prototype of a cardinal principle of twentieth-century natural and social sciences: the idea that "by its intervention science alters and refashions the object of investigation" (Heisenberg, *Physicist's Conception of Nature*, 29). (It is now understood, for example, that a physicist cannot glimpse an electron without deflecting it from its path; nor can an ethnographer observe a tribal community without altering it in the process.) If "The Philosophy of the Unconditioned" marks the first distinct statement of this thesis, as Hamilton's nineteenth-century commentators believed, this essay can be said to represent a fateful moment in the history of modern awareness.

Hamilton's and Mansel's critique of absolutist philosophy under the banner of Protagorean relativity (not that they acknowledge their affiliation with the great criminal of philosophy) is noteworthy for the precociousness which which it introduces trends of discourse and matrices of imagery we are likely to think of as the innovations of later intellectual generations. But it is also noteworthy and surprising for the moral that they derive from this critique. Their goal in subverting the category of the "absolute" is not to call for the development of relativistic philosophy, still less of relativistic science, but to argue that the inescapable defeat of the intellect by the refractions of relativity necessitates and is precisely the condition of religious faith. "Philosophy . . . is impossible" in a world where Difference is sovereign, they conclude, and the place of the defunct discipline is to be claimed by theology. They imply that the craving for religious assurance is necessarily exacerbated by the failure of philosophy, represented as the nightmarish spectacle of positive knowledge turning to water and running off into "the abyss of nothing." "By a wonderful revelation, we are thus, in the very consciousness of our inability to conceive aught above the relative and finite, inspired with a belief in the existence of something unconditioned beyond the sphere of all comprehensible reality," says Hamilton ("Philosophy," 22).

Abrupt and arbitrary as this argumentative turn may seem, Mansel in *The Limits of Religious Thought* carries it to a drastic set of conclusions, highlighting as he does so the treacherous instability that affects the whole subsequent history of relativity. On the one hand, this striking text by the future Dean of St. Paul's goes far to delineate the paradigm of a thoroughly skeptical and radicalized modern sensibility. Basing his theological speculations squarely on Protagorean principles, Mansel develops a sophisticated, uncompromising critique of the philosophy of absolutes; in this respect, he frames the paramount motif of the relativity movement ever afterward and the ground of its inevitable antagonism toward systems of authority. He sounds strong notes of what would later in the century be named "humanism" and pragmatism in philosophy and would serve as an important vehicle for Nietzschean intellectual modernism. No "complete system of scientific Theology" exists ready-made in Christian revelation, he declares, for example; "if it is to exist at all, [it] must be constructed out of it by human interpretation" (*Limits*, 48). In its own context of Anglican orthodoxy, Mansel's constructivist formula represents a transvaluation of values every bit as extreme as—and closely similar to—that proclaimed by Nietzsche several decades later. ("Facts are precisely what is lacking," says Nietzsche in *The Will to Power* [2:12], for example; "all that exists consists of *interpretations*.") Mansel does not fail to see that the project of interpretive construction in theology, necessitated as he claims by the absence of a transcendent signifier even in the holy scriptures, must contend at every moment with "an inextricable dilemma" (*Limits*, 79). The "absolute" things that it is the task of theology to represent lie forever outside the reach of human cognition, limited as our knowledge must always be to the domain of the mundane and the relative. The use of analytical reason in this form of investigation, therefore, is "on every side involved in inextricable confusion and contradiction" (*Limits*, 91); it leads inescapably to self-refutations, aporias, antinomies. "The Absolute cannot be conceived as conscious, neither can it be conceived as unconscious: it cannot be conceived as complex, neither can it be conceived as simple: it cannot be conceived by difference, neither can it be conceived by the absence of difference" (79).

As the rest of this chapter shows, Mansel's analysis of the self-contradictory character of theological absolutes furnishes an exact model for the critique of established doctrines that came to define avant-garde natural and social science several decades later. Similarly, his rules of method for "scientific Theology" prefigure language that was to become widely characteristic of modernistic movements across a range of sciences in later years. "In renouncing all knowledge of the Absolute," he remarkably says, for example, "[theology] renounces at the same time all attempts to construct *a priori* schemes of God's Providence as it ought to be . . . but confines itself to the actual course of that Providence as manifested in the world" (*Limits*, 132). It is precisely this rule, we may infer, that gives theology the right to call itself "scientific." Mansel's formulation of it prefigures Durkheim's disavowal of idealized preconceptions in the scientific study of sociological phenomena such as religion itself (Durkheim,

Elementary Forms, 38); or Lucien Lévy-Bruhl's disavowal of "general and abstract principles" in favor of localized positive science in the study of ethics (Lévy-Bruhl, *Ethics and Moral Science*, 155); or Ernst Mach's and then Einstein's disavowals of a priori notions like "absolute space" in physics in favor of a self-analytic science focused on the process of actual measurement of physical events; or Dewey's insistence that philosophical truths "are made, not a priori," and that a reconstructed philosophy needs to inquire rigorously into the dynamics of the "experimental" process of their making (Dewey, *Essays in Experimental Logic*, 320). Mansel's theology is closely continuous with all this trend of avant-garde secular speculation, of which he deserves to be named as one of the noteworthy pioneers. Most modernistic of all is Mansel's embrace of paradox as a necessary component of the scientific mode of knowledge—as, indeed, the identifying mark of intellectual rigor once the unsparing demystification of "the absolute" has been carried out. "It is our duty . . . to think of God as personal; and it is our duty to believe that He is infinite," he says, for example, adding, "it is true that we cannot reconcile these two representations with each other" (*Limits*, 106).

But *The Limits of Religious Thought* does not, as it might seem, develop these provocative arguments in order to emancipate "human interpretation" from the sway of orthodoxy and superstition or in order to banish the Absolute—that imponderable category that has disappeared "into the abyss of nothing," according to Hamilton—to the limbo of unmeaning metaphysics. Quite the contrary: for Mansel, relativistic critique in the Hamiltonian style leads simply to the discovery of a disabling flaw in all human speculation, and the lesson to be derived from it is the necessity of unquestioning submission to religious authorities in all matters both of belief and of conduct. The transcendent and binding character of Christian doctrine, including a strict principle of biblical infallibility, is taken for granted from the outset. In expounding this credo, Mansel shifts Hamilton's earlier declaration that one is "inspired" to religious faith by the relativistic critique of absolutes into a new register of obligation, coercion, and duty. Comparing the religious sensibility of the two parallel texts makes a vivid exhibit, indeed, of the cultural shift that had occurred in Britain between 1829 and 1858, by which time the influence of Calvinistic Victorian Puritanism had reached its high-water mark. Since reason is incapacitated in the field of religious speculation, we must abandon any attempt at critical evaluation of the dictates of "Faith," says Mansel. In religious matters, "it is a duty . . . to believe in that which we are unable to comprehend" (*Limits*, 110); nor is the smallest deviation from orthodoxy ever to be tolerated, since to "reject one jot or one tittle of the whole doctrine of Christ" is to asperse all Christian religion as fraudulent (215). In setting forth these precepts, Mansel emphasizes that he is concerned less with abstract theological issues than with the enforcement of codes of behavior following not from speculative but from "*regulative* ideas of the Deity"; from these, he declares, we derive, and must all obey, an "absolute standard of right and wrong," "an Absolute Morality" (132, 122). Mansel's Christianity is very much a system of rigid practical control appealing to an expressly authoritarian and absolutist ideology—an instance of precisely the mentality that Mill identifies as the bane of the age in *On Liberty*.

On the theological plane, the leading elements of Mansel's thinking are the primacy of the doctrine of "the Conviction of Sin" (*Limits*, 121) and a strong affirmative emphasis on the world of eternal torment reserved for the faithless after death. To readers who might be troubled by the idea of a supposedly benevolent deity visiting infinite punishment upon finite sins, Mansel asks, with the air of a philosophical inquirer exploring all reasonable hypotheses, "are we then so sure . . . that there can be no sin beyond the grave?" (196). We seem at this moment to tumble down a rabbit hole from a mode of sophisticated "scientific" analysis into another discursive mode altogether. It is no use protesting, however, against the eruption of the most bizarre fantasy in the midst of Mansel's reasoned inquiry, since as far as His function of prescribing moral laws goes, "God's judgments are unsearchable, and His ways past finding out" (202). It would be no less futile to protest against the sinister model offered to worldly governments by theological imagery of a giant agency for the infliction of torture upon any person who might infringe "one jot or one tittle" of official doctrine. Mansel's critique of the Absolute in the name of Protagorean relativity has at all events troped itself here in dizzying fashion into a celebration of profoundly mystified authoritarian absolutism; the law of Difference has become a law of subjection to an "Absolute Morality" that tolerates no difference whatever.

To the extent that Hamilton and Mansel deserve credit or blame for injecting the virus of Protagorean relativity into the bloodstream of modern discourse, their efforts can only be said to have fallen victim to the law of unintended consequences. The relativity principle they played notable roles in formulating soon became the instrument of a broad nineteenth-century insurgency against Christian religion, which for various critics (who could only dimly imagine the rise of totalitarian regimes on the twentieth-century model) stood as the defining instance of ideological tyranny and of the inherent tendency of absolutist systems to reach out from the realm of speculation, theory, and official mythologies into the exercise of practical power over human lives. But I am anticipating the story to be told in the next chapter.

If the critique of the Absolute in the name of the "relative" was thus initiated from the unexpected quarter of Anglican theology (and later, as we shall see, by Newman also), the same project was simultaneously being carried on, from the other side of what might seem to be an impassable ideological divide, by writers associated with Benthamite utilitarianism, a.k.a. "philosophical radicalism." The penchant for indeterminacy and paradox that marks modern relativity discourse makes it seem remote from utilitarianism, given the latter's overriding stress on practical reform and its commitment to what may seem almost parodically mechanical logic, tolerant of no ambiguity. (Bentham's "felicity calculus," which claimed to compute the quantitative value of pains and pleasures so as to make possible a scientific moral system, would be one index of this aspect of utilitarianism.) But a full history of relativity would need to emphasize strongly the role played in its propagation by the Benthamite tradition. One can think of dogmatic theology and philosophical radicalism as engaging at this early stage of relativity thinking in a contest for ownership of an ideological instrument in which each sensed the potential to exercise transforming influence over the modern mind.

Underlying the moral scandal of the utilitarian "happiness principle," which rated happiness higher than ascetic self-denial or nobility or other recognized virtues, was a philosophical scandal. Bentham and his disciples regarded happiness as an absolute, unqualified principle of value, for which they were sharply taken to task on relativistic grounds by Edward Westermarck in *Ethical Relativity* (see esp. 4–20). Westermarck failed to note that utilitarian ethics in fact rested on (and disseminated widely in the intellectual culture of the time) the deeply relativistic proposition that acts possess no inherent moral properties, such as those prescribed in codes of religious morality, but only relative properties that could be computed only indirectly, with reference to consequences.[5] Not only did philosophical radicalism disavow theological and other a priori absolutes in order to anchor values wholly in the field of human experience (though how to define "happiness" in terms suitable for rigorous philosophy proved an insurmountable difficulty); it clearly implied principles of potentially uncontrollable relativistic indeterminacy. T. H. Green, arguing against the utilitarians for the need for "a universal law" of ethics based on the principle of "an unconditional good" and on "the absolute imperative to seek the absolutely desirable" (Green, *Prolegomena*, 204, 207), makes just this point later in the century.[6] One and the same act can be computed good or bad, depending on circumstances; happiness means different things to different people. Moreover, the need to assess consequences implies in turn the need to follow them out in an endlessly expanding, ultimately incalculable reverberation that would seem to baffle any determination of a definite happiness quotient for a given act. Without factoring all the circumstances of an act into one's calculation, "it can never be known whether it is beneficial, or indifferent, or mischievous," says Bentham accordingly; but "of the circumstances of all kinds which . . . attach upon an event," he says without acknowledging the gravity of this problem for his would-be science, "it is only a very small number that can be discovered by the utmost exertion of the human faculties" (*Introduction*, 76, 79). The happiness doctrine, meant to function as an Archimedean leverage point for dislodging ancient systems of privilege and injustice and for reforming legal, governmental, and administrative institutions, proved for such reasons to be an unstable grounding, if too rigorously examined.

Bentham and his followers could only strive to keep the mischievous genie of relativism in its bottle by continuing to promote the happiness principle as what Feuerbach called the theologians' God: the "absolute measure" (*Essence*, 254). The deeply relativistic implications of utilitarian theory were made unmistakable in 1871, however, with the publication of W. S. Jevons's *Theory of Political Economy*, the founding text of the marginalist revolution in economics and one of the key works of Victorian intellectual modernism. Jevons proclaims himself a disciple of Bentham, eulogizing and quoting him repeatedly, and describing his own theoretical innovations as a rigorous attempt "to treat Economy as a Calculus of Pleasure and Pain" (*Political Economy*, vi). Jevons finds that this project means reconceiving economic science along radically relativistic lines. Focusing on the defining concept of utilitarian thinking, he takes as his central premise, for example, the rule that "there is no such thing as absolute utility, utility being purely a relation between a thing and a person" (xxxiii). Such is the fundamental implication of all Benthamite philosophy.

The theme of relativistic anti-absolutism was powerfully elaborated, too, in the associationist psychological theory that formed a crucial adjunct of utilitarian philosophy. This theory, associated with Locke, Hartley, and James Mill, carried distinct political valences and was invoked by John Stuart Mill and others with unmistakable militancy: its denial of the category of innate a priori ideas was directly continuous with the great Enlightenment critique of systems of society that justified themselves with reference to supposedly natural or divinely ordained and permanent principles of domination.[7] The world of things and ideas was not marked out in fixed, given, permanent ways, according to associationist psychology; rather, it was gradually *constructed*, inscribed on an initially blank slate, by each person out of the particulars of his or her experience. If this was so, then the world was open to being constructed both cognitively and politically according to other patterns than those currently established. Such was the would-be emancipatory message subtending classical associationist theory—and the reason why this theory, as expounded and given an extended range of application in the mid-nineteenth century by another author much cited by Jevons, Alexander Bain, in such works as *The Senses and the Intellect* (1855), *The Emotions and the Will* (1859), and *Logic* (1870), embraced and elaborated Hamiltonian relativity, repudiating "absolute" knowledge in the name of a newly radicalized principle of Difference. These are cardinal texts in the intellectual history I seek to trace.

Bain's attack on "the doctrine of innate ideas" (*Logic*, 2:391) rests, he says, on "a great mental law involved in the fundamental property of Discrimination . . . namely, the law of RELATIVITY," which tells us that "every mental experience is necessarily *twofold*" in the sense that "everything known to us is known in connexion with . . . the opposite or negation of itself" (*The Senses*, 8, 565). Lending support *avant la lettre* to Jacques Derrida's concept of "the trace" as it functions in the signification process Derrida calls *différance*, Bain asserts that "when we pass from one member of a contrast to the other . . . both members must be present. . . . The other member is still before us in a manner," absent yet crucially present (*The Senses*, 565). He makes clear that he considers this "principle of Universal Relativity, by which all objects of knowledge are two-sided, or go in couples" (*Logic*, 1:255), to form the basis of a radical and all-encompassing new scheme of thought, the significance of which he underlines typographically (in common with other writers of the day) by his frequent practice of capitalizing the revolutionary word RELATIVITY. Aristotle and others, he remarks, give the philosophical category of "relation" only a subordinate role, but in reality, "if Relation is recognized at all, it is fundamental and independent; everything comes under it, it comes under nothing" (*Logic*, 1:255). Thus "there is no escape from the principle of universal relativity," he says. "There is no possibility of mentioning a thing, so as to be intelligible, without implicating some other thing or things" (*Logic*, 1:61). In other words, to quote Bain's unacknowledged master, nothing is one thing just by itself.

What is clearest about the "great mental law" enunciated by Bain is that it introduces a profoundly equivocal character into the theory of knowledge, and sets relativity ever after under the sign of paradox, destabilizing the very rules of logic by which relativity thinking (which professes only to follow them with unprecedented rigor) accounts for itself. For Bain, the idea of a thing is always a conundrum: it always contains a self-contradiction, "the opposite or negation of itself." Every determination reached under the aegis of Difference thus is incurably dislocated and precarious, of two minds. According to "the Law of Relativity," says Bain, "whenever we have an object in our view, we have by implication the opposite," and "either every name must have a double meaning, or else for every meaning there must be two names," and "the simplest affirmation has two sides" (*The Senses*, 566; *Logic*, 1:54).[8] To imagine a mode of awareness in which such indeterminacy is not a disabling paradox after all, as Mansel claimed, but simply the necessary condition of genuine scientific knowledge is to move decisively into the orbit of the modern. From this point forward, epistemological insecurity, open-endedness, and existential self-fashioning will form defining characteristics of experience and "the moralization of objectivity" will play a reactionary role in European intellectual politics. Embracing this acutely destabilized intellectual condition is the cost (so runs Bain's unstated argument) of emancipation from the fallacy of innate ideas, from a priori absolutes and from the reign of traditional authorities. Yet Bain's conception of "the principle of Universal Relativity" seemed so extreme and subversive even to William James, the professed ardent anti-absolutist and professed disciple of J. S. Mill, that he decried it in 1890. If Bain's interpretation of the law of Difference were true, James declares, "the whole edifice of our knowledge would collapse" (*Principles of Psychology*, 1:12).[9]

Bain is careful to specify that logical propositions fall under the law of relativity fully as much as do names and "notions" (*Logic*, 1:79). Considered as an element of cognition, therefore, any proposition becomes vertiginously equivalent to its contrary. Bain illustrates this thesis with a striking astronomical metaphor. "According to Relativity," he says, "the simplest affirmation has two sides. . . . Thus the daily rotation of the starry sphere is either a real motion of the stars, the earth being at rest, or an apparent motion caused by the earth's rotation" (*Logic*, 2:391): the two propositions are logically equivalent, "according to Relativity." Bain's explicit identification of the new intellectual regime with the denial of an absolute "rest frame" in physics echoes the similar argument made by Spencer in *First Principles*, and reminds us again that Einstein's dramatic announcement of this principle in 1905 came not as "a Minerva-like creation" but as a confirmation of well-rehearsed themes of Victorian speculation.

Bain writes, however, not as a physicist but as the champion of a new movement of scientific psychology, and means in his advocacy of relativity theory to insist in particular on the impossibility of disentangling objective reality from the principles of human psychology (which distinctly includes, for Bain, the physiology of the nervous system). To claim as Bain does that *all properties of things arise from relations of difference* is at least implicitly to attribute an essential role to the human mind and to human interpretation in determining the very fabric of the natural world, since difference can scarcely be figured as a given property of things; it is an interpretive artifact, a construction of the intellect. Bain declares accordingly, with a strong sense of the historical mandate of the principles of analysis he is advocating, that all attempts to resolve the enigmas of science and philosophy "according to the spirit of modern thought, according to the modern laws of explanation" must begin by defining all things rigorously in relation to "our own sensibility or consciousness" (Bain, *Practical Essays*, 65). So, for instance, "the great leading notions called Time and Space are known to us only under the conditions of our own sensibility." Time is not a given of the natural world, nor is it, as Kant claimed, a necessary a priori category of thought. Rather, by the complex action of our senses, emotions, and thoughts, "we create time," says Bain. "And our notion of Time in general is exactly what these sensibilities make it, only enlarged by our constructive power" (*Practical Essays*, 59, 60).[10]

Bain's work marks a watershed in the emergence of the relativity movement: *The Senses and the Intellect*, in particular, takes a place in intellectual history as one of the earliest texts (along with Spencer's *Principles of Psychology* of the same year, 1855) to bestow the special name of "relativity" upon "the spirit of modern thought": from about this historical moment, the word becomes endowed with the dangerous mystique that it has never yet lost. I note for future reference that this is four years before the appearance of *The Origin of Species*, and that Relativity in a sense names itself as the great transformative idea of the age before Evolution does so. Distilling as it does the emancipationist and constructivist impulse of Benthamite philosophy, Bain's writing, in which the correlation of advanced scientific research and radical politics is easy to discern, also epitomizes the dimension of militancy and freethought that is the birthright of the relativity movement. Yet the juxtaposition with Bain's contemporary Mansel, his political opposite, shows that the ideological valences of relativity were open to conflicting interpretations in the mid-nineteenth century. The evil reputation of Protagoras notwithstanding, relativism had not yet acquired the diabolical and "repugnant" aspect in the eyes of intellectual and other kinds of orthodoxy that has since engulfed it, and still could be exploited by spokesmen for authority and revealed religion as well as by utilitarians and agnostic freethinkers.

To some extent, in Bain's major writings, ideological militancy seems only latent or implicit, as though relativity had not yet fully grasped its own destiny of potentially radical cultural critique. In one essay, indeed, Bain analyzes pointedly the "rigorous fixity of institutions," the "hatred of change," the "unreasoning acquiescence in a state of things once established" that allegedly marks human societies; "incipient reformers," he remarks, "are at once immolated *pour encourager les autres*" (*Practical Essays*, 261–62). But this overtly polemical note, with its evident biting reference to Bain's own society and, in particular, to the predicament of Benthamite reformers in Victorian Britain, is absent elsewhere in his philosophical and scientific writing. It is quite otherwise in the writing of his friend and professional collaborator J. S. Mill, who bears witness to the significance of the relativity theme for utilitarian thinking in the two-volume treatise he devoted in 1865 to the study of Hamilton's philosophy, and in which reference to Mansel also figures importantly.

Mill explicates at length the Hamiltonian doctrine of "The Relativity of Human Knowledge" and, like Bain, gives unconditional assent to that version of it based on the principle "that we only know anything, by knowing it as distinguished from something else; that all consciousness is of difference; that two objects are the smallest number required to constitute consciousness" (Mill, *Sir William Hamilton*, 1:13, 14). The law of Difference, he declares, is "a fundamental truth" (1:65). And he follows Hamilton by endorsing uncompromisingly, as almost his cardinal principle of philosophic reasoning, the thesis of the complete unknowability of "things in themselves"—that is, those hypothetical aspects of things lying beyond the reach of relativity. But he does not treat Hamilton as the honored pioneer of this line of speculation, "which impresses a character on the whole mode of philosophical thinking of whoever receives it" (1:25) and which potentially could revolutionize modern inquiry. Rather, he paints him as a pusillanimous thinker who shied away from the consequences of his own fundamental discoveries. Hamilton, Mill concludes, only upholds the relativity principle "in the scantiest meaning of which the words are susceptible," applying his critique only to the category of things imagined as wholly infinite or absolute, whereas, Mill declares, the relativity principle applies in fact to all objects of knowledge whatsoever (1:76). In this Mill follows Bain's insistence that "the principle of Relativity applies to everything that we are capable of knowing" (*The Senses*, 9). Mill's analysis of Hamilton thus rehearses what would become a recurring moment in subsequent philosophical discourse: that in which a militant proponent of deconstructive critique identifies in the work of an earlier master a failure of nerve or a lapsing back to essentialist categories.[11]

The motives impelling Mill to embrace the relativity principle so strongly must have been exceptionally compelling, since it represented a mode of thinking and of discourse in many ways insuperably alien to him. Hence his critique of the famous Hamiltonian formula "*to think* is *to condition*" (Hamilton, "Philosophy," 21; quoted in Mill, *Sir William Hamilton*, 1:54). What can be the meaning, if any, Mill asks, of this "emphatic and oracular" declaration (*Sir William Hamilton*, 1:69)? He decides that Hamilton is not referring to a process of causality and yet that he means more than the innocuous claim that thinking an object "is to give it a correlative" (1:72). He concludes that Hamilton seems "to reckon as a condition of a thing, anything necessarily implied by it; and uses the word Conditioned almost interchangeably with Relative"; probably, says Mill, he has in mind a Kantian reference to inherent categories of understanding (1:70, 72–73). Whether or no, Hamilton's idea of the "conditioning" property of thought is afflicted, Mill declares, with "the most dangerous kind of ambiguity" (1:71). In pronouncing this verdict, he can be seen to recoil, with a twinge of intellectual panic, from full acceptance of the principle of relativity, just as he blames Hamilton for doing. Things cannot be "conditioned" by thought, Mill implies, because in order to be rationally intelligible they must be seen as having a determinate existence unto themselves, sheltered from the dizzying inflections of Difference. The alarming alternative is a state in which the inherent properties of things and the operations of human intellect become caught up in a thoroughgoing reciprocity, each dependent on the other—and in which scientific objectivity might collapse as a result. Mill's dilemma in confronting Hamilton's "dangerous" principle, which seems to him to carry relativity thinking to an outlandish conclusion, is later reenacted by his disciple William James—another philosopher of relativity and a vociferous opponent of "absolutism"—when he makes a point of declaring in 1890 that "our inveterate love of relating and comparing things does not alter the intrinsic qualities or nature of the things compared, or undo their absolute givenness" (*Principles of Psychology*, 1:12).

Other nineteenth-century writers less frightened of the destabilizing of logic by relativity, far from condemning Hamilton's "oracular" principle, seize on it as the very prerequisite of lucid philosophical thought. For J. B. Stallo (about whom more in chapter 2), curing "the metaphysical malady" means recognizing that a "judgment" necessarily "transforms both concepts which it brings into relation," so that, for example, the judgment "hydrogen is a metal" gives "new meanings" to both terms, "hydrogen" and "metal" (Stallo, *Concepts and Theories*, 6, 158). In the light of this principle, concepts are not static elements anchored in "absolute givenness," but are immersed in the relativity, thus in the "perpetual flux," of cognition

(Stallo, *Concepts and Theories*, 159), just as Derrida insists in his critique of Saussure long afterward that not just signifiers but signifieds, too, are functions of *différance* and thus inseparable from "the play and the coming into being of signs" (*Of Grammatology*, 72, 48). So, too, the pragmatic philosopher F. C. S. Schiller insists in 1907 (in an essay probably written in 1905, the year of the promulgation of special relativity) that "*mere knowing always alters reality*" (*Studies*, 439). Though the logical structure of the universe would seem to be thrown into turmoil by so extreme a principle of relativity, twentieth-century science has of course caused this recurring Victorian intuition to come true. Hamilton's, Stallo's, and Schiller's principle receives its almost inconceivable confirmation in the double-slit experiment that lies at the heart of quantum mechanics. If we identify a specific photon passing through the slits, the experiment demonstrates, it behaves like a particle; if we do not identify it, it behaves, unthinkably enough, like a wave.[12] If at such a point we apply the logic of "Universal Relativity" to intellectual history itself, we will be driven ultimately to explain these results by saying that the physical phenomena described by quantum mechanics are themselves fundamentally "conditioned" from afar by Sir William Hamilton's controversial text and by the intellectual tradition arising from it. Had no such tradition existed, we may confidently assume, photons would never have thought to begin playing these fantastic tricks.

Mill may have recoiled from the proposition "to think is to condition," but his ambivalent endorsement of Hamiltonian relativity, like Bain's development of a scientific psychology emancipated from the fallacy of innate ideas, bases itself none the less on a conviction of the power of mind to transform reality. He insists on the primacy of the law of Difference as a strategy for expunging metaphysical pseudoentities as far as possible from the vocabularies of philosophy and science; to do so, he imagines, is the precondition for fostering an emancipated mental condition aware of its capacity to construct its own world and thus favorable to the radical reform of contemporary social life that is the goal of all his intellectual labor. His antagonism toward Hamilton and Mansel is clearly proportionate to his sense of their betrayal of their own potentially liberating discoveries out of servitude to what he took to be the very worst of the agencies of mental subjugation, Christian religion. Hamilton, says Mill, "has established, more thoroughly perhaps than he intended, the futility of all speculation respecting those meaningless abstractions 'The Infinite' and 'The Absolute,' notions contradictory in themselves, and to which no corresponding realities do or can exist" (*Sir William Hamilton*, 1:74). That which is severed from its constitutive relation to the human mind and thus rendered unknowable *does not exist:* in drawing back from such a conclusion, says Mill, Hamilton not only contradicts himself but reveals his complicity with mighty forces of reaction and oppression. Mill cites in this connection Hamilton's approving quotation of a principle of St. Austin's which to one of Mill's political temper could only seem like the essence of diabolical disinformation: "We know, what rests upon *reason;* but believe, what rests upon *authority*" (1:77). The Absolute is for Mill just a euphemism, finally, for that dire category, "authority," and relativity is its philosophical antithesis. Mansel's argument that God is unknowable and that His dictates are therefore beyond critique expresses this equation unambiguously; it is denounced by Mill as just the latest of many reasons given by apologists of religion "why we may assert any absurdities and any moral monstrosities concerning God, and miscall them Goodness and Wisdom." Mansel's is "simply," Mill concludes, "the most morally pernicious doctrine now current" (1:115).[13]

This critique of Mill's is thus continuous with his analysis in *On Liberty* (1859) of the "atmosphere of mental slavery" pervading contemporary British life and with his protest in the same text against such abuses as the persecution and even imprisonment of modern-day English citizens for religious heresies (*On Liberty,* 25–31, 33). For Mill, the great revival of religion in mid-Victorian England signifies only "the revival of bigotry" and of an attitude of "submission to all authorities found established" (*On Liberty,* 30, 49). Reading *On Liberty* in conjunction with *An Examination of Sir William Hamilton's Philosophy,* we can readily draw the conclusion, though it is never stated explicitly by Mill, that his commitment to the relativity principle in spite of the intellectual discomfort it arouses in him expresses his sense of its liberating potential for a period in which the bigotry of absolutism seemed to him to be in the ascendant.

Unity

From the moment that the relativity theme appears with a flourish of rhetorical trumpets in Victorian writing, proclaiming itself the key to a new philosophical outlook, it thus identifies itself with the dismantling of theological and crypto-theological "absolutes." The nineteenth-century revival of Protagorean theory was by no means a purely subversive one, however. Its richest developments lie in the attempts of a coalition of authors to define the properties of the relativistic field—that is, of the intellectual domain that is brought into being by the proclamation of "Universal Relativity." How do the constituent entities of such a field, the differential signifiers that act as proxies for real things, behave? It turns out that they obey two great principles that seem contradictory but figure in the logic and poetics and in the ethical allegory of relativity as one and the same: Unity and Proliferation. Relativity means that all things are rigorously bound up together in a single indivisible world; it means also that this world is not one after all, but uncontrollably multiple.

This twofold argument was formulated decisively by Herbert Spencer, the importance of whose role as an innovator of modernist intellectual and scientific culture has never since his own day (when it was clear to all) been sufficiently recognized.[14] He was the author of several momentous discoveries, which prove to be so intimately interrelated in his work as to seem finally like the different phases of a single ramifying idea: "the law of Evolution"; the principle of Relativity; and the danger posed to contemporary European civilization, and to Britain in particular, by the spread of (in his formulation, the regression to) what he termed "militant" social forms. Spencer made none of these discoveries *ex nihilo*, unaided by earlier and by contemporary writers. His own insistence on the doctrine of the imperceptibly slow development of all natural formations (in which category he emphatically includes social and intellectual formations), not to mention his demystification of the Carlylean and Millean idea of the heroic Great Man as an autonomous force in history (Spencer, *Study of Sociology* [hereafter *SS*], 30–36; see Mill, *On Liberty*, 66–67), would prohibit us from thinking that independent invention could ever occur in the field of theoretical inquiry, where discovery is bound to be a gradual and a mysteriously collective process. But Spencer gave to these several emergent ideas such compelling statement, and illustrated them with such amplitude of materials, that he has only been excluded from the pantheon of the intellectual heroes of modernism (centered on such names as Darwin, Durkheim, Frazer, Freud, Einstein) by dint of something akin to a conspiracy against his memory.

In the standard history of ideas, relativity figures as the distinctive scientific concept of the twentieth century and evolution (as to which, Spencer's claim to priority of authorship over Darwin, or at least to shared authorship with him and Alfred Russel Wallace, seems fairly clear)[15] as that of the nineteenth; but to become aware of Spencer's leading role in the construction of relativity theory is to suspect that these two currents of thought should be seen historically as coeval, as moments or figurings of the "general movement of intellectual reconstruction" that marked the period (Dewey, *Influence of Darwin*, iv)—if indeed relativity did not actually precede evolution in the sequence of discovery. "The most fundamental things in our minds," said Veblen's teacher Richard T. Ely, describing the avant-garde intellectual world of the late 1880s, "were, on the one hand, the idea of evolution, and on the other hand, the idea of relativity" (Ely, *Ground under Our Feet,* 154).[16] That the two were inseparable manifestations of a single transformative concept was evident to nineteenth-century observers. The point is clearly made in Pater's 1865 comments on "the 'relative' spirit" as the keynote of intellectual modernism, and it is stated no less clearly by St. George Mivart in 1898. The theory of evolution by natural selection, declared this vehement anti-Darwinist in the specific context of an attack on Spencer, is part and parcel of that skeptical modern philosophy "which affirms the essential *relativity of knowledge*"—that is, the doctrine that knowledge can never be "absolutely and perfectly true" (*Groundwork of Science,* 279). (Mivart is an instance of the turn-of-the-century "recrudescence of absolutistic philosophies" to which Dewey drew attention [*Influence of Darwin,* 18] and which manifests itself also in that cult of "the moralization of objectivity" recounted by Daston and Galison—no simple emanation of the "Victorian" ethos, we can now see, but a reactionary response to trends of speculation that had grown to alarming proportions since the middle of the century.) I will return to Darwin shortly. But the clearest indication of the coeval bond between evolutionism and relativity is undoubtedly found in Spencer himself, who stands historically so near the head of both streams of thought. The two movements are not exactly coincident, however. Often they function in concert, performing similar cultural work, but their alliance is never free of potential contradictions.

No other text of comparably early date may more clearly delineate this complicated twofold movement than Spencer's once famous treatise *First Principles* [hereafter *FP*] (1862), which later Victorian freethinkers venerated as "the agnostic Bible" (B. Lightman, *Origins,* 82). The first half of Spencer's anti-Scripture is devoted to an exposition of relativity theory and to a sustained critique of the idea of the "Absolute"; the second half expounds his version of the theory of evolution and certain of its corollaries, such as the doctrines of the instability of the homogeneous and of the multiplication of effects. Implicit in the structure of this text, therefore, is not only the inseparability of its two main themes, but also the intimation (if true, a very noteworthy point for intellectual history) that relativity is both logically and genealogically prior to evolutionism.

The first of the two main motifs that prevail in Spencerian relativity is *the radical interconnectedness of all things.*[17] It is an idea with Romantic antecedents. When a version of it is set forth by Wordsworth in the 1805 *Prelude* (published in 1850) as the foundation of a specifically modern intellectual and spiritual revolution, it takes the form of a mystical epiphany, laden with Christian overtones, in which the most disparate and contradictory-seeming natural phenomena are seen to be

> all like workings of one mind, the features
> Of the same face, blossoms upon one tree;
> Characters of the great Apocalypse,
> The types and symbols of Eternity,
> Of first, and last, and midst, and without end.
>
> (VI.636–40)

The relativity movement in nineteenth-century science and philosophy can be seen in one of its aspects as the vehicle of a great cultural project aimed at giving definite content to Wordsworth's intuition. Ernst Mach, one of the prime movers of the relativity movement in physics, frames its basic postulate at the early date of 1865: "all phenomena are so connected that any one of them can be represented as a function of any other," he announces (quoted in Carus, *Principle of Relativity*, 38). The world seen through the lens of relativity is above all an inextricable unity: *all* phenomena are connected. This is the fundamental postulate also of the mathematician and philosopher Henri Poincaré, to whose work we shall return at the end of this chapter. "We need not . . . ask if Nature is one, but how she is one," he declares; "the true and only aim [of scientific inquiry] is unity" (*Science and Hypothesis*, 145, 177). In articulating this principle, these authors and others are echoing one of Spencer's most insistent themes.

By the deepest necessity of the natural world, all things are defined wholly by their relationships with other things: this is the doctrine Spencer teaches in one text after another. He states as an axiom of evolutionary biology, for instance, that "organisms in their totality are mutually dependent, and in that sense integrated" (*FP*, 253); an independent organism would be a contradiction in terms. He insists similarly on the "consensus" and "mutual dependence of parts" that exists, he says, among the phenomena of any society or in the internal organization of any organism, in either of which "the changes in the parts are mutually determined, and the changed actions of the parts are mutually dependent" (Spencer, *Principles of Sociology* [hereafter *PS*], 1:431, 580, 439). This great principle of being, or great mythological construct, in which hierarchies of dominance and subordination are replaced by an all-encompassing reciprocity so radical that different organisms must be seen as "integrated" and "mutually dependent," almost as functions or aspects of one another, applies not just to the flora and fauna of a particular local habitat, and not even just to organisms and societies, but to every animate and inanimate item in the universe. "From the centre of our system down to a microbe, each aggregate is subject to incident forces derived from other aggregates large or small: even the Sun being affected by the planets" (*FP*, 451).

From this primary postulate of the interconnectedness of things flows Spencerian relativity—though the reasoning process at this juncture can just as well be expressed in reversed form, from relativity to the doctrine of the integration of nature, since each of its constituent elements derives from the other. Once all of nature is conceived as an inextricable system of interrelations, in any case, then, as Spencer says, "every thought expresses a relation," "thinking is relationing," and "every thought involves a whole system of thoughts and ceases to exist if severed from its various correlatives" (*FP*, 94, 106).

In expounding this conception of relativity in *The Principles of Psychology* [hereafter *PP*] (1855), Spencer focuses on "the Relativity of Feelings," or the lack of any determinate equivalence between objective realities and perceptions of them, and, further, on "the Relativity of Relations Between Feelings." As he says, "the whole question of the relativity of relations among feelings is reducible to the question of the relativity of the relation of Difference" (*PP*, 1:224). The mighty law of Difference, which some writers treat as essentially coextensive with relativity, turns out, says Spencer, to be subject to it after all. Two different observers, or the same observer at different moments, may construct differences differently, in other words. Like Bain, Spencer thus insists that relativity applies exhaustively to everything we are capable of knowing and that all knowledge of things in their independent existence is foreclosed to us as a result. "What we are conscious of as properties of matter . . . are but subjective affections produced by objective agencies that are unknown and unknowable" (*PP*, 1:206). External objects, as far as we can know them, are purely and simply relativistic mental phenomena. "It becomes impossible to suppose any identity between [an] objective connexion [between things] and some one of the multitudinous subjective relations answering to it" (*PP*, 1:215): no objective check on our knowledge by direct reference to things in the outer world is possible, that is, and no theory can give an accurate account of nature. "What we know as a relation is qualitatively and quantitatively determined by our own nature, and does not resemble any order or nexus beyond consciousness" (*PP*, 1:225).

From his reversible main assumption (thinking is relationing; all things are inseparably integrated), Spencer derives another consequence that will run throughout subsequent relativity literature: an insistence on the vital and dynamic character of reality, on the active intervention of each thing upon the being of each other thing—which is to say, on the impossibility of static, isolated "things in themselves." "Matter cannot be conceived except as manifesting forces of attraction and repulsion," says Spencer, for example (*FP*, 182): on relativistic principles, a particle of matter not affecting and being affected by others would have no being.[18] He expands significantly on this point in a passage in the last volume of *The Principles of Sociology*, hinting at the mode of interchange by which material objects in effect bring one another into being. He observes here that in modern physics, what earlier was conceived as inert "brute matter" now is seen as animated by "internal energy" releasing "an infinity of vibrations" passing in all directions through every point in space. "Dead matter," he says, turns out to be "everywhere alive" and, in the new scientific perspective, "gives rather a spiritualistic than a materialistic aspect to the Universe" (*PS* 3:172, 173). In such formulations, early relativity theory collapses the distinction between "thinking is relationing" and "Nature is one" by moving toward a conception of a universe in which all things are found to be enmeshed in a network not of physical collisions and gravitational forces, as the mechanical model would have it, but of *intelligent communication* with all other things, particularly with human beings. "What we term the properties of an object, are the powers it exerts of producing sensations in our consciousness," says Mill, implying such a formula (*Sir William Hamilton*, 15). To put it in another form: nothing occurs in nature but the transmission of information, and we know natural things only by virtue of their propensity for sending messages. No notion is woven more deeply into the poetics of Victorian relativity than this one. Every sensation, says the mathematician W. K. Clifford, consists principally in "a message that comes to us somehow . . . from the external world" (*Lectures*, 1:308); the evolutionist, socialist, and statistician Karl Pearson expresses the same idea in figuring the relation between the mind and reality as a telephone exchange through which "messages . . . come flowing in from that 'outside world'" (*Grammar of Science*, 61–62). The natural scientist in this model is cast in some sense as the medium of a spiritualist séance, the interceptor and interpreter of mysterious communications passing from one object to another and from the world of objects to us. When the anthropologist E. B. Tylor analyzed in 1871 "the primitive fancy that inert things are alive and conscious" (*Primitive Culture*, 1:287), he was obliquely making reference, consciously or not, to advanced speculative trends of natural science in his own day and hinting at how radically these trends had begun to diverge from scientific common sense.

In *First Principles*, Spencer, taking his lead from Hamilton's and Mansel's statements of the unknowability of the "Absolute," and quoting from them liberally (ensuring them thereby their rightful if remote places in the intellectual lineage arching from Protagoras to Poincaré, Einstein, Heisenberg, and Planck), focuses a trenchant relativistic critique upon a wide range of the supposedly self-evident constituent elements of modern physics: time and space, motion, force, matter, the "luminiferous ether." In all these categories, Spencer declares, as Mansel had with reference to the categories of theology, rigorous analysis leads to insurmountable contradictions; the Unknowable confronts us at every turn of scientific inquiry—necessarily so, since "Ultimate Scientific Ideas . . . are all representative of realities that cannot be comprehended" (*FP*, 48). The distinctive outcome of the doctrine of "the Relativity of All Knowledge," he thus argues, is that "of necessity . . . explanation must eventually bring us down to the inexplicable" (*FP*, 54). "While it is impossible to form any idea of Force in itself," for example, "it is equally impossible to comprehend its mode of exercise" (*FP*, 45). In the case of the ether, we are compelled by the logic of classical mechanics to hypothesize its existence to account for the interrelation of objects and the exercise of force across empty space, yet the concept is logically incoherent, Spencer argues, requiring as it does the very supposition of force acting at a distance that it was meant to avoid; the ether, the basis of nineteenth-century physics, is "an insoluble enigma" and a scientific monstrosity (*FP*, 182–83, 44–45, 49). Spencer's theoretical deconstruction of the ether in 1862—apparently taken note of by no latter-day historian of relativity—received experimental confirmation in the 1880s in Albert Michelson's and Edward Morley's epochal and totally unanticipated failures to detect "ether drag" experimentally by means of a special apparatus (the interferometer) designed for the purpose, and it forecasts almost verbatim, as I have previously noted, Einstein's repudiation of the ether in 1905. It is for making such arguments against the fundamental building blocks of classical physics that Spencer is singled out as late as 1913 by the antirelativist Paul Carus for having plunged modern science "into the bottomless abyss of the incomprehensibility of existence" by his evil work of preaching "a universal and intrinsic relativity" (Carus, *Principle of Relativity*, 2–4).

To his discredit in the eyes of various critics (among them Mill, Bain, Clifford, T. H. Huxley, Arthur James Balfour, F. C. S. Schiller, and Dewey), Spencer at this point in his argument follows Hamilton and Mansel one step further, asserting that a theory based on relativistic and evolutionary principles inescapably implies the reality of objective things in themselves, even though these things must remain forever and utterly unknowable. "The existence of a non-relative is unavoidably asserted in every chain of reasoning by which relativity is proved" (*PS*, 1:209). Radical as Spencer's relativity was, he recoiled from the prospect of a system of knowledge wholly untethered from an objective natural reality. For Schiller, this move on Spencer's part signifies a "final surrender" to the prejudices of the old metaphysics of fixed absolutes (Schiller, *Studies*, 225). Dewey gives the same disparaging analysis: that "the habit of seeking justification for ideal values in the remote and transcendent" causes even Spencer, the sworn enemy of these habits of thought, to hypothesize "an unknowable absolute" equivalent to God (Dewey, *Influence of Darwin*, 16). Spencer in his final pact with the Absolute does seem to betray the rigor of his own account and to abandon relativity itself. He finds no way to maintain his relativity principle, in any case, but by recourse to an already obsolescent and intellectually incoherent model of mechanical causality in which an effect logically presupposes a cause, even though the nature of this cause and its mode of action may lie forever outside the scope of human thought. The disavowal of causality itself that would soon be proposed by relativity theorists like Clifford, Stallo, and Poincaré (following the lead of Hume, of course) was beyond the ken of Herbert Spencer, who was pioneering new intellectual territory without having burned all his bridges to the old.[19] Yet in view of his proclamation of a sweeping, exhaustive relativity in the field of human cognition, where "the relativity of the relation of Difference" is the sole governing principle and where "multitudinous subjective relations" form the sole content of knowledge, his act of obeisance to unknowable objective realities amounts to little more than a placatory formula. And one must underline clearly that Spencer's endorsement of the Absolute was not designed to bring much comfort to the party of metaphysical and religious orthodoxy.

The declared purpose of *First Principles*, as Spencer says after delivering a long preamble calling for the extension of individual freedoms "against the invasions of State power" (*FP*, 6) and thus strikingly underlining the political vectoring of much Victorian relativity theory, is to effect a reconciliation between science and religion. Contrary to the indictments of the moral depravities of religion, especially Christian religion, that run through his writings, and utterly unlike the evolutionary ethnography of *The Principles of Sociology* (1876–96), where he argues scandalously that "ghost-propitiation is the origin of all religions" (*PS*, 3:7), Spencer in *First Principles* adopts a conciliatory stance toward the powerful forces of mid-Victorian piety. The near-universality and consistency of religious ideas are evidence, he says here, that they express "some essential verity" (*FP*, 9). He identifies this verity with the very proposition that forms the chief corollary of his relativity doctrine: namely, the unknowability of anything lying outside the sphere of relativity. Religion and its basic conception of a nonrelative First Cause, which "must be Absolute" (*FP*, 28), is in effect that about which human beings are incapable of knowing or saying anything at all.

So Spencer's concession to such writers as Mansel and Hamilton to the effect that "in the very denial of our power to learn *what* the Absolute is, there lies hidden the assumption *that* it is" (*FP*, 65) is extended at the cost of abolishing all theology and all specific religious belief whatever, for he offers no support to their position that "duty" requires the acceptance of established doctrines. As far as religious orthodoxy goes, Spencer carries out a transvaluation of all values that piety, could it make sense of the sophisticated rhetorical machinery set in motion in *First Principles*, could only regard as a nihilism quite as devastating to theology as Feuerbach's own. Religion, says Spencer, has always been "more or less irreligious," since "it has all along professed to have some knowledge of that which transcends knowledge, and has so contradicted its own teachings" (*FP*, 74, 75). "In the devoutest faith as we commonly see it, there lies hidden a core of scepticism"—namely, the conviction that the Incomprehensible is to some degree comprehensible after all (*FP*, 75). Religious belief turns out in this antic formulation to be inherently skeptical and irreligious, and authentic religion is in effect vested in relativistic modern science: such is Spencer's professed reconciliation of Religion and Science.

Thus the great Victorian agnostic harshly criticizes Mansel (whom he has cited respectfully hitherto) for the lack of "reverence" he displays in upholding such superstitious notions as that of an intelligent, personal, morally perfect deity, toward whom human beings may entertain sympathy. True reverence, Spencer announces, will insist on nothing less than the elimination of all religious doctrine. "Volumes might be written," he piously declares, "upon the impiety of the pious" (*FP*, 82). And in spite of his professions of respect for religion even in its theologically conservative manifestations, he makes more than plain the depth of his revulsion from Victorian religious orthodoxy. "It is hard to listen calmly to the futile arguments used in support of irrational doctrines, and . . . to bear the display of that pride of ignorance which so far exceeds the pride of science." A rational person, he declares, will hardly be able "to conceal his repugnance to a creed which tacitly ascribes to The Unknowable a love of adulation such as would be despised in a human being," which teaches "that divine vengeance is eternal," or which condemns actions of unselfish sympathy as "intrinsically sinful" when detached from religious faith (*FP*, 89–90). Far from succumbing to conventional religious apologetics, Spencer traces in *First Principles* exactly the polemical line used by Mill in his own fierce critique of Mansel (Mill, *Sir William Hamilton*, 1:113 ff.) or by Clifford in his equally fierce critique of the 1875 theistico-scientific volume by P. G. Tait and Balfour Stewart, *The Unseen Universe* (Clifford, *Lectures*, 1:268–300). His embrace of the Absolute is a suavely diabolical strategy of demolition of orthodox belief rather than a "surrender" to it. As Feuerbach said in reference to just the sort of position taken by Spencer, to deny positive predicates to God "is simply a subtle, disguised atheism" (Feuerbach, *Essence*, 15).[20]

In the course of the second half of the nineteenth century and in the early years of the twentieth, the Spencerian theme of the radical interrelatedness of things presided over advanced thinking in every area and produced a continuing intellectual upheaval.

It presides in especially distinct form over what proves to be one of the cardinal texts of modern relativity and one of the important influences on Spencer himself: Darwin's *The Origin of Species* [hereafter OS] (1859). To trace this pattern in Darwin's book also serves to underline again the intimate historical bond linking the principles of evolution and relativity. For Darwin, a biological species seen in evolutionary context is not so much an autonomous thing defined by an independent identity as it is a phenomenon expressing in every detail of its physical form and its instinctual codes the whole relational structure of the universe. Given any two life forms simultaneously in existence, an interpreter ought in principle to be able to extrapolate by Darwinian reasoning the entire system of nature. To my knowledge, Darwin never frames his argument in such hyperbolic terms, but he does repeat with almost obsessive insistency his proposition that "the form of each [organism] depends on an infinitude of complex relations," or, as he memorably puts it at another point, that "the structure of every organic being is related, in the most essential yet often hidden manner, to that of all . . . other organic beings" (OS, 131, 77). According to the interpretive paradigm of Darwinian biology, to describe an organism in evolutionary terms is thus to describe it wholly and exhaustively in its relations to every other element in its ecosystem, which ultimately includes nothing less than "the whole economy of nature" (64). "I am tempted to give one more instance showing how plants and animals, remote in the scale of nature, are bound together by a web of complex relations," says Darwin at another typical point in *Origin of Species* (OS, 74), stressing the concept that pervades his work, with revolutionary effect.[21]

The revolutionary character of this concept is highlighted in 1892 by Karl Pearson. "Many things pass in the universe for absolutely independent, which a finer power of analysis or observation would demonstrate to be associated," he says. In the traditional scientific conception, "all things are . . . either independent or causative," but, he says, "the newer, and I think truer, view of the universe is that all existences are associated in a higher or lower degree." "The intellectual attitude which sees between all existences diverse degrees of association . . . conceptualises the universe under a new category" (Pearson, *Grammar of Science*, 163, 166). Pearson applies this dictum to physics as Darwin had to evolutionary biology and as Spencer had to sociology and other fields, and as Saussure then applies it to linguistics in the first decade of the twentieth century. Saussure's fundamental tenet might have come verbatim from Spencer: "language is a system of interdependent terms in which the value of each term results solely from the simultaneous presence of the others," a "complex equilibrium of terms that mutually condition each other" and that therefore teaches scientists above all never to "isolate the term from its system" (*Course in General Linguistics*, 113, 122). Just as Darwinian biology posits that "the form of each [organism] depends on an infinitude of complex relations" (*OS*, 131), Saussure postulates in linguistics that the "content" of a signifier "is really fixed only by the concurrence of everything that exists outside it" and therefore that "*to explain a word is to relate it to other words*" (*Course in General Linguistics*, 115, 189). Modern linguistics manifestly crystallizes from a rich nineteenth-century imaginative tradition. Saussure thus makes clear, as Spencer did in such an intensive fashion, the necessarily relativistic character of any system of thought based on the principle of all-encompassing interconnectedness. "In language," he declares, in his own rendering of the Protagorean slogan of the nineteenth-century relativity movement, "there are only differences *without positive terms*"; the scientific study of such a phenomenon is, he says, necessarily "governed by the . . . paradoxical principle" that "values remain entirely relative" (120, 115, 113).

The advent of the relativistic paradigm of interdependency in the nineteenth century can only seem a riddle from the point of view of a rationalistic history of ideas. The new model was dictated by no empirical findings and seemed, as Saussure notes, to render paradoxical any analysis carried out under its auspices. Its irresistible appeal for scientific thinkers will only make sense if we take into account its powerful convergence with themes and susceptibilities of Victorian imagination at large. As I have suggested, the conception of fields or systems in which "all existences are associated in a higher or lower degree" must have been evocative for nineteenth-century thinkers of an ideal world of harmonious relations in which conflict and isolation had been abolished and reciprocity reigned. Transposed into another idiom, this idea forms a governing theme in the work of contemporary writers such as Dickens, Elizabeth Gaskell, and Spencer's close friend George Eliot, where it is given the powerfully moralized interpretation that can never be more than covert in its scientific formulations.

In *The Mill on the Floss* (1860), for example, Eliot's narrator, expressly invoking the advanced scientific thought of her day, calls for "a large vision of relations" as the sovereign corrective for what she diagnoses as the characteristic moral pathology of the Victorian middle classes—for the vindictive moral attitude that mistakenly imagines, for one thing, that the effects of punishment can possibly be confined to a single intended victim (*Mill on the Floss*, 215) or that lowly-seeming forms of human life can complacently be thought of as dissociated from ourselves:

> For does not science tell us that its highest striving is after the ascertainment of a unity which shall bind the smallest things with the greatest? In natural science, I have understood, there is nothing petty to the mind that has a large vision of relations, and to which every single object suggests a vast sum of conditions. It is surely the same with the observation of human life. (239)

Much the same code of values, based on the commandment that "the true and only aim is unity" (the principle that Freud allegorizes as Eros, calling it a fundamental human drive), is at least subliminally present in Spencer's declarations that "organisms in their totality are mutually dependent, and in that sense integrated" and, as we shall see in chapter 2, in J. B. Stallo's impassioned preaching of the lesson of "the relativity and consequent mutual dependence of natural phenomena" (*Concepts and Theories*, 146). For these and other writers of the early relativity tradition, it must have seemed clear that their announcement of this principle marked not only a movement into a new theoretical episteme but an assertion of a fundamental framework of values. As Poincaré revealingly puts it, a science based on the study of "isolated facts" would "no longer be to us of any value, since it would not satisfy our need of order and harmony" (*Science and Hypothesis*, 130). A science of interdependency signified, by contrast with that of isolated facts, the intellectual cognate of the nineteenth-century dream of constructing a socialist society in which unity and cooperation would replace "warlike aggression of the middle classes against the workers" (Engels, *Condition of the Working Class*, 320), or of bringing all the heterogeneous populations of the world into the unified, supposedly harmonious fold of Christian religion. Colin Falck's attack on Saussurean linguistics and on relativistic theory in general as irreligious is to this extent seriously misleading. Even Darwin suggests at times that his conception of nature as a "complex equilibrium" in which each element is exquisitely and reciprocally fitted to all the others represents a kind of utopia. For example, he says at one point that "natural selection tends to make each organic being as perfect as, or slightly more perfect than, the other inhabitants of the same country with which it comes into competition" (*OS*, 212)—a formula in which evolution seems for a moment to yield a wonderful and egalitarian improvement of life for all beings (though the concept of "perfection" in an evolving relativistic system is of course nonsensical).

For many nineteenth-century theorists, the imaginative appeal of the relativistic principle of unity and coherent interrelatedness was powerful enough to overcome the grave difficulty that the conceptual field created by this principle is inherently paradoxical and unstable. The paradox in one of its forms is the same one stamped on all formulations of the law of Difference, which says that the concept or perception of any thing incorporates a plurality of things, *including the antagonistic opposite* of the one in question.[22] How then can harmonious cooperation and potentially violent conflict be stably distinguished from each other, as any morally coherent theory is bound to require? I need hardly say that this dilemma is played out in its most vivid form in *The Origin of Species*. In Darwin's model of the natural world, the system of providential unity, in which "the structure of every organic being is related . . . to that of all . . . other organic beings" (*OS*, 77), is of course not at all one of harmonious coexistence and "consensus," but rather one of fierce genocidal combat waged by each organism against "all the other organic beings, with which it comes into competition for food or residence, or from which it has to escape, or on which it preys" (*OS*, 77). The unity that defines the conceptual and at some level the moral ideal of relativistic imagination thus proves to be in Darwin a form of that "absolutely unbridled freedom of competition" that Engels identified as the core of the bourgeois system of values (Engels, *Condition of the Working Class*, 313): seamless unity is identical here to a spectacle of conflict among a multitude of antagonistic separate individuals.

A version of the same dizzying paradox inhabits that later expression of the relativistic imagination, Saussurean linguistics. On the one hand, the rhetorical stress in the *Course in General Linguistics* falls constantly on the theme of the interdependency of signifiers and the harmonious equilibrium that reigns in the kingdom of any language as the precondition of the establishment of intelligible meanings (Saussure, *Course in General Linguistics*, 88). On the other hand, Saussure's military metaphor of chess and the economic metaphor of "value," and the soft-pedaled recognition that a language has an evolutionary or diachronic dimension as well as a relativistic synchronic one, all express a differently valenced idea of language "as a system based entirely on the opposition of its concrete units" (Saussure, 107). "Within the same language, all words used to express related ideas limit each other reciprocally" and "mutually condition each other"—they compete for niches in the system, in other words, just as biological organisms do in Darwin's model (Saussure, 116, 122). Such is existence in any realm to which the law of Difference (which states that nothing can exist in isolation from its opposite) applies. One may figure this state of affairs as one ideologically inflected allegory or another, but the inescapable point in all versions is that the powerful impulse of unity in relativistic thinking is at the same time necessarily a powerful assertion of multiplicity. From the point of view of multiplicity, the insistent, rhetorically resonant language of unity invoked by theorists is likely to seem either like an ideological ruse (as in Darwin's praise of the "perfection" attained by evolutionary competitors as a result of natural selection, clearly meant to disguise the morally repugnant aspects of such a system) or like the arbitrary imposition of uniformity upon a scene in which diverse constituent entities strive to proclaim and to defend their individuality. This is the "paradoxical principle" (Saussure, 115) to which all relativistic thinking is keyed.[23]

On another plane, perhaps even more pressingly, the paradigm of all-encompassing interrelatedness brings with it a set of purely intellectual liabilities and even a threat of extreme epistemological disorder. It professes to emancipate scientific inquiry from the influence of illusory metaphysical absolutes, but emancipation, for many in the nineteenth century, felt like a frightening crisis that threatened the annihilation of knowledge. As we have seen, and as Spencer particularly emphasizes, relativity places all rigorous scientific investigation under the sign of "the Unknowable" and brings it at last "down to the inexplicable." It teaches the doctrine that instability, "ceaseless flux" rather than permanence, is the mode of reality (Spencer, *SS*, 118), that "change is universal and unceasing" (*FP*, 451). In Spencerism, there is no room whatever for participation in the philosophical conspiracy to idealize current social arrangements as natural, necessary, and permanent;[24] but neither is there room for the conviction that the truth on any question has been or ever could be found. To the extent that the concept of "truth" maintains its usefulness in the domain of "the new outlook," it is only on condition of giving up its claim to timeless permanency. "Truth must submit to the common doom and undergo change," declares the shrewd spokesman of relativity theory Samuel Butler in the mid-1880s (*Collected Essays*, 1:139). At this philosophical juncture, relativistic indeterminacy (according to which "science alters and refashions the object of investigation") and evolutionary flux coincide.

Most dismayingly of all, the interdependency thesis, in which "every single object suggests a vast sum of conditions," entails an irreparable breakdown of the explanatory mechanism upon which modern scientific understanding had always depended: the concept of causality. Particularly with regard to the analytical interpretation of complex phenomena, talk of determinate causes is essentially meaningless once one assumes that, as Alfred North Whitehead put it, "in a certain sense, everything is everywhere at all times" (*Science and the Modern World*, 128). "It is not in acted, as it is in written History," writes Carlyle as early as 1830. "Actual events are nowise so simply related to each other as parent and offspring are; every single event is the offspring not of one, but of all other events, prior or contemporaneous, and will in its turn combine with all others to give birth to new." Hence, says Carlyle, the dubious intellectual status of the "'chains,' or chainlets, of 'causes and effects'" that historians labor to construct in defiance of the truth that in human history "each atom is 'chained' and complected with all!" ("On History," 40). This principle becomes a leitmotif of theoretical literature in the later Victorian years. In 1870, for example, W. K. Clifford states it as a basic doctrine of physics: "The hypothesis of continuity [of space and time] . . . involves such an interdependence of the facts of the universe as forbids us to speak of one fact or set of facts as the *cause* of another fact or set of facts" (Clifford, *Lectures*, 1:122). Poincaré offers much the same principle in a different formulation in 1902, arguing that "all things are interdependent" to a degree that even throws into doubt the sacrosanct "simplicity criterion" (that is, the rule that simplicity is a criterion of scientific truth): "for if all things are interdependent, the relations in which so many different objects intervene can no longer be simple" (Poincaré, *Science and Hypothesis*, 145, 177, 146). Relativistic interdependency throws the very possibility of scientific reason into doubt, as we will see further in chapter 3.

Proliferation

If relativity is taken as identical to a theory of the radical unity of things, and thus gives rise to the conceptual problems we have noted, it also turns out to be identical to a contrary-seeming theory: the theory that things uncontrollably multiply themselves within the relativistic field.

As we have observed, it is the idea that relativity promiscuously sanctions differing perspectives and points of view, even differing "truths," that has always represented its most scandalous aspect in the eyes of a tradition based on the dogma that truth is and must be absolutely single. The intellectual terror haunting the Western tradition, says Foucault (if I may quote this passage once again), is "the cancerous and dangerous proliferation of significations" (Foucault, "What is an Author?" 118). If truth should be robbed of its absolute monopoly and exclusivity, cry out a host of writers, then nihilism reigns in the form of "[the] doctrine that all views are equally good," nothing can be believed, and the world is plunged into the "abyss" of meaninglessness. The model underlying this apocalyptic scenario coincides closely with the civic rationale invoked by post-ancien-régime authoritarian governments for the suppression of free thought and speech: some regime or other proclaims every day that public order depends on the principle that there is one truth, and only one. Indeed, so close is the rapport between the philosophical and the political in this respect that one may wonder whether they may be identical from the start, and whether all the polemics of the past century and a half on behalf of the cause of "absolute truth"—for example, the theistic philosopher T. H. Green's insistence in 1883 on the principle of "a single and unalterable order of relations" in the universe, and then on its supposed corollary, the "unconditionally binding" character of "every one . . . of the duties which the law of the state or the law of opinion recognises" (Green, *Prolegomena*, 17, 207)—amount to more than the translation of political imperatives into the abstract language of philosophy. Is it possible in the real world of discourse to maintain a theory of "absolute truth" that is unallied to motives of coercive authority and to the intended suppression of specific people and ideas? Does the bitter hostility to relativistic theory spring finally from anything but outrage at the idea that this theory denies one the right to impose one's thinking by compulsion upon others? Such questions are injected with increasing distinctness into the Victorian intellectual arena.

The alarming concept of relativistic multiplicity was articulated by Spencer as early as 1855. If between entities in the outer world and their symbolic representations in our sense impressions "there is no such correlation as that which the physicist calls equivalence," says Spencer, stating what for him is an unquestionable axiom, then the closest we can come to objective truth will be "the multitudinous subjective relations answering to it," each as justifiable as any other (Spencer, *PP*, 1:194, 215). Spencer emphasizes this conclusion, giving it a considerably more radical interpretation than does such a latter-day relativist as Saussure. Saussure follows the Spencerian paradigm in declaring the lack of determinate relations between signifiers and signifieds and in proclaiming that "in language there are only differences *without positive terms*" (*Course in General Linguistics*, 120). He evidently assumes, however, that the differences themselves are determinate and that linguistic "value" as determined by a given set of correlates is unambiguous. Spencer, with his stress on "the relativity of the relation of Difference" (*PP*, 1:224), argues that the main property of the symbolic forms by which we know reality is that they are indeterminate and subject to constant variation. "Besides concluding that in no two species are the subjective effects produced by given objective actions absolutely alike, qualitatively and quantitatively; we may conclude that they are absolutely alike in no two individuals of the same species," he says. Indeed, "even in the same individual the quantity, if not quality, of the feeling excited by an external agent constant in kind and degree, varies according to the constitutional state"; "very possibly the ratio [between object and perception] is never twice the same" (*PP*, 1:197–98). This head-spinning theme is developed by a line of subsequent writers in the nineteenth century and made into an axiom of progressive scientific thinking—thinking governed, that is, not by any dogmatic idea of "truth" but by "a wide vision of relations." Arthur James Balfour argues in 1879, for example, in the course of a skeptical analysis of scientific arguments based on the (for Balfour, always illegitimate) concept of causality, that "if two or more explanations of the universe are barely possible, they must, for anything we can say to the contrary, be equally probable; which is as much as to say, that one version of history need not be less likely than another, merely because it seems in comparison unnatural and extravagant" (Balfour, *Defence*, 283).

This radical implication, like much else in the work of Spencer and his Victorian disciples, is elaborated in a text that I have cited once or twice already and that should be better recognized than it is as one of the definitive works of modernist literature: Henri Poincaré's *Science and Hypothesis* [hereafter *SH*] (1902; first English translation 1905). The young Einstein, still a patent clerk in Bern, read this book with intense interest along with his friends in the Akademie Olympia: "*Science and Hypothesis* engrossed us and held us spellbound for weeks," reported one of the members of this circle, Maurice Solovine (Solovine, "Introduction," 9). No less spellbound, Henry Adams read Poincaré's book feeling "green with horror"; it seemed to him to plunge modern thought into a vortex of indeterminacy and "endless displacement" from which it might never escape (Adams, *Education*, 455). The center of the dangerous allure of Poincaré's book is precisely its argument that scientific representations of reality are bound by the logic of relativity to be, as Spencer said, "multitudinous," and that one can deny this principle only at the cost of sacrificing the scientific spirit itself.

Poincaré states in his preface the central theme of the book to come. "The aim of science is not things themselves, as dogmatists in their simplicity imagine," he says here, "but the relations between things; outside those relations there is no reality knowable" (*SH*, xxiv). In announcing this credo with such polemical fanfare, Poincaré is staking out for himself a radically modern line of thought, but we know by now that it is one with a long Victorian pedigree that reaches back, in particular, to Spencer. Poincaré occupies indeed an equivocal historical position that is curiously similar to Spencer's. Just as Spencer could and did claim coauthorship of the theory of evolution, so Poincaré could (though he did not) claim coauthorship of the special theory of relativity—and has in fact been credited with it more than once.[25] Already in this text of 1902, he not only rejects as untenable the concepts of absolute space and absolute time (as had Spencer and others), but plainly states the crucial scientific and philosophical doctrine of Einsteinian special relativity, the inconceivability of simultaneity in regard to events occurring at a distance from one another (*SH*, 90)—a doctrine that sounds the death knell of ordinary scientific rationality. For the moment, however, it is Poincaré's relation to Spencer that concerns us. To a degree sure to surprise those who have not questioned the legend of the latter's irrelevance, Poincaré duplicates in *Science and Hypothesis* many of the particulars of his argument, from the proclaiming of "the law of relativity" (*SH*, 76) as the supreme principle of science (obviously as challenging a thesis in 1902 as it had been forty years before) to tracing the manifold consequences of this law by means of deconstructive critiques of common-sense ideas of space, time, force, matter, and that metaphysical emanation "the luminiferous ether," the main physical property of which was not to transmit electromagnetic waves but to absorb the intellectual energies of many eminent nineteenth-century physicists for several decades.

Like Spencer's *First Principles*, Poincaré's daring book, in its systematic indictment of "the absolute" in all its forms, has the potential appearance of a purely skeptical tract, and must therefore perform the ritual disavowal of nihilism that is obligatory for all exponents of relativity. Poincaré performs it by declaring his conviction that "absolute scepticism is not admissible" (*SH*, 188). The English edition of *Science and Hypothesis* both reinforces and subtly contradicts this message by means of a very nervous introduction by the physicist Sir Joseph Larmor urging the reader not to take as a denigration of "the wonderful fabric of human knowledge" Poincaré's argument that the laws of science, rather than embodying infallible truths, are merely "convention[s] of language" (*SH*, 90) that we institute heuristically for the sake of "convenience" (Larmor, xx). *Science and Hypothesis* is faithful, too, to the paradoxical deep structure of *First Principles* in evoking two contradictory-seeming tropes as the joint foundations of emancipated scientific thinking: the trope of interconnection and unity on the one hand, and the trope of proliferation on the other. Rather than carefully separating the two, as logic would dictate, both writers exemplify the paradigm shift of which they are among the founders by seeming to take these opposite and antagonistic themes as equivalent.

Poincaré thus declares that it follows directly from "*the law of relativity*" that interpretations of physical experiments based on non-Euclidean geometries will always be possible in any particular case, and that such interpretations will be "quite as legitimate" as Euclidean ones (*SH*, 76, 90). "What, then," asks Poincaré, "are we to think of the question: Is Euclidean geometry true? It has no meaning. We might as well ask if the metric system is true. . . . One geometry cannot be more true than another; it can only be more convenient" (*SH*, 50).[26] With this formulation, the line of nineteenth-century avant-garde speculation returns in a sense to its Benthamite origins by proclaiming "truth" to be a name for a sheer utilitarian value, presumably justifying itself at last by a magnification of human happiness, "convenience." What holds true for alternative geometries in physics holds true more broadly, says Poincaré, for all explanatory theories of natural phenomena, particularly in electrodynamics and in ether theory. In such areas of investigation, "it may be shown that we can explain everything in an unlimited number of ways"; for "if . . . a phenomenon allows of a complete mechanical explanation, it allows of an unlimited number of others, which will equally take into account all the particulars revealed by experiment" (*SH*, 168, 222). Poincaré dwells on this heretical principle with the same insistence that marked Spencer's treatment of it. Thus he cites various modern theories of dispersion that have been proposed in the aftermath of Helmholtz's and that obtain the same equations as his. "I venture to say," proclaims Poincaré, "that these theories are all simultaneously true" (*SH*, 162). In this text, the cancerous proliferation of signification breaks out in its most extreme form: this it surely was that caused Henry Adams to go "green with horror" in reading it.

In *Science and Hypothesis*, this theme is invoked by Poincaré, as it had been by Balfour, almost entirely with destructive intent. To imagine that this great scientist intends in thus using it to undermine science would be preposterous. He does, however, passionately argue that an acceptance of "what we shall call, for the sake of abbreviation, *the law of relativity*" (*SH*, 76), means a banishment of the category of "truth" from the philosophy of science, if by "truth" one means any formula with an exclusive claim to validity. Scientific truths "are conventions and definitions in disguise" (*SH*, 138)—"it is *by definition* that force is equal to the product of the mass and the acceleration," for example (*SH*, 104)—and such constructions are by their nature potentially "unlimited" in number. The many controversialists who propagate the mythology in which a "wicked race of deceivers" has abducted Truth and "hewed her lovely form into a thousand pieces" would need to include Poincaré in the wicked race.[27]

But beyond his searching critique of absolutist notions of truth, Poincaré urges a fundamental break with the mode of science based on the longing for explanation itself. He cannot at this pre–quantum-mechanics stage imagine a form of scientific discourse that would be able, in fact, wholly to divest itself of the fetish of causal explanation. "That day has not yet come," he says in the peroration of *Science and Hypothesis;* "man does not so easily resign himself to remaining forever ignorant of the causes of things" (*SH*, 223). Lacking a perfected example of a scientific model emancipated from mechanistic metaphysics, Poincaré holds up nonetheless the image of Clerk Maxwell, on whose magnificent achievements all later work in optics and electrodynamics depends. He portrays Maxwell as a protopostmodernist who, departing radically from the traditional ideal of absolute explanatory mastery, used scientific research as the vehicle for the creation of a new intellectual sensibility. "[Maxwell] does not try to erect a unique, definitive, and well-arranged building," says Poincaré; "he seems to raise rather a large number of provisional and independent constructions, between which communication is difficult and sometimes impossible" (*SH*, 215). The ideal of flawless coherence is replaced here, it seems, by an ideal of imperfect, precarious, improvisational logic; unique and definitive truth is replaced by a congeries of provisional constructions. Obviously, Poincaré's own great dictum that "the true and only aim [of science] is unity" proves equivocal and problematic when once fully immersed in the destructive element of relativity.

In reading Poincaré's philosophical treatments of relativity theory in physics, one cannot help being struck by the close intellectual relation linking him not only to Spencer but to the new structural linguistics, and thus by extension to all the theoretical developments that have since flowed from the latter. The writings of Poincaré and of Saussure read often like translations of one master text into two specialized but closely parallel languages, an effect eloquently exemplifying T. S. Kuhn's thesis that scientific innovation is fundamentally a process not of experimental discovery but of constructing data in accord with powerful new themes of thought—a "purely fictitious" process yielding "free inventions of the human intellect," as Einstein himself often insisted (Einstein, *World*, 34, 33). Thus Poincaré, explaining physical relativity, stresses that a rigorously rational science begins with the rule that no experiment can possibly reveal the "absolute . . . orientation" of the bodies in a material system (in other words, their relation to space itself), but only "the state of the different parts of the universe, and their mutual distances," since "experiments only teach us the relations of bodies to one another" (*SH*, 76, 77, 79). Saussure declares, for his part, that "just as the game of chess is entirely in the combination of the different chesspieces, language is characterized as a system based entirely on the opposition of its concrete units" (*Course in General Linguistics*, 107). Expressing the supremacy of "the law of relativity" in other terms, modern thought imagines the physical world as "form nearly devoid of matter," says Poincaré (*SH*, 224; see also 20); Saussure asserts similarly that in his new science, language "*is a form and not a substance*" (*Course in General Linguistics*, 122). Such are the parallel outcomes in different fields of modernistic inquiry once they have come to be "governed by the . . . paradoxical principle" that "values remain entirely relative" (Saussure). In embracing this principle, fully articulated already, as we have seen, in a speculative literature extending back fifty years and more, linguistics and physics (not to mention other fields) become homologous expressions of one all-encompassing scientific outlook.

The official interpretation of the special theory of relativity draws back sharply from Poincaré's radicalism, as we have seen. It insists on "the basic premise that there exists an invisible and invariant truth which cannot be perceived in any single earthly manifestation alone but only through a comparative series of manifestations," and that behind "the welter of [relativistic] particulars" lies "some abstract, categorical rule or regularity" (Ermarth, *Realism*, 31, 37; see Cassirer, *Einstein's Theory*, 381, 393, 447). The pungent ideological flavor of this recuperative rhetoric would have been anathema to Paul Feyerabend, who insisted that "proliferation of theories is beneficial for science" and is "an essential part of a humanitarian outlook" (Feyerabend, *Against Method*, 24, 38). He insisted by the same token that we incorporate into philosophy an inquiry into the aggressive motives that, so he alleged, underlie the repudiation of theories that lend credence to the terrible heresy of coexisting and equally valid systems of truth.

One nineteenth-century writer who carried out just this inquiry is Ludwig Feuerbach. His commentary on monotheism and polytheism in *The Essence of Christianity* [hereafter *EC*] (1841) will remind us once again of the persistent infusion of religious implications into the history of modern relativity thinking. For Feuerbach, the institution of the monotheistic principle in ancient Judaism was at heart purely a strategy of political aggression by the nation of Israel, a declaration of supremacy over other peoples. "Jehovah . . . troubles himself about nothing but Israel, . . . is nothing but the personified selfishness of the Israelitish people, to the exclusion of all other nations"; he embodies, in a word, "absolute intolerance, the secret essence of monotheism" (*EC*, 113–14).[28] The cultural invention of monotheism is for Feuerbach the very expression of intellectual narrowness and dogmatism, of "political vindictiveness" and of the claim to be able to impose one's will upon others as a result of self-identification with "absolute divine power" (*EC*, 120, 121). "Hence science, like art, arises only out of polytheism," says Feuerbach, "for polytheism is the frank, open, unenvying sense of all that is beautiful and good without distinction, the sense of the world, of the universe" (*EC*, 114). Polytheism is the religious affirmation of Feuerbach's own fundamental commitment to what he calls "plurality and difference" (*EC*, 85). Neither Spencer nor Poincaré nor Feyerabend invokes the illustrious example of Feuerbach, and none of them describes the concept of the proliferation of theoretical possibilities in science as a revived polytheism, but plainly the four writers constitute a continuous lineage of thought and argument.

The twin principles of unity and of "unlimited" multiplicity are thus linked in an indissoluble nexus at the center of nineteenth-century relativity theory. This nexus can be described as paradox, incoherence, or self-contradiction; alternatively, to invoke the religious themes that have run throughout the present chapter, one could describe it as a theological mystery like that of the Trinity, or as a scientific one like the particle/wave duality of matter. The various descriptions are, no doubt, "all simultaneously true." We need not try to adjudicate among them to see the importance of this twofold pattern for the growth of Victorian relativity thinking.

CHAPTER TWO

Relativity and Authority

> In the republic of the sciences sedition and even anarchy are beneficial in the long run to the greatest happiness of the greatest number.
>
> —W. S. Jevons

If, as I argue in this chapter, the nineteenth-century relativity movement served from the first as the vehicle of a radical moral and political ideology and owed to this fact much of the transformative leverage it exercised upon modern inquiry, it may seem surprising that Victorian expositions of relativity consistently adopt the languages of almost ostentatiously esoteric philosophical and scientific investigation. To some extent, this method of discourse served as a strategic means of making arguments potentially chargeable with promoting "sedition and even anarchy" from the relative safety of innocuous-seeming intellectual locales. (In this age, as I will note again, one could go to prison in Britain, and many did, for "blasphemous" attacks on Christian doctrine, for example.) At the same time, Victorian relativity constituted too inclusive a movement to be restricted to its political significance, even construing "political" very broadly. I have stressed that relativity was regarded by its advocates as a principle that, like a religious revelation, transforms consciousness, that "impresses a character on the whole mode of philosophical thinking of whoever receives it" (Mill), that "conceptualises the universe under a new category" (Pearson), that "from the beginning was associated with a new philosophical outlook" (Eddington). The goal of nineteenth-century relativity was not just that of establishing new protocols of analysis, in other words, but the apocalyptic one of bringing forth a new world of thought; transforming the imagination, the scientific intellect, morality, and the world of political institutions were diverse facets of one and the same project.

It is thus revealing to think of early relativity pronouncements in relation to Victorian literature at large and, in particular, to the dominant form of writing of the age, the popular novel. Similar themes run crucially through both. But from the perspective of the polemicists of relativity, the moral and intellectual complacencies of the mainline Victorian novel prior to (say) *The Way of All Flesh* and *Jude the Obscure* stand out in high relief. Victorian popular novelists did not by any means ignore the social crisis of their day, a crisis of which the most unmistakable symptom was the persistence of desperate poverty in the midst of a gigantically wealthy society; it is largely thanks to the detailed testimony of these writers that we are able today to imagine it as fully as we do and, among other things, to understand the ramified pathologies of the Victorian middle-class subject as forming determining features of the modern environment. (Every significant Victorian novel suggests, for example, that the rampant sexual disorders of the modern bourgeoisie and the miseries endured by the poor are linked in some profound historical nexus, if only we could learn to decipher it.) At the same time, the Victorian novel takes as its rhetorical principle a fundamentally optimistic and melioristic appeal to the best (most humane, most morally earnest) qualities of its readership. The Victorian public is animated by altruistic good will and by responsiveness to revelations of iniquities and suffering; the thralldom of ideology is not impermeable; the status quo is salvageable: these are the premises from which the complex rhetorical systems of Victorian mainstream novelists derive. On this basis, these writers are able to maintain a fully serious art that remains moderate, decorous, uncynical, communitarian, respectably Christian.

What runs through relativity literature and immediately collateral writing of the same period is, by contrast, a note of feeling not often allowed expression in other contemporary literary fora: a note of rage, disaffection, nausea. These writers refuse to express comforting solidarity with conventional thinking; rather, they proclaim an urgent need to dismantle ideology to its foundation and to install in its place a revolutionary new literary, moral, and intellectual regime presided over not by Jesus Christ but in effect by the pariah of philosophy, Protagoras. The wicked Sophist is rarely invoked by name, but there could be no other guiding figure for the movement that Matthew Arnold described in 1864 as being based on "the baneful notion that there is no such thing as a high correct standard in intellectual matters; that every one may as well take his own way." If no discursive choice is innocent, the predilection of nineteenth-century Protagoreans for highly abstract and technical issues of theory such as definitions of mass and velocity or of economic "value" or of the nature of cognition needs in part to be read as a reaction to the overly rich humanitarian sentiment that dominates the typical stylistic register of Victorian fiction and perhaps muddles its thought. But the texts of the relativity movement are not so barren of affective values as they may often seem at first. Learning to read them with their cultural valences unimpaired means sensitizing ourselves to the polemical import and to the note of rage given off, at sometimes almost inaudibly high frequencies, by works bearing such titles as *The Concepts and Theories of Modern Physics* or "On the Electrodynamics of Moving Bodies."

I am referring here to institutions of discourse and their stylistic properties, not necessarily to the psychology of authors. Very possibly the author of "Electrodynamics of Moving Bodies" was unconscious of expressing any affective values whatever or even of calling for "a new philosophical outlook" as he set down the derivation of the momentous equations of special relativity, demonstrating that there are as many systems of time as there are inertial frames of reference. Yet there is an abundance of psychological and biographical data to invoke in the interpretation even of so abstractly mathematical a text as this one, if one wished to do so. If, for example, one accepts at least provisionally the historical model here proposed, in which the trajectory from Herbert Spencer's work to Einstein's defines a main axis of nineteenth-century relativity, then one cannot put off for long inquiring into the significance of the striking likeness of the autobiographical personae these writers project.

Each describes himself as formed from childhood by an almost morbid hatred of authoritarianism in any form. Spencer belabors this supposedly dominant motive of his life's work throughout his *Autobiography* [hereafter A], by way of illustrating his belief—an axiom, I think, of any sound history of ideas—that in the elaboration of any significant system of thought, "the emotional factor is . . . perhaps as large a factor as the intellectual nature" (A, 1:vii). From his Huguenot ancestors and their longstanding "resistances to religious authority," he derived an "ingrained nonconformity of nature," he says (A, 1:7). From the Wesleyanism of his immediate family, he similarly inherited a "repugnance to priestly rule and priestly ceremonies" and more generally a "lack of regard for certain of the established authorities, and readiness to dissent from accepted opinions" (A, 1:171, 12); hence "the disregard of authority, political, religious, or social" he declares to be his predominant character trait (A, 1:13). Already in his school days he was ruled, he says, by "an aversion to everything purely dogmatic" and by a "nature . . . prone to resist coercion" and "to carry individual freedom as far as possible" (A, 1:123, 139, 238). His instinctive sympathy with Mill's anti-establishmentarian character is matched by his scorn of Carlyle for "his despotic temper and resulting love of despotic rule"—the same unforgivable perversions with which he identifies that irascible reactionary, the Christian God (A, 1:265, 171–72). Spencer does not in the *Autobiography*, or for that matter elsewhere, draw an explicit link between his all-encompassing personal antagonism toward "authority" and his systematic and career-long critique of "absolute" values in physics, biology, and other fields.

In his 1949 "Autobiographical Notes" [hereafter AN], Einstein presents an uncannily similar self-portrait, describing what seems to have been the formative period of his boyhood as "a positively fanatic [orgy of] freethinking coupled with the impression that youth is intentionally being deceived by the state through lies." Out of this "crushing impression" grew, he says, as though quoting Spencer, an attitude of "suspicion against every kind of authority . . . an attitude which has never again left me" (AN, 5). His later scientific research took specifically the form of a struggle against the "dogmatic rigidity [which] prevailed in matters of principles" in the physics of his youth; by the same token, as he insists here and often elsewhere, the ideal scientific work consists not in the application of rigorous rules of deduction to experimental evidence, but rather in an anarchic-seeming "free play with concepts"—a doctrine more in line with the libertarian intellectual politics of writers like Mill, Jevons, and their twentieth-century disciple Feyerabend than with traditional ideas of scientific reasoning (AN, 19, 7), and one that even seems to hint that the scientist is at liberty to conceptualize the universe according to his own pleasure or (to use Samuel Butler's and Poincaré's term) convenience. In his later years, Einstein the public figure became widely known for his forceful denunciations of "arbitrary authority" (Pais, *Subtle Is the Lord*, 38).[1] In all this, his profound temperamental affinity with Spencer makes itself manifest and draws attention in a new way to the intimate connections linking the two writers' theoretical constructions.

Like Spencer, Einstein fails to analyze the relation of his iconoclastic work in relativity physics and his idealization of free intellectual creation, on the one hand, to his phobic personal hatred of authoritarianism, on the other. Yet throughout the work of these two overshadowing figures, and throughout the literature of the nineteenth-century relativity movement, one of the clearest thematic patterns is the mirroring of theoretical expositions of the relativity principle by polemical witnessings against repressive authoritarian agencies. The law of Difference served as the intellectual medium or at least as the signpost of a movement of political liberation—a truth that is only made more apparent by the examples of a few writers like Mansel who sought to enlist the relativity idea in the service of Authority itself. This movement defines the political sphere broadly, directing its emancipatory energies not only against authoritarian official institutions such as governments, but against an ideology of authority and domination that these writers saw to be profoundly institutionalized in Victorian culture, and to be inseparably entwined with the philosophical cult of the "absolute." To dispel the aura of this one fetishized word would be to cleanse European thought of its mystified character and to open the way to an era of intellectual and political liberty: such was their creed. Given the tremendous upsurge of evangelical Christianity in this period, particularly in Britain, it is not surprising that superstitious religion figures in polemical relativity literature as the prime manifestation of absolutism and "arbitrary authority" and of its inescapably tyrannical and violent nature—even as this literature manifests its own deep strain of evangelical moral purpose.

Relativity and the Image of Tyranny

In their concerted critique of the religious mentality and of abusive authority generally, writers in the orbit of Victorian relativity offer radicalized versions of an argument made more cautiously by other contemporary observers. If there is a single prevailing theme of the nineteenth-century popular novel in Britain, for example, it is the theme of the cruelties of tyrannical authority. Often in these novels the focus falls on imagery of tyrannical officialdom: the various sadistic public agencies of *Oliver Twist*, the dictatorial Lord Steyne in *Vanity Fair*, the court of Chancery in *Bleak House*, Kurtz and the whole macabre apparatus of imperialism in *Heart of Darkness*, or, in a late version of this Victorian motif, the sinister psychiatrist Sir William Bradshaw and his medicalized system of tyranny in Virginia Woolf's *Mrs. Dalloway*. Given the focus of the novel in this period on the narration of family life, the theme expresses itself here still more often in images of domestic tyranny, which percipient readers are urged to see not merely as an effect of individual personality distortions, but as a constituent institution of modern society. Victorian fiction is thus a long gallery of portraits of dictatorial husbands asserting the right and obligation to exercise unlimited authority over wives and families: Quilp and Murdstone and Dombey, Hindley Earnshaw, Rochester, Old Osborne in *Vanity Fair*, Casaubon, Grandcourt, Melmotte, Theobald Pontifex, Louis Trevelyan in Trollope's *He Knew He Was Right*, Gilbert Osmond, Angel Clare, and a host of others, all individualized but all recognizable as instances of the same recurring social identity. Sometimes domestic tyranny is usurped, doubly disturbingly, by women: Mrs. Clennam and the lesbian Miss Wade in *Little Dorrit*, Mrs. Proudie, Aunt Glegg in *The Mill on the Floss*, Miss Havisham. As seen from the perspective of the Victorian novel, respectable middle-class life discloses itself as a nightmare of abusive authoritarianism, one frequently tinged with (or expressly based upon) perverse religiosity. That humane domestic orders are commonly and comfortingly reestablished by the end of the usual Victorian narrative, and that the literary formula of the domestic tyrant may function culturally to normalize and to exempt from critique subtler forms of domination exercised in real-life households, does not dispel the proposition so insistently put forth in contemporary novels that the tendency toward despotic relations represents a systemic syndrome of nineteenth-century life. When Carlyle proclaims the "divine right" of aristocrats to rule over the lower classes, "being themselves very truly . . . BRAVEST, BEST; and conquering generally a confused rabble of WORST, or at lowest . . . of WORSE" (*Past and Present*, 205), or when his echo Nietzsche declares that the moral antithesis of good and bad originates in "the chronic and despotic *esprit de corps* and fundamental instinct of a higher dominant race coming into association with a meaner race" (*Genealogy of Morals*, 20), they are simply idealizing, promoting into the source of moral values (and in Nietzsche's case, as we shall see, immersing in a rabid anti-Semitism not absent from Carlyle) a prevalent trend of nineteenth-century existence, at least as novels represent it.

One could interpret the overwhelming focus of Victorian popular literature on the theme of domestic politics as an ideological ruse designed to divert attention from issues of power in political institutions as such, and particularly from the great issue of the oppression and exploitation of the underclass in the age of the "industrial revolution." Is the critique of middle-class marriage the stylistic sign of complicity in larger structures of exploitation? I will not pause to argue this problem, but I will note recurring imagery in the novel and in other modes of literary discourse, especially as the century moves toward its end, of which the theme is the prospect of widespread calamity as the result of a rise of authoritarian cults. Some of these dire fables are allegorical only. An instance would be Ernest Pontifex's brush with the charismatic (and sexually perverse) Pryer and his sinister new religious organization, the college of Spiritual Pathology, in *The Way of All Flesh*. The ability of the clergy to effect spiritual improvement among the laity will be regrettably limited, says Pryer, "until we have a discipline which we can enforce with pains and penalties" (*Way of All Flesh*, 286). More explicit is Conrad's *The Secret Agent* (1907), a novel in which the pursuit of fanatical and violent authoritarian politics is linked intriguingly to the influence of evangelical religion,[2] and in which a wise old diplomat cries out on his deathbed, "Unhappy Europe! Thou shalt perish by the moral insanity of thy children." Giving substance to the warning, the novel ends with the Professor's rapturous prophetic vision of a vast death camp, "a world like shambles, where the weak would be taken in hand for utter extermination." "Exterminate, exterminate!" he cries. "That is the only way of progress" (*Secret Agent*, 36, 246).

No nineteenth-century commentator, even one of Conrad's uncanny prescience, could fully have foreseen those developments in which the cult of absolute values could mutate into cults of absolute leaders and into modern totalitarianism, and could produce a series of calamities beyond anyone's capacity to imagine. But there was no need to await the horrors of twentieth-century history in order to recognize the despotic tendencies that the collapse of the *ancien régime* in the previous century had left in its wake. The European political scene in the later nineteenth century was dominated by a triumvirate of notorious autocrats: Napoleon III, Bismarck, and (a late survival of the *ancien régime*) the tsar. England had no such absolutist state as the ones presided over by these Continental rulers, but the climate of authoritarianism and political repression in the country was none the less intensely felt throughout the century, from the Peterloo massacre in 1819 to the suppression of the Jamaica uprising by Governor Eyre in 1865 to the brutal police suppression of free-speech demonstrations in Trafalgar Square in the "Bloody Sunday" riots of 1888. Mill's protest in *On Liberty* (1859) against the systematic official (and unofficial) persecution of unorthodox beliefs in England was extreme but not excessive: as Joss Marsh reminds us in her recent study *Word Crimes*, two hundred prosecutions for the crime of blasphemy were carried out in England in the nineteenth century (Marsh, *Word Crimes*, 15). Spencer catalogued a long list of similar instances of authoritarian repression in Victorian Britain. When the philosopher of the "absolute" T. H. Green proclaimed the doctrine of "a single and unalterable order of relations" in the universe and derived from it the principle of the "unconditionally binding" power of all established laws and opinions (Green, *Prolegomena*, 17, 207), he was using philosophical arguments with obvious and immediate political, cultural, and legal correlatives in the Victorian world.

Only at a late stage of the relativity movement was its implicit logic linking the philosophical critique of the "absolute" to the cause of resistance to political absolutism—and, in particular, to its penchant for physical violence—made explicit. This is the theme running, sometimes subliminally, sometimes in unmistakable form, through the writings of the leading school of relativistic philosophy at the turn of the century, the pragmatic school. It is implicit in William James's criticisms of "absolutist" writers like F. H. Bradley, dogmatists who reject the possibility of "a plurality of reals" and who reason in terms of what James pointedly calls "violent extremes" (James, *Pluralistic Universe* [hereafter *PU*], 75, 74). But it was Hegel, says James in 1909, who was the premier exemplar of a mode of thought "dominated by the notion of a truth that should prove incontrovertible, binding on every one, and certain, which should be *the* truth, one, indivisible, eternal, objective, and necessary": and this, says James, is "the dogmatic ideal" that pragmatic philosophy, as its primary mission, seeks to overthrow (*PU*, 100). Despite the striking overtones of some of his evocations of "the logical machinery and technical apparatus of absolutism" (*PU*, 314), which seem to arise from the same intuition as does the Professor's glowing vision of efficiently engineered extermination camps two years previously, James is in fact an inconsistent anti-absolutist: in *The Principles of Psychology* (1890) he disavows "the doctrine of universal relativity" and denies that it proves anything about the "absolute qualities" of objects of knowledge; in *The Meaning of Truth* (1909) he goes so far as to declare that pragmatists, like every "relativist who ever actually walked the earth," adhere faithfully to "the notion of absolute truth" (*Principles*, 1:12; *Meaning*, 309). Not surprisingly, James does not give the same sharp anti-authoritarian valences to relativistic arguments that his fellow pragmatists, notably F. C. S. Schiller and John Dewey, often do.

For Schiller, absolutism as a philosophical doctrine harks back ultimately to the fatal influence of Plato, but for practical purposes is to be seen as a culturally specific phenomenon of the late nineteenth century in England. It was imported, he says, from Germany and from Hegel in the service of embattled British theology; its commission was to roll back "the great scientific movement of the nineteenth century," which between 1850 and 1870 threatened to submerge orthodoxy in "the logic of Mill, the philosophy of Evolution, the faith in democracy, in freedom" (Schiller, *Studies*, 277, 278). The leading characteristics of this crypto-theological and politically fraught absolutism, apart from "the fanaticism with which it is held," are its "hostility to individual liberty" and its "insistence on absolute conformity," says Schiller with his usual belligerence (*Studies*, 289, 297). He describes the wide-ranging new school of "pragmatic" philosophy as a countermovement that has arisen to oppose "the intensity of intellectualistic prejudice and the intolerance of Absolutism"; among its leading lights are not only Dewey, Peirce, and James, but, he shrewdly asserts, Poincaré and Ernst Mach (*Studies*, xiii, xiv–xv). Pragmatism in its struggle against a reactionary absolutist theology is historically part and parcel of the construction of relativity theory in physics, declares Schiller, contradicting Einstein's assurances to the archbishop and Toulmin's "coincidence" hypothesis. (Schiller seems not yet to have heard of Einstein in 1907, however.)[3]

Schiller's identification of pragmatism—or, as he calls his version of it, "humanism"—with the principle of relativity is emphatic. "The 'essence' . . . of a thing is relative to the point of view from which it is regarded," and philosophical "truths" are inescapably relative to human purposes and activities, he teaches (*Studies*, 12). These doctrines of sound philosophy all originate, Schiller provocatively insists, in the work of Protagoras himself, "the great teacher" slandered unpardonably by Plato as a skeptic and a nihilist. Protagoras not only developed the one satisfactory philosophical account of "truth," according to Schiller, but is to be regarded, his evil reputation notwithstanding, as a great moralist who seems in the *Protagoras* "as clearly to excel Socrates in nobility of moral sentiment as he falls short of him in dialectical quibbling" and whose teachings, in fact, "very distinctly savour of the moral fervour of St. Paul" (*Studies*, xvii, 113, 299, 36). The modern-day revival of the tradition of Protagorean relativism is impelled by the same moral militancy, says Schiller. Its basis is the radical defense of "liberty of thought"; absolutism, by contrast, is "autocratic in authority," has an instinctive aversion to "toleration, mutual respect, and practical co-operation," and—here Schiller echoes Mill and William James—is driven above all by "the tyrannous demand for rigid uniformity" (*Studies*, 139, 7, 18). Schiller thus portrays contemporary relativism as the response to alarming cultural trends in the direction of autocracy and tyranny, though he illustrates them not by reference to contemporary instances but to Protagoras himself, who was assailed on the level of argument by Plato and on the political and practical levels by the Athenian state, which, Schiller says, feared Protagoras's theory so much that it set the state executioner to burn his book on Truth, depriving us of "one of the great monuments of Greek genius" (*Studies*, 37). It would not be an overstatement to say that for Schiller, the philosophy of absolute values is inescapably allied with the forces of politically motivated repression; it is fundamentally ideological.

Dewey describes the new pragmatic philosophy in similar terms; its primary imperative, he says, is to "[forswear] inquiry after absolute origins and absolute finalities," including the search in the moral field "for *the* final good, and for *the* single moral force" (Dewey, *Darwin*, 13, 48). He goes beyond Schiller in his indictment of "absolutistic philosophies"—in a phase of alarming recrudescence in 1910, he declares (*Darwin*, 18)—for their historic complicity in many oppressions, particularly those carried out in the name of religion. The pretense to be able to define "absolute" values by means of abstract reasoning has, Dewey charges, long served to justify "class distinctions of superiority and inferiority as between man and woman, master and slave" and serves ultimately no other function than the "idealization of the existent." Greek philosophy on the Aristotelian model thus provided a rationale for a social system based on "harsh intolerance"; absolutist moral theory in the age of feudalism similarly dictated a rigid distinction between "those possessed of sacred rule and those whose sole excellence was obedience" (*Darwin*, 50, 53). Absolutist philosophy cannot abide differences of opinion or interpretation, says Dewey; rather, it indoctrinates its public in the conviction that "truth is . . . so important that it must be authoritatively imparted and enforced" (*Darwin*, 54). Its great principle is "the deification of hereditary prejudice . . . masquerading in the guise of absolute and eternal truth." The absolutist reason everywhere promotes the ascendancy "of control, of system, of order and authority," and everywhere helps "to establish inequality and injustice among men," says Dewey (*Darwin*, 291, 294, 75).

In such writings, the pragmatists articulate the political animus of relativity theory more distinctly than had previously been done, and yet the texts I have been citing here are in some ways less sharply dissident and less inflammatory rhetorically than the nineteenth-century tradition they interpret and maintain. For one thing, they emphasize less sharply than do some earlier writers the specific political manifestations of the absolutist code of "intolerance" and "authority": in particular, its craving to express itself in the form of *violence*—violence inflicted not only upon books and ideas but upon human bodies. For nineteenth-century advocates of a revived Protagorean philosophy, this seemed a vital distinction, for the insisting upon which no illustrative imagery was too vivid or too out of keeping with the decorum of polite intellectual debate. They tell us insistently that the Absolute, when one peels away its veneer of reasoned demonstration, is the name of a world of sanguinary ideological violence. They suggest that its philosophical incoherence (so plainly manifest in the theory of classical physics, for example) links it in some profound way to the brutal use of force, as though in fury at its own ultimately hopeless inarticulateness.

Spencer, for example, develops this theme with great emphasis, though only by a kind of metonymic implication. Two large arguments run on parallel tracks through his work: on the one hand, the argument asserting the universal relativity of knowledge, denying all access to the "Absolute" and denouncing as impostors those such as Mansel who claim to reveal it; and on the other, a sustained critique of the violent system of repression that is institutionalized, he declares, not just in the openly authoritarian nations of continental Europe, but at home in Britain. The correlation between these two themes, the one philosophical and abstract, the other specifically political, is not made explicit in Spencer's writing, as far as I know; yet it forms the fundamental logic of much of his work.

In one text after another, particularly in those concerned with social science, Spencer thus denounces what he portrays as an ever more vindictive assault on freedom by contemporary cultural and political institutions. From the point of view of his theory of social evolution, one crux of which is the supposedly natural, inexorable transition from early "militant" to modern "industrial" society, this alleged trend in Victorian Britain is wholly aberrant and pathological. In "militant" society, discipline and regimentation are the dominant principles, and they are enforced by "a despotic controlling agency" (*Principles of Sociology* [hereafter *PS*], 2:608). In such a society, Spencer explains, "absolute subordination . . . is the supreme virtue, and disobedience the crime for which eternal torture is threatened" (*PS*, 1:548). The echoing of his attacks elsewhere on the mystificatory philosophical and scientific category of "the Absolute" in this analysis of the social code of "absolute subordination" is not an accidental or an incidental effect. A mode of speculative thought based on theories of absolutes forms the natural symbolic expression and the ideological substructure of an authoritarian social order, he audaciously suggests—too audaciously indeed for him to put this insight in so many words, as his intellectual heir Thorstein Veblen does soon after.[4] No less audacious, for a Victorian public, is Spencer's direct identification of coercive state power with religious dogma, as in the analysis here of the maintenance of quasimilitary social controls by the threat of "eternal torture" of the disobedient. One is reminded of the same author's bitter descriptions elsewhere of representative cruelties performed in the name of theology. Remarking "how close has been the connexion between killing enemies and pleasing deities," for example, he declares that often "it seems that Christian priests agree with the priests of the Comanches" in their bloodthirsty code of revenge (*PS*, 3:109, 113). In Spencer's analysis, the psychological cognate of the religious cult of obedience and savage retributive violence, and a principal characteristic of the contemporary British citizenry, is the "awe of embodied power"—the reverential worship of authorities and the glorifying of absolute loyalty to them—that he identifies as a key syndrome of "militancy" (*Study of Sociology* [hereafter *SS*], 175). By contrast, the evolutionary transition to "industrial" society is marked in the Spencerian model by increased personal liberty and individual initiative, and by the ascendancy of voluntary self-government as opposed to state coercion.

(Michel Foucault, Spencer's latest intellectual descendant, offers a closely parallel analysis in *Discipline and Punish*, as does Freud in *Civilization and Its Discontents*, though both take a considerably more pessimistic view than Spencer does of the growth of internalized self-government.)

Contemporary European history, not as unidirectional as his evolutionary model would otherwise suggest, seems to Spencer rife with evidence of an evolutionary anomaly, a slide back into the grim darkness of militancy, with its agencies of dictatorial control, its passionate devotion to the ethic of revenge, its glorification of national supremacy and of predatory military aggression (all the features of the rise of the ferocious Serbian state of Slobodan Milosevic that is unfolding as I write this chapter in April 1999). As signs of this trend, Spencer cites, for example, the "diabolical cruelties" inflicted by European settlers upon American Indians, the sanguinary career of Napoleon Bonaparte—"all this slaughter, all this suffering, all this devastation, was gone through because one man had a restless desire to be despot over all men"—and the atrocities committed by Governer Eyre in Jamaica, which seemed to Spencer, as they did to Mill, signs of "a return to barbaric principles of government" (*SS*, 212, 157–58, 244).[5] The war years from 1775 to 1815 were a period of many major and minor "aggressions upon the individual by the State" (*PS*, 2:626), and subsequent decades, he says, have witnessed a "return towards the militant type in our institutions generally—the extension of centralized administration and of compulsory regulation," as typified in the tyrannical Contagious Diseases Acts of 1868, the despotic administration of the Poor Laws, predatory empire-building, "usurpations of civil functions by military men," and other similarly alarming developments (*PS*, 1:570, 2:591; *SS*, 278–80). Since 1860, he reports, fifty-nine acts have been passed in Parliament for "regulating the conduct of citizens," and since 1884, "coercive legislation affecting men's lives has greatly extended" (*PS*, 3:591). In Victorian England, brutal generals are honored more than great scientists; missionaries are in league with ruthless colonialists (*SS*, 138–44).[6] Spencer diagnoses in Victorian public affairs the same movement toward harshly repressive authoritarianism that novelists of the day transcribe so intensively in the area of marital and familial relations.

While other Victorian sages (Carlyle, Pugin, Arnold) pointed with alarm to the seeming spread of disorder in their society and indulged in reveries of therapeutic medievalism or called for new institutions to monitor and regulate national standards of thought, Spencer thus relentlessly protested against the widening sovereignty of centralized government and its accompanying authoritarian ideological apparatuses (of which Protestant religion was in his eyes one of the most virulent). It may have been in significant part thanks to Spencer's prophetic hectoring that the terrifying spiral into modern revolutionary totalitarianism, when it came, occurred in Russia, Germany, Spain, Italy, and elsewhere, rather than in Britain. The point here is not to canonize Spencer as the conscience of his nation, though this might only be fair recompense for his service to it, but merely to make clear that his mission as dissident social critic and his pioneering role as relativistic theorist of science were intimately interconnected—at some level, identical—enterprises.

In the remainder of this chapter, I propose to situate the central case of Spencer by reference to two bracketing texts by non-British writers: Feuerbach's *The Essence of Christianity* [hereafter *EC*], published in 1841 and translated into English by Spencer's close friend George Eliot in 1854; and J. B. Stallo's *The Concepts and Theories of Modern Physics* [hereafter *CTMP*] (1881). The triangulation of these three writers makes a telling exhibit of the way Protagorean relativity worked in the nineteenth century to propel simultaneous movements of political emancipation and of scientific speculation and discovery.

Feuerbach and the Theology of the Absolute

In his polemical analysis of theological anthropomorphism, Feuerbach does not refer to Sir William Hamilton (or, of course, to Mansel, who comes later), but he closely echoes Hamilton in deploying systematically, as the basis of his philosophy, the theory of relativity and of Difference. "Difference is a positive condition of the reason," Feuerbach asserts, "an original concept, a *ne plus ultra* of my thought, a necessity, a truth" (*EC*, 85). Things possess their defining properties not as inalienable belongings, he explains, but by virtue of differential relations with other things in our cognitive field. He illustrates this foundational principle by means of a radically relativistic allegory of astronomy:

> The Sun is the common object of the planets, but it is an object to Mercury, to Venus, to Saturn, to Uranus, under other conditions than to the Earth. Each planet has its own sun. . . . It really is *another* sun on Uranus than on the Earth. The relation of the Sun to the Earth is therefore at the same time a relation of the Earth to itself. . . . Hence each planet has in its sun the mirror of its own nature. (*EC*, 4–5)

All things are fundamentally "conditioned," as Hamilton would say, by their relations with other things, and in turn impress their own points of view narcissistically upon the universe. There are as many different Suns as there are planets standing in relation to it. Needless to say, were one to take this outlandish-seeming proposition not as a philosophical allegory but as a rigorous law of science, it would be necessary to reconstitute physics along lines that would seem wholly contrary to reason.

Feuerbach's project of intellectual reconstruction is, however, aimed in the first instance not at physics but at religion. From this perspective, one immediate consequence of the proposition is that the apparent one-way relationship of dominance and subordination obtaining between, say, Sun and planets becomes more ambiguous than it may seem. (According to Feuerbach's allegory, in which the Sun is in fact a creation of the planets that it subjugates, could a planet not wishing to imagine itself as a helpless satellite reconceive this relationship? Could mankind emancipate itself from superstitious ideas of an omnipotent divinity?) As the above passage suggests, in any case, relativity for Feuerbach is not simply the logical precondition of knowledge, as it is for Hamilton and Mansel and subsequently for Einstein. Rather, it is a condition dictated by a supposed psychological compulsion. One strives to attain self-awareness not by introspection but by projecting one's image upon external objects and thus, since "man is nothing without an object" and "to know oneself is to distinguish oneself from another" (*EC*, 4, 81), to see oneself as separate from these same objects—this coincidence of separateness and identity being what gives relativistic theory in all its manifestations the appearance of "insurmountable paradox," as Schiller later said (*Studies*, 428). What Feuerbach's schema implies on the psychological plane is a disjointed state of being in which subjectivity is forever specular, forever estranged from itself. On the epistemological plane, it teaches us to regard what are called "facts" not as permanent realities external to us, but as functions of the imperious needs of human feeling: they are in fact "just as relative, as various, as subjective, as the ideas of the different religions" (*EC*, 205). "A fact," says Feuerbach in 1841, long before Nietzsche professed similarly heretical views, "is every possibility which passes for a reality, every conception which, for the age wherein it is held to be a fact, expresses a want, and is for that reason an impassable limit of the mind. A fact is every wish that projects itself on reality" (*EC*, 205–6).[7]

From this dizzying set of premises, Feuerbach carries out his indictment of Christian theology. The key device and the original sin of theological reason is for Feuerbach the representation of the deity (properly understood anthropomorphically as a projection of human imagination) in the guise of an external and autonomous reality; God as misconceived by theology, in other words, "is no longer a merely relative, but a noumenal being *(Ding an sich)*" (*EC*, 199). Correspondingly, the Bible, "an historical book" necessarily conditioned by all the historical circumstances of its composition, is pronounced for theological purposes to be "an eternal, absolute, universally authoritative word" (*EC*, 209). Religious superstition is nothing other than the denial of relativity by theology, says Feuerbach. This is the intellectual act that contains "the noxious source of religious fanaticism, the chief metaphysical principle of human sacrifices, in a word, the *prima materia* of all the atrocities, all the horrible scenes, in the tragedy of religious history" (*EC*, 197).

Violence is the essence of religion in Feuerbach's account because God is endowed, as human wish-fulfillment would have it, with supreme authority over the universe. This invention of a deified tyrant is analyzed by Feuerbach in remarkable proto-Freudian terms. Religion, he says, denies and seeks to root out egoism in human beings, but, as though in obedience to a kind of law of the conservation of psychic energy, only to restore the suppressed impulse "unconsciously" to the deity, who is idealized in the image of a monomaniacal tyrant, "a selfish, egoistical being, who in all things seeks only himself, his own honour, his own ends" (*EC*, 27). Christians worship in God the selfishness that they are taught to condemn in human beings, and worship in this way the deep truth of their own nature. From this perverse mechanism arises the deep strain of moral perversion in Christianity that Feuerbach anatomizes in his most inflammatory polemical prose. Alongside the benign God corresponding to the tenet "God is love," the Christian deity appears also, inevitably, "in another form besides that of love; in the form of omnipotence, of a severe power not bound by love; a power in which . . . the devils participate." This is "the God—the evil being—of religious fanaticism," "an unloving monster, a diabolical being, whose personality, separable and actually separated from love, delights in the blood of heretics and unbelievers" (*EC*, 52, 53).[8] In its next phase of perverse displacements, religious imagery reacts back upon its creators, says Feuerbach, producing a characteristic complex of sadomasochistic impulses: for "how should not he who has always the image of the crucified one in his mind, at length contract the desire to crucify either himself or another?" (*EC*, 62).

The idea of God the Father as "the absolute being" rather than "a merely relative . . . being" and the establishment of a regime of "dictatorial" theological tyranny based on "absolute intolerance" and on idealization of "absolute divine power" thus go inseparably hand in hand, declares Feuerbach (*EC*, 46, 199, 115, 114, 121). The denial of relativity in the name of "truth, absolute truth" is no philosophical nicety for Feuerbach, but an unleashing of cruelty and terror: the metaphysical violence of this denial in effect translates itself directly into the practical violence of persecution by ecclesiastical authorities and into "all the horrors of Christian religious history" (*EC*, 211, 257). In *The Essence of Christianity*, the fetishizing of the philosophical concept of the Absolute is thus treated very distinctly as the founding ideological operation of what already in 1841 went by the name of "absolutist" systems of political power. Later philosophical expositions of the absolute in whatever field are placed by Feuerbach's epoch-making text under the burden of assessing the implications of any such argument for the maintenance of civil liberties and freedom of thought amid potentially hostile social environments, under pain of being open to a charge of bad faith—a burden that few if any champions of the absolute have ever been willing to assume.

Moral depravity in religion is inescapable, says Feuerbach, since all that matters to the Church is loyalty to its dogma, which is to say, first and foremost, belief in the God of the theologians. "Whether under this God thou conceivest a really divine being or a monster, a Nero or a Caligula, an image of thy passions, thy revenge, or ambition, it is all one, —the main point is that thou be not an atheist" (*EC*, 202). Religious faith is on principle utterly intolerant of all who do not believe; as Dewey later emphasizes, it automatically institutes social inequality by declaring believers a superior caste (*EC*, 249–50); and of course it declares as its supreme rule, says Feuerbach, the insistence on a single truth, on the "One True Vision" that for Ernest Gellner and other antirelativists is the sacred principle of modern science:

> Faith discriminates thus: This is true, that is false. And it claims truth to itself alone. Faith has for its object a definite, specific truth, which is necessarily united with negation. Faith is in its nature exclusive. One thing alone is truth, one alone is God, one alone has the monopoly of being the Son of God; all else is nothing, error, delusion. Jehovah alone is the true God; all other Gods are vain idols. (*EC*, 248)

The idea that truth is necessarily *one* is for Feuerbach the basis of the violent practical outcome of religious philosophy in the real world, where any who refuse to conform to the reigning code of belief are anathematized and made subject to persecution by religious power acting on "divine authority" claiming "unconditional, undivided sovereignty" (*EC*, 274, 256). The assertion of knowing the one and exclusive truth and the exercise of unrestrained violent power are, again, the two sides of a single coin: the one implies the other, for the concept of a single truth "is necessarily united with negation." Among other implacable passages from scripture cited by Feuerbach to illustrate this principle is 2 Thessalonians 1:7–10, which promises that "the Lord Jesus shall be revealed from heaven with his mighty angels, in flaming fire taking vengeance on them that know not God, and that obey not the Gospel of our Lord Jesus Christ" (*EC*, 258). The *essence* of religious faith is vengeance. It is in fact what Conrad's Professor imagines as the outcome of his own political faith: extermination.

> What God condemns, faith condemns, and *vice versâ*. Faith is a consuming fire to its opposite. This fire of faith regarded objectively is the anger of God, or what is the same thing, hell; for hell evidently has its foundation in the anger of God. But this hell lies in faith itself, in its sentence of damnation. The flames of hell are only the flashings of the exterminating, vindictive glance which faith casts on unbelievers. (*EC*, 255)

Feuerbach would quickly have identified the exterminating flames kindled for the Jews and other undesirables in twentieth-century Europe as not a terrible aberration explainable somehow by the Treaty of Versailles or other political circumstances, but as expressing the fundamental organizing principle of Christian society and as just one of the most enormous among "all the horrors of Christian religious history."

Contrary to Einstein's reassuring formula for the benefit of the Archbishop of Canterbury, and contrary to Hamilton's and Mansel's discovery of "a wonderful revelation" of Christian faith in relativity theory, *The Essence of Christianity* thus presents what may be the original declaration of relativity as the antithesis of the ideology of tyranny. Ultimately, Feuerbach argues, the only defense against the menace of divinized leaders in the shape of "a Nero or a Caligula" and of the genocidal impulse that they embody is to dismantle the very principle that "one thing alone is truth," that it is possible to say categorically "this is true, that is false." Diverging responses to this extreme proposition among the next several generations of writers go far to define the ideological dynamic of the Victorian public sphere.

On the one hand, W. K. Clifford echoes the Feuerbachian paradigm in a slightly more optimistic register. Clifford was the exponent of a radical evolutionary relativity that held all knowledge and all systems of value to be necessarily impermanent since human nature itself is forever in flux (Clifford, *Lectures*, 2:278, 282). "Instead of contemplating an eternal order, and absolute right," he says in a vivid formula, "we find only a changing property of a shifting organism" (*Lectures*, 2:279). In a universe thus emptied of "immutable and eternal verities," the one legitimate source of systematic understanding, according to Clifford, is science—not a science claiming to discover absolute truths, but one that brings itself into consonance with the natural world by insisting that its body of knowledge "must be regarded as only provisional and not final, as waiting revision" and as having always "a character of incompleteness about it" (*Lectures*, 2:283, 278). Religion in Clifford's Feuerbachian analysis sets itself up as the contrary of enlightened science not only in its claims to possess absolute truth but—the necessary concomitant of such claims—in its habitual reliance upon various kinds of coercive machinery. The fundamental unnaturalness of such a system manifests itself, in effect, in the terroristic character of religious ideology. In the Feuerbachian manner, Clifford illustrates this character by citing Christian writings: for example, "the two famous texts: 'He that believeth not shall be damned,' and 'Blessed are they that have not seen and yet have believed'" (*Lectures*, 2:214)—tenets expressing the dogmatic and antiscientific attitude in its purest, most explicitly violent form. Against this mentality, Clifford sets an idealized portrait of scientific reason. "This sleepless vengeance of fire upon them that have not seen and have not believed, what has it to do," he asks, "with the gentle patience of the investigator?" (*Lectures*, 1:299).

Matthew Arnold evinces a more ambivalent response to the trends of thought issuing from Feuerbach. He professed the same doctrine as did Clifford or Mill with regard to the supreme value of what he called "openness of mind and flexibility of intelligence," of a freely critical, undogmatic attitude ("Literary Influence of Academies," 263). Yet Arnold saw these prized attributes of culture as being favored by, as almost dependent upon, agencies of "severe discipline" in thought, such as "a sovereign organ of opinion" like the French Academy ("Literary Influence," 267, 261). Nothing was more repulsive to Arnold than "the baneful notion that there is no such thing as a high correct standard in intellectual matters; that every one may as well take his own way"; he firmly upheld the creed that in intellectual matters, "there is a right and a wrong" ("Literary Influence," 267, 262). As he revered the principle of authority in the realm of culture, so too his politics were marked by a distinct strain of authoritarianism tinged with that visceral love of violence that Feuerbach and Clifford diagnosed as the core and the essence of the authoritarian psyche. In *Culture and Anarchy* [hereafter *C&A*], he declares the necessary allegiance of men of culture to "a State in which law is authoritative and sovereign" in its repression of civic disorders (*C&A*, 204). In support of the cult of an all-powerful law, he tropes directly into political discourse the religious and philosophical formula of the uniqueness of "truth, absolute truth" and of the manifold nature of error, calling the former category by such names as "right reason" or "our best self." The political state is thus ideally, he says, the expression "of our best self, which is not manifold, and vulgar, and unstable, and contentious, and ever-varying, but one, and noble, and secure, and peaceful, and the same for all mankind" (C&A, 204). Arnold dogmatically denies in such a statement the whole progressive intellectual program associated with evolution and relativity, according to which truth is necessarily multiplex and in perpetual flux: this is the ideology, he says, of "anarchy." Nothing that is "unstable" or "contentious" can be tolerated in an Arnoldian social order. Right-minded people will "steadily and with undivided heart support [whoever may be in power] in repressing anarchy and disorder," says Arnold, even in the case of demonstrations in support of such a worthy-seeming cause as the abolition of the slave trade (C&A, 203).

Arnold makes perfectly clear that he is calling for the swift repression of even relatively minor civic disorder by the application of harsh physical violence. "Lovers of culture . . . prize and employ fire and strength," he says; in cases of such disturbances as the Hyde Park riots of 1866, he accordingly calls for the authorities to "flog the rank and file" and summarily put the leaders to death, in "the old Roman way" (C&A, 205, 203). It may be disconcerting to observe the idealistic Arnold imitating in such passages the protofascistic ranting of Carlyle at his coarsest. (In bad historic times threatened by "lawless anarchy," says Carlyle, "it is a dire necessity of Nature's to bring in her ARISTOCRACIES, her BEST, even by forcible methods," for "despotism is essential in most operations"—*Past and Present*, 254, 207, 271.) Feuerbach and Clifford would see Arnold's glorification of a politics of summary executions as expressing the inherent practical logic of all crypto-theological codes of "truth, absolute truth," however expertly a talented publicist might mystify them with an urbane rhetoric of disinterestedness and liberality that professes to hold itself aloof from the uncouth sphere of "public life and direct political action" (C&A, 207). The example of Arnold helps us in any case to understand the cultural environment within which the nineteenth-century relativity movement sought to articulate its mission. A setting in which a spokesman for an ethic of flexible open-mindedness could define "sweetness and light" as "fire and strength" is one in which an ideology of violence is both deeply ingrained and deeply dissimulated, as Spencer never tired of warning his contemporaries.

Stallo's New Physics

Johann Bernhard Stallo (1823–1900), an autodidact in the field of physics, produced what may be the definitive expression of the complex of late-Victorian argument we are tracing. He is not a well-known figure in the history of modernist thought, still less in the history of scientific discovery, so a few biographical details may be useful. He emigrated from Germany to Cincinnati in 1839 at the age of fifteen, wrote an influential textbook on the teaching of German, attended law school, and had a distinguished career as a lawyer, eventually becoming a judge and American ambassador to Italy. Like Feuerbach, Mill, Spencer, and Clifford in his own day, and then like Einstein, he was a passionate champion of the cause of freedom in what seemed to him a repressive age; he was a staunch opponent of protective tariffs and, on the slavery question, an ardent abolitionist who abandoned his original allegiance to the Democratic Party on this ground and became an early supporter of Abraham Lincoln. He became locally famous for organizing "the Stallo Regiment" of Cincinnati Germans to fight for the Union in the Civil War. This was a man deeply embroiled in the great issues of contemporary politics and always opposed to anything he regarded as "arbitrary authority." In 1870, he won on appeal to the Ohio Supreme Court a celebrated and still influential case against the Protestant churches of Cincinnati, which had sought to block the removal of Bible study and hymn singing from the curriculum of the public schools. In the course of these proceedings, following a campaign of sermons preached from Cincinnati pulpits in favor of school Bible education, Stallo delivered a lecture titled "State Creeds and Their Modern Apostles" [hereafter SC], which was published as a pamphlet two years later to coincide with a convention of Christian activists urging a Constitutional amendment acknowledging "God as the author of our nation's existence and the source of its authority . . . Jesus Christ as its ruler, and the Bible as the fountain of its laws and supreme rule of its conduct" (*SC*, iii). This lecture gives a vivid idea of Stallo's political commitments (and of his formidable polemical eloquence), and sheds light indirectly on his amazing contribution to the early history of the relativity movement in physical science.

For Stallo, the insertion of religious dogmatism into public affairs in such a form as the proposed amendment was no small or merely symbolic matter; he claims that it would represent an act of ideological tyranny "threatening to revolutionize all our political institutions and to subvert the most precious of our liberties, the liberty of conscience" (*SC*, iv). No doubt mindful of the frequent prosecutions for blasphemy in England at the time, Stallo in his lecture imagines the institution in the United States of "future blasphemy laws" under which Jews and other non-Christians could be "sent to the penitentiary" for failure to practice the approved state religion. And lest his readers think him guilty of reckless exaggeration in warning of the holocaust of persecution the innocuous-seeming amendment could ignite, he reminds them that the Protestant divines most eager to institute the new religious code were men who not long before had justified human slavery in the United States with an appeal to "the 'special revelation' of

Biblical moral law" (*SC*, v, vi). For Stallo, the debate about relations of church and state can only be serious and responsible if it is keyed at every moment to the propensity of the state to impose direct physical force upon citizens—a propensity which religious influences, he says, can only inflame, with incalculably dangerous possible consequences.

In the rhetorical manner characteristic of the Feuerbachian lineage of nineteenth-century freethought, Stallo portrays the history of Christian belief as a spectacle of harsh coercion aimed above all at suppressing ideological deviance—at the enforcement by violence of the "One True Vision" in religious matters. Whenever a state comes to imagine that it possesses "the absolute truth [or] the absolute right," Stallo warns his readers, it becomes without fail "an abominable engine of tyranny and oppression." To be an *absolutist* state in this philosophical sense is automatically to tend to the exercise of unrestrained terroristic force, in other words; historically, says Stallo, the official dogmas of such states have been "practically interpreted by the rack and the stake on which heretics were tortured and burned" (*SC*, 25, 35, viii). Stallo thus imagines his own American version of the slide back into barbaric "militancy" and evokes the prospect of a modern era obsessed with the violent repression of heresy, blasphemy, and Jews in the name of "the orthodox State religion" (*SC*, 24). No illusion could be more dangerous than the idea that our modern society is exempt from fanatical absolutist ideology, says Stallo, at the risk of sounding unreasonably alarmist. "The logic of the Spanish inquisitor is in every respect identical with that of the zealot in the frock or gown of a Unitarian or Methodist preacher, or under the ermine of the judge" (*SC*, ix)—or also, we may say (reading this strain of Stallo's work into his work in another field, philosophy of science), in the academic robes of another group of fanatical absolutists, contemporary professors of physics. Stallo's interventions in this other field exemplify as plainly as does the work of any other writer of the day the principal theme of this chapter: the intimate affiliation between the development of relativistic science and the motive of moral and social emancipation—the motive that made relativity physics, as "purely [a] scientific matter" as it ostensibly was, an unpardonable ideological crime for Lenin and Hitler alike.

As early as the early 1870s, Stallo the scientific amateur began publishing portions of the remarkable book that appeared as *The Concepts and Theories of Modern Physics* in 1881 (the year of the Michelson-Morley experiment that failed to detect the ether and is often said to have triggered the development of modern scientific relativity). Stallo's book, "today almost completely forgotten" (Bridgman, "Introduction," ix), deeply impressed Ernst Mach, who wrote a long laudatory introduction for its 1901 German translation. Well might it have impressed him, for it offers a searching, thoroughly Machian critique of the whole range of prerelativity physical theory and a forecast of many revolutionary aspects of Poincaresque and Einsteinian science, essentially all of which, according to the physicist Percy Bridgman, Stallo appears to have discovered independently on the strength of his own "prophetic vision" (Bridgman, "Introduction," xix, xxviii). Mach praised Stallo in striking language for being instrumental in "the gradual complete emancipation of science from . . . traditional, often primitive and barbaric, modes of thought" (quoted in Drake, "J. B. Stallo," 25). Any reader can see that Stallo conceived his book in precisely these terms, as a militant act equivalent at some level to his many and varied acts of insurgency against "tyranny and oppression" in the sphere of politics. *The Concepts and Theories of Modern Physics* performs this emancipatory mission by standing as perhaps the original example (or, let us say, the second, the first being Protagoras's lost treatise *On Truth*) of a work of philosophy in which the principle of relativity is given the character of a radical presiding dogma and serves as the basis for an entire, integrated view of the world. Though Stallo's book deals solely with a range of technical issues in physics, it invokes relativity as the transformative principle of a self-consciously modernist rationality implicitly applicable to all areas of thought—for example, to the defense of civil liberties against potential religious tyranny. The secret of Stallo's genius lies just at the unstated point of nexus between "State Creeds and Their Modern Apostles" and *The Concepts and Theories of Modern Physics*—the point at which moral and political passion transmutes itself into intellectual method and scientific discovery. As attentive readers of Poincaré, we will not embark on the fool's errand of seeking to construct a mechanical or a psychological model of this process, but we can seek to describe the streams of discourse that flow from it.

It would be hard to imagine a book more intensively, more almost monomaniacally devoted to the exposition of a single philosophical proposition than Stallo's. In the course of *Concepts and Theories*, he proclaims his cardinal axiom of "the twofold relativity of all physical phenomena" (*CTMP*, 12) scores of times, invoking it as the basis of each new analysis he undertakes. As we saw, relativity is twofold in that it encompasses the differential connections of any thing both with other things and with the mind and point of view of the observer. "Objects are known only through their relations to other objects," Stallo explains. "They have, and can have, no properties . . . save these relations, or rather, our mental representations of them. Indeed, an object can not be known or conceived otherwise than as a complex of such relations. . . . Things and their properties are known only as functions of other things and properties. . . . Relativity is a necessary predicate of all objects of cognition" (*CTMP*, 156). Not to multiply parallel citations from Stallo's book or to belabor the obvious continuities between this language and that of writers as far back and as different as Sir William Hamilton and Mansel, I will quote at length one of the most inclusive of such passages. It reaches its climax in the promulgation of a set of ten commandments, each in the form of a radical negative, in which the absolutism of those given to Moses on stone tablets is (in the allegorical idiom of physical science) systematically abolished:

> The real existence of things is coextensive with their qualitative and quantitative determinations. And both are in their nature relations, quality resulting from mutual action, and quantity being simply a ratio between terms neither of which is absolute. Every objectively real thing is thus a term in numberless series of mutual implications, and forms of reality beyond these implications are as unknown to experience as to thought. There is no absolute material quality, no absolute material substance, no absolute physical unit, no absolutely simple physical entity, no absolute physical constant, no absolute standard, either of quantity or quality, no absolute motion, no absolute rest, no absolute time, no absolute space. There is no form of material existence which is either its own support or its own measure, and which abides . . . otherwise than in perpetual change, in an unceasing flow of mutations. (*CTMP*, 201)

Even intellectually sophisticated readers at the turn of the twenty-first century, for whom the denial of "absolute truth" is supposed to be old hat, are likely to find such a passage startling for the boldness (not to mention the very early date) of its rejection of prerelativistic forms of thought. As for Stallo's original readers, none of them could possibly have failed to see that such a text heralds not just a revision of a particular set of technical doctrines in physics, but the revelation of a new heaven and a new earth, and of a transformed consciousness. The arguments presented in sharply compressed form in this passage are of course not Stallo's own independent inventions, any more than any fundamental theoretical principle ever could spring from a single brain: by the 1870s, they already had achieved a degree of currency in advanced intellectual circles through the works of writers like Spencer, Bain, Jevons, and Helmholtz (whom Stallo cites at length—*CTMP*, 200–201). Stallo's originality in the development of the idea of relativity lies in the uncompromising militancy and eloquence of his exposition, and in the forcefulness with which he puts this idea to work in a wide-ranging reconstruction of contemporary physics. In its devotion to mechanical and atomistic theory, physics seems to him to embody not just a series of mistaken beliefs but an entire superstitious and fanatical mode of thought—the same one at bottom as that anatomized in "State Creeds and Their Modern Apostles," with its prophetic vision of Jews being rounded up en masse and sent to the penitentiaries.

For Stallo, the essential fallacy of classical physics lies in its unacknowledged pervasive devotion to its irrational antiprinciple, to "metaphysics"; he structures *Concepts and Theories* therefore as a systematic unmasking of "latent metaphysical elements" in one component of physical theory after another (*CTMP*, 4). Both in the main text and in a preface where (once again performing the ritual self-cleansing disavowal of nihilism) he seeks to defend himself against "the charge of being a mere destructionist" (*CTMP*, 15), Stallo concedes readily that it is legitimate to rely in science upon provisional "working hypotheses" not intended to stand as final doctrines and claims to call into question only fundamental and unquestioned articles of the modern scientific creed. His deconstructive interrogation of them is meant to show that the vocabulary of physics, infested as it is with metaphysical parasites, results in a starkly irrational science plagued with insoluble contradictions and unable to give even a minimally coherent account of itself. Modern physics, he claims, demands unconditional allegiance to a set of absolutist postulates such as that of "the absolutely independent and passive existence of matter," no matter how illogical the explanations of these postulates may prove to be (*CTMP*, 20).

It may at first glance seem odd for Stallo to have shifted his focus from burning public issues such as slavery and school prayer to abstruse theoretical issues in physics. But it is plain that for this writer, no disjunction exists between political and scientific work, no matter how we may try to partition the two spheres. His urgency in challenging (as an uncredentialed amateur in Cincinnati, Ohio) the seemingly impregnable fortress of modern science comes from his view of scientific theory as epitomizing the native discursive mode of tyranny, in its cloaking in mystificatory mumbo-jumbo of "absolutes" a system of thought unable to bear the light of rigorous inspection and in its insistence on the duty of belief in that which we are unable to understand or even to express intelligibly. He notes pointedly in the preface to the second edition of *Concepts and Theories* that the prestige of modern physical theory has been invoked to prop up university chairs of metaphysics, to lend credence to the smuggling of religious ideology into such pseudoscientific venues as P. G. Tait's and Balfour Stewart's *The Unseen Universe* (1875), and to legitimate "dogmatic theology" preached from church pulpits (*CTMP*, 20). Physics is an influential ideological apparatus, says Stallo plainly. When he speaks of seeking by means of a rigorously relativistic critique of physics to "emancipate" his public from "thralldom" to "the insidious intrusion . . . of the old metaphysical spirit" into the realm of science (*CTMP*, 11, 4), we should not doubt that he thinks of this thralldom to absolutism as identical with other variants of that mental condition that Feuerbach calls "fanaticism," variants ultimately enforced by physical violence as well as by the sort of polemical "violence" with which, he reports, *Concepts and Theories* has been denounced (*CTMP*, 19–20). And when he states, in terms that Paul Feyerabend later echoes closely, that his goal is to establish a conception of science that will "foster and not . . . repress the spirit of experimental investigation" (*CTMP*, 4), he implicitly links the ethos of contemporary laboratory science to the wider-ranging repression of individual creativity that Mill diagnoses in *On Liberty* as a principle of modern middle-class culture. Stallo's book all by itself should be enough to refute the notion that the simultaneous development in the late nineteenth century of relativity theory in physics and the critique of absolutist morality was merely a historical "coincidence."

I cannot undertake a detailed study of *Concepts and Theories*, but I need to highlight a few of the concepts and theories Stallo brings before the judgment bar of relativity—and finds guilty.

One leading theme in the book centers on the definition of the prime category of classical mechanics, "matter." Stallo quotes P. G. Tait and other leading physicists of the day to show that the "metaphysical" conception of matter as an isolated thing in itself, such that a body can be said to exist independently of its relations to other bodies and to cognition, involves contemporary scientists in drastic contradictions. In obedience to one of the most frequently proclaimed principles of mechanics, these authorities declare that "matter is simply passive," but at the same time, the all-important concept of inertia requires them to claim that "matter has an innate power of resisting influences"—two absolutist conceptions that not only contradict one another fatally, rendering the entire system of mechanics irrational, but in themselves (for Stallo, no less fatally) fly in the face of the principle of universal relativity. Indeed, no other concept so arouses Stallo's abhorrence as that of a material body as being in itself, until impressed with movement by the impact of another body, "absolutely passive, *dead*," "absolutely dead" (*CTMP*, 13, 183). This conception is impossible to square, Stallo insists, with the twofold principle that the properties of things arise wholly from their relations with other things and that to speak of the being of a thing apart from its observable properties is to leave science and plunge into an abyss of metaphysics. "When a body is considered by itself—conceptually detached from the relations which give rise to its attributes—it is indeed inert, and all its action comes from without," says Stallo. "But this isolated existence of a body is a pure fiction of the intellect. Bodies exist solely in virtue of their relations; their reality lies in their mutual action. Inert matter . . . is as unknown to experience as it is inconceivable in thought" (*CTMP*, 182). Once physics has been radically reconceptualized in the light of "the twofold relativity of all physical phenomena," and the impossible superstition of isolated, self-existent entities abolished, the principle that matter is *active*—we might say that matter is action—becomes central. Since, as Stallo says, "mass reveals its presence or evinces its reality only by its action," since "the properties of things are nothing else than their interactions and mutual relations" and "all physical existence resolves itself into action and reaction," he proposes the formula, "a body is where it acts" (*CTMP*, 180, 200, 220, 166). In conformity with this line of thought, he launches a critique of another primary doctrine of classical physics: the impossibility of action at a distance. "The very presence of a body in space and time, as well as its motion," he declares, "implies interaction with other bodies, and therefore *actio in distans*" (*CTMP*, 182). By following out the implications of the law of Difference, Stallo thus claims to revolutionize our notion of "material presence"—or, as he puts it elsewhere, in what seems to be a theological pun, "real presence" (*CTMP*, 166, 181). His intellectual heir, Jacques Derrida, follows the same logic a century later in his claim to overturn the "metaphysics of presence" by the disclosure of the law of *différance*—the law that the components of the linguistic sign can only be constructed as what Stallo calls "[terms] in numberless series of mutual implications."[9]

Stallo carries this insistence upon the necessarily interactive and interdependent character of material objects farther than any other writer of his day known to me, and gives it more philosophical importance, but it forms a recurrent theme in the literature of the nineteenth-century relativity movement. I mentioned in chapter 1 the increasing use in this literature of metaphors of a traffic of information to describe the active interrelations of things. This trend of thinking, so powerfully developed by Stallo, lies behind Schiller's argument in 1907 that the principle of a "relation of mutual implication of self and world" requires us to reject "the notion that 'matter' must be denounced as 'dead' in order that 'spirit' may live." The same principle compels, he says, the philosophical recognition that cognition always entails *a reciprocating flow of messages*, and that even inanimate objects in this sense must be vested with intelligent awareness. A stone, Schiller goes so far as to declare, "is aware of us and affected by us on the plane on which its existence is passed, and quite capable of making us effectively aware of its existence in our transactions with it" (*Studies*, 443, 470, 442). We may take such a claim in the name of the principle of relativity as a philosophical extravagance; but in developing it as he does, Schiller is only giving voice to the radical moral and ideological implications of the theoretical trend we are tracing. "That [physical objects] do not respond more intelligently, and so are condemned by us as 'inanimate' " he says, "is due to their immense spiritual remoteness from us, or perhaps to our inability to understand them, and the clumsiness and lack of insight of our manipulations, which afford them no opportunity to display their spiritual nature" (Schiller, *Studies*, 444). A new physics will constitute a moral transformation, says Schiller. His protest against arrogantly denying intelligence and spirituality to objects unlike ourselves, and condemning them to "death" as a consequence, makes a vivid parable, for example, of the scientific and theological denial of full humanity to "primitive" peoples (often declared by well-informed contemporary authorities not to be so sensitive to pain as Europeans, to be incapable of compassionate moral feeling, and to be naturally unintelligent). It is a parable, too, of the inevitable genocidal violence visited upon these lower peoples as a result. The relativistic definition of "matter" in Victorian and early twentieth-century theory is not so far removed from issues of tyrannical persecution and of ideologically motivated violence as one might think.

Stallo's commitment to a demystified (that is, relativistic) definition of matter expresses itself in particularly dramatic form in his critique of C. G. Neumann's thought experiment designed to salvage absolute space and absolute motion by imagining the effect upon the shape of a star if all other celestial bodies were suddenly annihilated. (Ellipsoid as a result of centrifugal forces produced by its rotation on its axis, the star would, according to the theory of physical relativity as interpreted by Neumann, suddenly be at rest in the absence of other bodies and ought instantly to become spherical—obviously an "intolerable" supposition, in Neumann's view.) In his critique of Neumann, Stallo does not avail himself of the chance to observe how readily the absolutist imagination moves in this case to reveries of mass annihilation; he does, however, peremptorily dismiss Neumann's speculation on the ground that it "is forbidden by the universal principle of relativity," according to which "the very presence of a body in space and time . . . implies interaction with other bodies" (*CTMP*, 215). In a universe containing only a single body, says Stallo, the law of Difference would lapse, and the universe itself would vanish. "The annihilation of all bodies but one would not only destroy the *motion* of this one remaining body and bring it to rest, as Professor Neumann sees, but would also destroy its very *existence* and bring it to naught, as he does not see. A body can not survive the system of relations in which alone it has its being" (*CTMP*, 215). Possibly because this text of radical relativity in physics reads so much like a thinly veiled moral allegory, even so sympathetic a commentator as Bridgman (whose own philosophy of science is deeply allied to Stallo's) complains about the "primarily verbal" character of the latter's relativism and refers to the refutation of Neumann as a "paradoxical remark" (Bridgman, "Introduction," xxvi).[10] It is not intended as paradox, however, but as a straightforward expression of the inviolable rule that "nothing is one thing just by itself."

Seen in the perspective I am suggesting, Stallo's refutation of Neumann offers at the same time an early instance of what becomes a recurring figure, and one evidently charged with exceptional imaginative potency, in Victorian and post-Victorian relativity discourse. In this figure, the establishment of the reign of the absolute and thus the abolition of relativity is conceived not in the guise of a logical or philosophical error but as nothing less than *the annihilation of the world*. The introduction of the smallest particle of "absolute knowledge" into the world, says Samuel Butler, framing the idea in its starkest and most elliptical terms, would lead infallibly to "the destruction of all life and consciousness whatever" (*Collected Essays*, 1:126, 125). The career of this striking, hyperbolic-seeming leitmotif suggests again that the emergence of the relativity principle in nineteenth-century philosophy and science is bound up closely with the appearance of a new register of moral awareness keyed prophetically to the possibility of catastrophe on an unthinkable scale, a catastrophe that would eclipse the whole existing universe of moral reference. Let me digress to cite a few further instances.

In *The Will to Power*, Nietzsche, for instance, duplicates Stallo in a series of propositions that exemplifies the whole course of Victorian Protagoreanism from Spencer and Bain onward (and exemplifies, of course, his indebtedness to it):

> The qualities of a thing are its effects upon other "things."
>
> If one imagines other "things" to be non-existent, a thing has no qualities.
>
> That is to say: *there is nothing without other things*. (*Will to Power*, 2:66)

Nietzsche does not dwell here on the sort of moral insanity that would be implicit in imagining the nonexistence of "other 'things,'" but he does suggest apocalyptic consequences in any attempt to suppress the relativistic principle. "It is *the point of view* . . . which accounts for the character of 'appearance,'" he declares. "As if a world could remain over, when the point of view is cancelled! By such means *relativity* would also be cancelled!" (*Will to Power*, 2:70).

W. K. Clifford transcribes Stallo's and Nietzsche's *idée fixe* into a more particularized set of references. The denial of relativity by monolithic dogmatism, and the calamitous consequences of such an act, are exemplified by him in his prophecy of a possible resurgence of sacerdotalism—that is, the rule of a priesthood "which claims," as Clifford says, picking up the Feuerbachian phraseology, "to declare with infallible authority what is right and what is wrong" (Clifford, *Lectures*, 2:224). Such an authoritarian movement, says Clifford, would render it "quite possible that the moral and intellectual culture of Europe, the light and the right . . . may be clear swept away by a revival of superstition"; under such a regime, "civilisation [would be] perverted to the service of evil" and "the wreck of civilised Europe would be darker than the darkest of past ages" (*Lectures*, 2:233, 234, 256). Clifford's imagination fails him at this point; the abstractness of his language is the sign of his difficulty in trying to picture the annihilation of European society by the abolition of relativistic values in the modern era. Possibly it is not until 1907, when Conrad's Professor (the image not of an old-fashioned despotism but of a nihilistic authoritarianism on a new model altogether) describes his genocidal utopia of extermination camps, that a truly post-Victorian image of the reign of the Absolute comes into focus. Here the mad idea, broached by an earlier professor, of annihilating all physical bodies in the universe except one as a scientific proof of its absoluteness takes its true, essential form: that of the annihilation in effect of all *human* bodies but that of the annihilating tyrant himself.

Reflecting on these issues in the aftermath of Hitler's translation of professorial fantasy into practical action (but prior to Pol Pot's, the Rwandan Hutu's, or Slobodan Milosevic's), Hannah Arendt reprises these themes, finally, in a form that reads at times like a direct gloss of Stallo and Butler. Her wonderful essay "On Humanity in Dark Times" seems designed, in particular, to confirm the underlying equivalence of Stallo's critique of physics and his critique of political despotism. She quotes her hero Lessing to the effect that society incurs "much harm from those who wish to subject all men's ways of thinking to the yoke of their own" ("On Humanity," 26), and sets up Kant as Lessing's philosophical antithesis, as Schiller, for the same polemical purpose, sets up Plato as Protagoras's. "Kant argued that an absolute exists, the duty of the categorical imperative which stands above men, is decisive in all human affairs, and cannot be infringed even for the sake of humanity"—just as Matthew Arnold declares that the duty to preserve the British social order against disturbance is sacred and must never be compromised, even in the name of such a cause as the abolition of slavery. "The categorical imperative is postulated as absolute and in its absoluteness introduces into the interhuman realm—which by its nature consists of relationships—something that runs counter to its fundamental relativity" ("On Humanity," 27). Any such absolutist postulate, says Arendt, expresses "the inhumanity which is bound up with the concept of one single truth" (27). Absolutism is inhuman, she says, invoking anew the apocalyptic trope that resurfaces time and again in this literature, in the sense that the establishment of "a single absolute truth" would abolish discursive interactions among people and would in this way spell "the end of humanity" (27). Were that impossible thing, an absolute verity, ever to be discovered, it would be a catastrophe, says Arendt,

> because it might have the result that all men would suddenly unite in a single opinion, so that out of many opinions one would emerge, as though not men in their infinite plurality but man in the singular . . . were to inhabit the earth. Should that happen, the world, which can form only in the interspaces between men in all their variety, would vanish altogether. (31)

This, I take it, is precisely the moral allegory subtending Stallo's "paradoxical remark" about Neumann's thought experiment on behalf of absolute values in physics, and subtending Butler's remark about the annihilating effect of "absolute knowledge."

Stallo's demonstration of the irrational and "metaphysical" character of classical mechanics covers almost the whole gamut of its central doctrines and offers a clairvoyant glimpse, for the most part, of the revolutionary scientific trends of the succeeding century. He thus criticizes the atomic hypothesis, exposing fatal flaws in the concept of absolutely hard, inert, unchanging, determinate elementary particles of mass, flaws that remained insoluble until the advent of quantum mechanics. He denounces the concept of force as "an independent, substantial entity," asserting instead that "all force is essentially a stress—an action between two bodies," which is to say, "a relation between two terms at least" (*CTMP*, 36, 37, 188). The relation called "force" runs both ways, of course. "Without its relation to and union with force or motion, [mass] has no existence, just as force or motion has no existence without its relation to and union with inertia" (*CTMP*, 170). He (mistakenly) repudiates the kinetic theory of gases on the grounds that its idea of particles moving in straight lines until colliding with other particles "is based on a total disregard of the relativity and consequent mutual dependence of natural phenomena." Bodies "which . . . move independently without mutual attraction or repulsion or any sort of mutual action . . . are unheard-of strangers in the wide domain of sensible experience," he declares (146). He rejects the idea of space as "an absolute, self-measuring, objective entity" (116), and he proclaims the relativity of time in terms almost identical to those used by Einstein more than two decades later in the opening of his 1905 special-relativity paper. Time, says Stallo, is "measured by the recurrence of certain relative positions of objects or points in space"—time and space are as inseparable from each other as are force and mass, in other words—and "the periods of this recurrence are variable, depending upon variable physical conditions" (219). Different systems of time, says Stallo, are produced by different physical conditions! No wonder Bridgman opines that Stallo "would probably have found Einstein's special theory of relativity completely congenial" (Bridgman, "Introduction," xxvi).[11]

In common with other early proponents of relativity, Stallo never quite articulates the conceptual hinge of his philosophy, though he implies it by stating in the first sentence of his original preface that he sees *Concepts and Theories of Modern Physics* as "a contribution, not to physics, nor, certainly, to metaphysics, but to the theory of cognition." The origin of the modern principle of relativity, as we have seen, is that body of nineteenth-century speculation, formulated by writers like Hamilton, Mansel, Bain, and Spencer, that goes under the name of "the Relativity of Human Knowledge"; Stallo shows his indebtedness to this lineage by arguing that in the realm of scientific theory, *human cognition is primary* and, if one may use the term that for him was anathema, *absolute*. The dogma tacitly underlying all of his analysis of physical concepts—and underlying the relativity movement in all its nineteenth-century unfolding—is the positivistic rule that proclaims that scientific reality and the field of actual or potential human knowledge are strictly coextensive. All that we can possibly know of the external world is our own knowledge of that world. That which cannot manifest itself as knowledge in observable and intelligible form, such as "force" or "things in themselves," has consequently no physical reality that could ever pertain to science; it is "metaphysical" or a "fetish." Out of this founding postulate, rarely explicitly stated and sometimes disavowed, comes the powerful tendency of all the different strains of relativity thinking from Protagoras onward to collapse together the categories of the natural and the mental, and to render the objective and the anthropomorphic identical—a tendency we explore in chapter 4.[12]

Stallo by no means goes so far in this philosophical direction as, say, Clifford, who declares that "the subject of science is the human universe; that is to say, everything that is, or has been, or may be related to man," and that "the universe . . . consists entirely of mind-stuff" (Clifford, *Lectures*, 1:141, 2:72). But he does place great emphasis on the principle that "thought deals, not with things as they are, or are supposed to be, in themselves, but with our mental representations of them," which is to say, with a world of *symbols*—a doctrine he traces historically from Leibniz through Hamilton and Spencer (*CTMP*, 156–57). Since, according to this view, all we can possibly know of nature is forms of our own thought, *to state a law of human cognition—a law of symbolism, in other words—is to state a law of nature*. No distinction could be drawn in a scientifically rigorous way between "the relativity of human knowledge" and relativity as a principle of "the universal order of nature" (*CTMP*, 18): these, for Stallo, are two verbal forms for one and the same indivisible category. This conception abolishes "absolutes" while yielding what for Stallo is the only authentically rational basis of scientific objectivity, the only one that does not rely on mystification and does not lead to tangles of logical contradictions like the ones he shows to proliferate in classical physical theory. It leads, however, to a world of knowledge that must be forever indeterminate, a world bound to appall those devoted to fallacious ideals of perfection and of what Stallo derisively calls "absolute permanence and immutability" (*CTMP*, 12):

> The concepts of a given object are terms or links in numberless series or chains of abstractions varying in kind and diverging in direction with the comparisons instituted between it and other objects. . . . All thoughts of things are fragmentary and symbolic representations of realities whose thorough comprehension in any single mental act, or series of acts, is impossible. And this is true, *a fortiori*, because the relations of which any object of cognition is the entirety, besides being endless in number, are also variable—because, in the language of Herakleitos, all things are in a perpetual flux. (Stallo, *CTMP*, 158–59)

The delusive mental operation philosophers and scientists perform in order to escape this condition of incompleteness and flux is, according to Stallo, the very one that Feuerbach identifies as the theological and that Marx subsequently identifies as the ideological mechanism of capitalism: it consists in reifying a manufacture of human thought into an estranged, independent, permanent, "absolute" reality external to ourselves and endowed with dictatorial authority. Just as liberation from religious superstition can only come, according to Feuerbach, from the recognition that God is "a mere figment devised by man" (*Essence*, 75), so Stallo's project of emancipation from the thralldom of "metaphysical" science rests on the principle that "[the] isolated existence of a body is a pure fiction of the intellect" (*CTMP*, 182): he displaces Feuerbachian critique from theology to physics simply by writing "a body" where Feuerbach would have written

"God." Stallo claims to cleanse physical theory of such mystifications purely by means of rigorous logical analysis, and in so doing he traces a path of reasoning that subsequent generations of scientists have been compelled to follow. But I hope it is clear by now that fully understanding a text like *The Concepts and Theories of Modern Physics* would be impossible if we were to perform an estranging maneuver of our own by abstracting its apparatus of logical analysis from its motive of ideological critique, specifically that of "dogmatic theology" in the broad sense of the term. In dismantling the metaphysics of absolute value in one field of modern science and replacing it with a system founded on the joint principles of Difference and universal reciprocity, and in seeking by the same token to make possible a scientific establishment likely "to foster and not to repress the spirit of experimental investigation," Stallo makes clear his conception of himself as a liberator of human thought from a broad spectrum of institutions of authority—institutions whose ultimate recourse, as he often reminds his readers, is always to violence.

High-minded idealist and activist that he was, Stallo does not provide good evidence for the old libel that relativism, by destroying "truth," destroys moral fiber and leads to "cynicism," "callous indifference," and "apathy." He would be more vulnerable to the opposite charge, that of creating a new "dogmatic theology" of his own based on the sacred axiom of "the twofold relativity of all physical phenomena," an axiom infallible everywhere in the universe for all time and marking precisely the divide between the wicked and the righteous in intellectual matters. His admirer Stillman Drake suggests this problem in referring to Stallo's championship of "what might be called, for want of a less paradoxical term, an absolute relativism" (Drake, "J. B. Stallo," 27). That Stallo is himself an absolutist, and that his near obsessive criticism of absolutism is to some degree a signal of intellectual discomfort on his part, is the clearest thing about him. The emergence of the principle of relativity in the nineteenth century may have rendered modern methods of analysis immensely more sophisticated and precise, and this principle may even correspond, as its partisans never cease affirming, to a genuinely revolutionary moral ideal; but it can lay claim to no more logical coherence, in the last analysis, than can the absolutism that it both opposes and seems destined to reproduce.

To scholars accustomed to thinking of knowledge in every domain as a phenomenon of cultural imagination, the discovery that early relativity science and radical political critique are linked intimately as this chapter has argued they are may seem an unsurprising conclusion hardly calling for so much demonstrative apparatus, even though this line of analysis can scarcely be taken for granted, considering how vehemently it has been ruled out of court by spokesmen for the immaculacy of scientific reasoning. The typical pattern among relativity theorists is the one exemplified by Stallo or, say, by Einstein, in which a writer sharply partitions the two registers of argument, the moral/political and the scientific, into parallel but separate texts like "State Creeds and their Modern Apostles" and *The Concepts and Theories of Modern Physics*. One could insist that any collapsing together of the two registers in the absence of definite evidence of causal linkage is untenable. But no such evidence can ever be forthcoming. Bereft of it and faced with the sorts of textual materials we have surveyed, the best a historian can do is to tell the most coherent possible story about them that is consistent with the assumption that the scientific imagination has a specific cultural history—or that at least does not rule out this possibility ahead of time. To isolate "State Creeds" from *Concepts and Theories* so as to divorce Stallo's clairvoyant Einsteinian science from his lifelong political militancy would seem from this point of view arbitrary and bound to do violence to historical understanding. One makes an admittedly speculative and inconclusive argument based on the principle of interpretive coherence as the alternative to this sort of mystifying ideological violence. "Whatever we do," as Samuel Butler says, "we should not . . . try to remove our actions from the category of the speculative when in reality we speculate with every step we take" (*Collected Essays*, 1:140).

CHAPTER THREE

The Relativity of Logic

> Irrationalism . . . rules unnoticed and uncontested in the defense of "logic."
>
> —Heidegger, "Letter on Humanism"

Throughout its vexed career from Protagoras to Einstein to the present day, the theory of which the central principle is that nothing is one thing just by itself has been identified with "irrationalism." This view of relativity thinking was particularly characteristic of the second half of the nineteenth century, given the idolization of science that was central to Victorian intellectual culture and given the popular idea of the method of scientific investigation—its supposedly rigid system of true/false logic and its strict protocols for analysis of factual data—as identical to rationality itself. The current creed of advanced thought in every domain, says Arthur James Balfour in 1879, is "that Science is the one thing certain," a state of affairs he regarded as a recent cultural phenomenon, the result of the appearance of a new class of literature in which experts, in mediating scientific technicalities to the public at large, indoctrinate nonspecialist readers into a cult of the perfection of scientific knowledge (Balfour, *Defence*, 308–9). Well might Poincaré begin *Science and Hypothesis* [hereafter *SH*] in 1902 with the threateningly hedged statement, "to the superficial observer scientific truth is unassailable, the logic of Science is infallible." Scientists are popularly supposed, says Poincaré, to attain certainty thanks to the "chain of flawless reasonings" mathematics allows them to construct on the basis of direct observations of nature (*SH*, xxi). Isaiah Berlin has subsequently argued that the hegemony of the ideology of logical reason in the period we are considering was every bit as unqualified as such statements suggest. "The basis of all progressive thought in the nineteenth century," he incautiously asserts, was the concept that scientific knowledge formed "a great harmonious system, connected by unbreakable logical links and capable of being formulated in precise—that is, mathematical—terms" (Berlin, *Crooked Timber*, 5). The component principles of this concept in fields from ethics to natural science were threefold, according to Berlin:

> In the first place, that . . . all genuine questions must have one true answer and one only, all the rest being necessarily errors; in the second place, that there must be a dependable path towards the discovery of these truths; in the third place, that the true answers . . . must necessarily be compatible with one another and form a single whole, for one truth cannot be incompatible with another. (*Crooked Timber*, 5–6)

The cultural dominance of this credo, in which epistemological absolutism (one true answer and one only) is equated with the infallibility of logical reasoning, suffered a shock from which it has yet to recover with the publication of the special theory of relativity in 1905. Einstein's theory was widely viewed by critics as not merely scientifically indefensible but an affront to rationality itself. The 1905 paper was arguably less significant as the establishment of a perfected system of measurement of relativistic physical phenomena—Lorentz's transformation equations and FitzGerald's contraction hypothesis were perfectly adequate for this purpose—than as the proclamation of *a new mode of scientific logic* which by contemporary standards was stupefyingly irrational. The astounding principle of the constant velocity of light, in particular, was declared by many critics of special relativity to be manifestly "a logical contradiction" that "[offended] *a priori* self-evident judgments" and thus seemed "absurd" from the point of view of conventional scientific reason (Wenzel, "Einstein's Theory," 587).[1] Sir Arthur Eddington, an early interpreter of relativity theory, emphasizes its seeming irrationality—as latter-day canonizing interpreters of it rarely remember to do. For instance: an electrically charged body at rest on the earth has no associated magnetic field; but the same body viewed from a distant nebula is in motion relative to the nebula, and thus *generates a magnetic field*. "How can the same body both give and not give a magnetic field?" asks Eddington. This may seem starkly illogical, a violation of the all-important logical law of noncontradiction, but in relativity theory, both effects are simultaneously true: magnetic fields, like every other physical phenomenon, "are relative" (Eddington, *Nature of the Physical World*, 22). In the intellectual culture of 1928, at least among theoretical physicists, statements that to an earlier generation would have seemed well nigh unintelligible had come to count as rational. Gaston Bachelard, stressing the shift, thus portrays special relativity as "a systematic revolution of basic concepts" involving first and foremost the establishment of "a relativism of the rational and the empirical," which is to say, "a Nietzschean transmutation of rational values" (Bachelard, "The Philosophical Dialectic," 565, 567). What greater offense against reason could there be than the assertion that the rules of rationality are not fixed and immutable, but are *themselves relative to empirical circumstances?*

It would be misleading, however, to think of the relativistic revolution in logic as occurring in the form of a sudden epistemic break in 1905. This is the view one might well get from a strict history-of-science perspective, especially one deeply invested in the cult of the genius of Einstein. In fact, over the course of several previous decades, a party of theorists had set in motion a "general movement of intellectual reconstruction" that sought nothing less than to abolish the scientific model of "a fixed universe hung on a cast-iron framework of fixed, necessary, and universal laws" (Dewey, *Influence of Darwin*, iv, 13, 73)—a universe, that is, suited to rigorous logical analysis. Anxious late-Victorian testimonies accumulated accordingly to the effect that making "the principle of Universal Relativity" (Bain, *Logic*, 1:255) and its cognate, the evolutionary principle, central to scientific thought necessarily entailed a radical reevaluation of the laws of logic. Hence the polemics of the biologist St. George Mivart, who confirms Berlin's image of the nineteenth century by squarely identifying the defense of science with that of the principles of formal logic, which he portrays in 1889 as being under subversive attack. "If we deny or doubt about the law of contradiction," he asserts, for example, "we are thereby landed in absolute scepticism, which . . . is absurd" (*On Truth*, 42).[2] His special anathema falls on "the doctrine of the relativity of knowledge," which is, he declares, logically self-refuting (Mivart, *Groundwork of Science*, 280–81). Science can rely with absolute confidence on the established principles of logic, he declares, since it has become obvious "that something analogous to reason . . . pervades the great whole [of nature]" (*Groundwork*, 288).[3] Not even the anti-absolutist William James was immune to anxiety about the seeming irrationality of relativity. How can one speak of determinate relations between terms if the terms themselves have no free-standing reality? Writing in 1890, he quotes Bain's "doctrine of relativity" as meaning that we cannot know "any one thing by itself, but only the difference between it and another thing" and protests that "if this were true the whole edifice of our knowledge would collapse" (James, *Principles of Psychology*, 12).[4] Nietzsche himself, in *The Genealogy of Morals* (1887), one of the radical statements of relativistic philosophy of its day ("there is only a seeing from a perspective, only a 'knowing' from a perspective") and a text that eulogizes contemporary "ascetic" freethinkers for their "determined reversals of the ordinary perspectives and values," inconsistently also reviles these same thinkers for perversely delighting in "an infliction of violence and cruelty on *reason*" (*Genealogy*, 153, 152, 151).

The multiplication of such comments bears witness to the gradual emergence of the intuition that logical incoherence or absurdity might after all form an ineradicable element of the natural order, and that modern-style scientific thinking might need to incorporate it. To trace this process, as I do in this chapter, is to document the gradual formation of the "thought style" required to ensure the successful propagation of a paradigm-altering new idea. I speculate accordingly that the development of a body of subversive Victorian commentary on the rules of logic formed the historical precondition for the advent of the special theory of relativity. This is not to say that Einstein was influenced directly or indirectly by any of the writings studied below, but just that this iconoclastic discursive tradition enabled the "absurd" propositions of special relativity to seem not only plausible to advanced thinkers when they were put forth, but so irresistibly persuasive that they were rapidly accepted despite their irrational character, even in the face of angry denunciation and what seemed initially to be decisive experimental disproof.[5] Somehow the "apparently absurd state of affairs" (Rindler, *Introduction to Special Relativity*, 8) created by Einstein's law of light propagation had become by 1905, not grounds for rejecting special relativity outright, but the reverse, a mark of intellectual legitimacy.

I argue, then, that mutinous dissidence with regard to the "laws" of logical reasoning appears not as a byproduct of relativity, but as one of its governing motives. No other aspect of the relativity movement expresses more decisively the furious drive of modernist intellectuals to emancipate themselves from traditional structures. Much textual evidence inclines one to speculate, indeed, that the relativity principle was constructed expressly for throwing the regime of a priori logic into a state of constitutional crisis in which it would be compelled, if not to abdicate its office, at least to share power with new forms of thought that give only conditional validity to the law of contradiction and other supposedly absolute dictates of reasoning. In other words, the emergence of a coordinated radical critique of scientific reason on relativistic and Darwinian grounds must be understood as the historical *Doppelgänger* of the extravagant Victorian reverence for science. The relations of the two trends were complexly reflexive all along: on the one hand, the critique of reason invariably took the form of hyper-reason, of scientific thought reflecting pitilessly upon itself; on the other, the strong reassertion of the ideology of classical science and "the moralization of objectivity" was itself in part a response to the subversive new movement.

The relativistic insurrection against the sovereignty of logic can be narrated as a counterpoint of two phases. What can be called the first phase (though it has more an intuitive than a historical primacy) is the militant emancipatory phase, in which the rhetorical stress is on expansiveness and creativity and the assertion of freedom. The second, more rueful phase expresses itself in attitudes of philosophical pessimism or resignation. Relativistic knowledge in this phase is defined by its necessary failure. In place of Napoleon, the object of idolatry for Nietzsche (the consummate figure of the militant first phase), another symbolic protagonist is invoked by a series of second-phase authors: Sisyphus. This phase of relativity thinking returns in a sense to its origins in Spencer's mythology of "the Unknowable" and in his insistence that the incalculability of consequences renders radical political and social action unacceptably risky. However, my project is not to try to resolve these contrasting vectors of discourse into a single historical pattern, but to unpack their manifestations in specific texts and to try to situate them within the larger drama of modernist intellectual and spiritual self-construction.

Mill and Schiller on Logical Necessity

The way to a relativistic critique of logic was in fact opened by the definitive Victorian textbook of logic, J. S. Mill's *System of Logic* [hereafter *SL*] (1843). Mill has little to say about the principle of relativity in this work (in contrast, say, to his study of Sir William Hamilton, where he identifies it as the key to philosophy), but he does devote himself to a systematic overturning of the supposedly absolute and necessary character of general logical principles such as the law of noncontradiction in favor of the theory that these principles are "only empirical laws, resting upon induction by simple enumeration merely" (*SL*, 350). They do not have the character of revealed truths, nor do they have the capacity to impose themselves upon the mind by their undeniable self-evidence. Mill's goal is thus to destroy "that mystical tie [between natural things and the rules of logical thought] which the word *necessity* seems to involve, and the existence of which, even in the case of inorganic matter, is but an illusion produced by language" (*SL*, 340). This disillusioning analysis formed a key part of Mill's broad campaign against the appeal to intuitive, self-evident a priori principles (as exemplified, for example, by his philosophical *bête noire* William Whewell), which, he believed, served as the chief theoretical instrument of the forces of orthodoxy and domination in his society. Demystifying the theory of the necessity of logical principles was thus inseparable, for Mill, from the critique of the status quo in nineteenth-century society. The law of noncontradiction was no more a necessity of the order of nature than was the subjection of women. Yet it is important to insist that his advocacy of the supremacy of induction allows no scope for willful departure from logic, and that logical necessity prevails after all in Mill's universe. Induction from experience—inference from particulars to particulars—is genuinely creative, in contrast to the pseudoinferences of syllogistic reasoning; but its operation is as strictly dictated by factual reality as is Whewell's intuitionism by the supposed self-evidence of general propositions.

Yet the possibility of radical relativity peeps out dangerously like a Satanic hoof from Mill's analysis of scientific logic. For Mill, the fundamental law of logical thought is that of uniform causation, which he asserts to be equal in certainty to the first principles of mathematics, despite its "merely" empirical character (*SL*, 343). Indeed, inductive logic is for Mill primarily constituted by causality (215). But he makes the striking reservation that we cannot assume the law of causation to be universally valid. "In distant parts of the stellar regions, where the phenomena may be entirely unlike those with which we are acquainted, it would be folly to affirm confidently that this general law prevails," he remarks. "The uniformity in the succession of events, otherwise called the law of causation, must be received not as a law of the universe, but of that portion of it only which is within the range of our means of sure observation" (342). Elsewhere, logic, geometry, and arithmetic could all assume different forms. All the laws of thought, being "merely" empirical, are limited in validity, says Mill, to the "time, place, and circumstance" in which they are determined (349). In other words, Mill expressly postulates in this famous passage[6] that "relativism of the rational and the empirical"—that variability of logic according to circumstances—that Bachelard identifies with Einstein.[7] The passage may amount only to a passing qualification in the *System of Logic*, but it points to an array of radical developments in subsequent theory. Most obviously, it seems to prophesy the eventual scientific discovery of the breakdown of causality and logic not in other galaxies after all, but in the atomic structure of matter. And it points to modernist speculations that the unimaginably remote regions where logic becomes illogical are simply those of other human societies. The validity of the law of noncontradiction may "depend upon social conditions," says Durkheim in 1912, for example. "The rules which seem to govern our present logic . . . far from being engraven through all eternity upon the mental constitution of men . . . depend, at least in part, upon factors that are historical and consequently social" (Durkheim, *Elementary Forms of the Religious Life*, 25). Such is the potential outcome of having undone "that mystical tie" between logic and reality that Mill exposes as purely a poetic effect, "an illusion produced by language."

We may assess this outcome still more comprehensively with reference to another text of 1912, F. C. S. Schiller's *Formal Logic: A Scientific and Social Problem* [hereafter *FL*]. This book, Schiller's magnum opus, can be said to give codified, official form to the nineteenth-century relativistic critique of rationality as it derives from Mill (eulogized as "our leader" by Schiller's close associate William James [James, *Pragmatism*, 3]); it closely echoes in many points, notably, Nietzsche's assault on logic in book 3 of *The Will to Power*.[8] Schiller, the professed disciple of Protagoras, surveys the whole system of formal logic, doctrine by doctrine, exposing each point in the system as being riddled with "self-contradictions and absurdities" (*FL*, 11). Denying the feasibility of defining terms in logic in accordance with their "essential" properties—an impossibility, since "everything is always related in some way or other to other things" (28)—is only the beginning. Schiller ridicules the three sacrosanct "laws of thought" (the laws of Identity, Contradiction, and Excluded Middle); denies the reality of self-evident first principles; denies the possibility of determinate causal analysis; and performs many other startling acts of philosophical rebellion along the way. For one thing, he follows both Nietzsche and Dewey in laying great stress on the "upheaval in logic" (56) which results necessarily from evolutionary theory, since formal logic, he argues, presumes fixed and stable conceptual entities and loses its validity in a world in which all things are in flux. All three great laws of thought are thus nullified by "the paradox of Change" (117). (Nietzsche: "Logic only deals with formulae for things which are constant. . . . *Knowledge* and the process of *evolution* exclude each other"—*Will to Power*, 2:33–34.) Particularly noteworthy is Schiller's destructive application of this principle to the law of noncontradiction (*FL*, 121–23), which traditionally, Nietzsche says, figures in logical theory as "the most certain of all principles . . . the most ultimate of all, and the basis of every demonstration" (*Will to Power*, 2:31).

The fundamental error of formal logic, which is replicated, Schiller says, in each of its specific doctrines, is to consider reason as an abstract structure disconnected from real processes of thought; thus logic dogmatically erects a system of false "absolutes" in place of the wholly relative constituents of actual reasoning. Logic condemns itself to the status of "an unmeaning pseudo-science" (*FL*, 384) by failing to see, in particular, that rational arguments and the materials on which they operate are only to be understood and evaluated in relation to particular human purposes. In place of the insistence of professors of formal logic that "Truth, being *absolute*, is true without regard to circumstances" (398), Schiller proposes a pragmatic scheme in which logical truths are understood to be limited to local, temporary, provisional, relative validity. He presents this conceptual shift not as a compromise or a mere adjustment in established doctrines, but as a "transmutation of rational values" in which, among other drastic consequences, unity dissolves into uncontrollable multiplicity. This effect is apparent with regard to the logical category of Definition, for example. If relevance to an inquirer's purpose is the primary requisite of definition, "it follows that the essences and definitions of things are necessarily plural, variable, and 'relative,' and never 'absolute,' " says Schiller; "there may have to be as many definitions as there are purposes, and the neat finality of the Aristotelian scheme is shattered beyond repair" (70, 69).

Unlike logicians, whose models of thinking are wholly formalistic and abstract (and thus, he claims, vacuous), Schiller imagines reasoners as *actors* embedded in particular circumstances and endowed with psychology. By insisting on such factors, he gives his relativistic science of thought a powerfully existential character with unmistakable ideological overtones. Reasoning for Schiller is above all *a free activity governed by the will and intelligence of the reasoner*. It is not subject to the authority of rigid "necessary" laws possessing "coercive" power, and there is no logical path to the attainment of the "absolutely certain truth" (*FL*, 192) sought vainly by logicians. In a relativistic, ambiguous, dangerously unpredictable world like the one we inhabit, thinking is not an affair of syllogistic rules but is at every point a "thoroughly purposive, selective, and personal process," an expression first and foremost of human freedom; even such rudimentary logical acts as postulations of identity "are adventures of thought and always involve a risk" (127). The "laws" of reasoning are not axioms, as logicians claim, but *postulates* we invent and freely use as far as they can help us achieve our purposes. We employ them not to seek to attain the chimera of "categorical certainty" (224), but to expand and perfect our knowledge in the midst of a condition in which eternal uncertainty is the only certitude.

Schiller thus in effect subjects logic and logicians to the same critique Feuerbach momentously formulated more than seventy years before in reference to theology and theologians. In both cases, a motivated human invention (the system of logic; God) is misconceived as a reality existing independently of mankind and as endowed with dictatorial powers. Absolutist logic is invoked by its advocates, Schiller shows, as if it were a system of infrangible divine ordinations, and as if to dissent from, say, the law of noncontradiction were to commit sacrilege. With regard to the scientific "facts" that supposedly form the substratum of logical inference, "relevance, purposiveness, and selection deprive scientific 'fact' of its absoluteness and 'independence,'" Schiller declares; the selecting of any fact for investigation is thus "an 'arbitrary,' artificial, human interference with the given"; "alike for science and for action *every 'fact' is man-made*, as a condition of its being a particular fact at all" (*FL*, 269, 282). (Nietzsche: "There is no such thing as a 'fact-in-itself,' *for a meaning must always be given to it before it can become a fact*"—*Will to Power*, 2:64–65.)

Despite telling hints along the way, Schiller's own compelling purposiveness in his long deconstructive study of formal logic is not allowed to transpire plainly until the end. The hints come in such forms as his analysis of the allegedly "necessary" quality of the rules of logic. The syllogism, he declares here, was in its origin a practical invention designed by Aristotle to ensure "that an opponent should be *forced* to confess himself beaten" in public disputation if he violated its formalized rules; it was a device to "*compel* him to surrender" by means of "a method so coercive that, once committed to it, there was no escape for any one" (*FL*, 189, 190). In the final pages of *Formal Logic*, the implications of this view are given unsparing polemical emphasis—at the cost of occasional slippage of the stylistic control that elsewhere marks Schiller's urbanely elegant, sardonic prose. It would be a grave error, he says, to regard formal logic, with its code of absolute truth and its claim to coerce the thinking of every citizen through the mechanisms of the "laws" of thought, as merely an innocuous academic activity without practical significance. By implication, he dissents from Mill's critique of syllogistic reasoning as essentially vacuous and, in disguising merely empirical proof as "necessary," as claiming more creative power than it possesses. In reality, says Schiller, the syllogism is a fundamental ideological formation calculated to foster every kind of tyranny in human society. Logic is a sociopolitical institution, in other words: a cognate of that trend toward "centralized administration and . . . compulsory regulation" that Spencer decried in Victorian society (Spencer, *Principles of Sociology*, 1:570). In arrogantly severing itself from the relativity of real intellectual life, logic "has become brutally and blindly dogmatic, and unaccustomed to argue reasonably," Schiller asserts (*FL*, 386). Given that "its ideal of formal perfection is *Fixity*," it naturally appeals to and tends to promote "the blindest and most intractable sort of conservatism" (397). (Nietzsche: "Reason is the road to a static state"—*Will to Power,* 2:88.) But more than this, formal logic is fueled by a desperate impulse *to abolish human freedom*.

> Formalism's "ideal" of the motion of Thought is that it should be, not *free*, but *compulsory*. Even as a slave's evidence was not good in Roman Law unless it had been given under torture, so a conclusion is worthless in Formal Logic unless it has been *forced* upon the mind. "Inference" is to be "logically necessary," all "proof" is to be "coercive." Its aim is to terrorize, and not to attract. Truth is to be believed, not because it is desirable and good to believe, and better than error, but because it imposes itself by sheer force on a mind that "cannot help" believing it. . . . Evidently this "ideal" has educational affinities with the barbarism of the old disciplinary methods; but is it calculated to promote a *love* of truth? (*FL*, 397)

"Necessity," Schiller adds, "is as evidently the tyrant's plea in logical as in political absolutism, and neither has any use for the freedom of human activity" (*FL*, 398).[9]

Schiller's argument against necessity is at bottom a critique of the central logical/metaphysical dogma of the singleness of Truth. In sharp distinction from commentators like Mivart, Gellner, or Gross and Levitt, who identify the authority of science with the rigid absolutism of One True Vision, Schiller idealizes science as a utopian field of fluid, humane experimentalism, a privileged domain of human freedom. "If Truth is formally *one*, and there can Formally be but one true theory of anything, it is clear that it leaves no room either for a plurality of sciences or for a plurality of theories within each science"; but in fact, he says, the sciences "praise, and largely practise, a Freedom of Thought, which involves difference of opinion, and a plurality of theories which are actually held to be true, and are treated with tolerance if they promise to promote the growth of science" (*FL*, 400). The sciences thus defy the intolerant regime of logic, which leads by its very nature, says Schiller, directly to demagoguery and totalitarianism.

> The absolute system of immutable Truth is *one*. No more than one view, therefore, can be true. You either have The Truth, or you have not. If you have it not, you are lost; if you have it, no one should dare to contradict you. You do right, therefore, to get angry with those who dispute the Truth. The Truth is yours, *nay, it is you*. . . . *La vérité c'est moi*, the Formal logician can then proudly say. (*FL*, 398–99)

Nor does the code of a single, logically demonstrable Truth authorize fierce combat merely on the intellectual level, as we can see with special clarity in the long history of religious terrorism, declares Schiller (who in this polemical turn echoes Feuerbach, Stallo, Clifford, and others, and whose fictional formal logician has an unmistakable affinity with Conrad's megalomaniac Professor, who dreamt several years before of global extermination of the unfit in the name of his own transcendent code of value). "To conceive The Truth as compulsory and coercive is in principle to authorize every form and measure of persecution," Schiller says. "It makes the sword and the stake the proper instruments for effecting religious conversions" (402). If Mill sought to strip the syllogism of its "mystical" authority in order to foster a Victorian intellectual culture more conducive to progressive reform, Schiller returns to the same project at the outset of the twentieth century from what must have seemed to contemporaries an exaggerated dread of the appearance of regimes of total persecution whose characteristic institution would be "a method so coercive that . . . there [would be] no escape for any one."

Schiller's manifesto against the cultic institution of logic makes a striking point of reference for a wide swath of postmodern critique.[10] But if *Formal Logic* looks forward in this way to later critical movements that have forgotten its author's name, it also looks back to its own enabling tradition—the tradition constituted by a broad Victorian critique of scientific logic in the name of "the law of relativity." This critique, gingerly touched on by Mill, was prosecuted with great force by a series of nineteenth-century writers of divergent theoretical and political orientations. Briefly tracing this series will not render the paradoxical ideological system of nineteenth-century discourse any less perplexing; if anything, it will do the reverse. But it will enable us at least to gauge at what a deep level of the nineteenth-century imaginary these themes originate.

Newman's Grammar of Relativity

The same year that saw publication of John Henry Newman's *Grammar of Assent*, 1870, also marked the publication of Bain's *Logic*, which set up "the principle of Universal Relativity" as the fundamental rule of philosophy (Bain, *Logic*, 1:255) and argued on this basis that any element of thought was logically equivalent to its antitheses. Logical inference became in this model a dizzying play of Difference from which absolutes and essences were banished, and in which the law of noncontradiction was utterly confounded. Rational thought in Bain was conceived to be not a structure of rigid component parts (definite identities, innate ideas, interlocking laws of logical construction) but a mercurial element in which instability or "change of impression" (1:8) was all. Newman's central category of "certitude," a settled conviction reached by a course of logical analysis, has no relevance whatever to such a scheme as Bain's; certitude in this perspective is just another name for the moment of the "change of impression" into one of its negative opposites.

Bain's assault on "the doctrine of the Absolute" (Bain, *The Senses*, 9) in the name of Universal Relativity formed a counterpoint to Spencer's. Without naming him, Bain condemns Spencer's argument that the doctrine of relativity logically implies the existence of the nonrelative (Bain, *Logic*, 2:392); more generally, he develops his own uncompromisingly relativistic theory ("the principle of Relativity, if true at all, must be true without reservation or exception," he declares—*Logic*, 1:66) as an alternative to that strain of philosophical pessimism in Spencer that is so shockingly dissonant with the prevailing optimism of Victorian public discourse—and, for that matter, with the nominal optimism of Spencer's own progressivist evolutionary scheme. For Bain, dismantling absolutist logic is not a matter of questioning the possibility of knowledge, but rather, of imagining and fostering an invigorated mode of thought freed from stasis and rigidity. Logical indeterminacy in Bain's system is inseparable from a condition of vitality and unimpaired humanity: so we may infer from *The Senses and the Intellect*, his work of fifteen years earlier. "As change of impression is an indispensable condition of our being conscious, or of being mentally alive either to feeling or to thought," says Bain here, "every mental experience is necessarily *twofold*" (*Senses*, 8). To abridge the contradictory doubleness of any experience in the name of Aristotelian logic, Bain implies, is to diminish mental life itself.

Spencer's relativity, by contrast, is based on an overriding intuition of the necessary failure of thought in its longing to grasp the truth of the world in "absolute," unconditional terms. His version of the relativity principle—"thinking being relationing, no thought can ever express more than relations" (*First Principles* [hereafter *FP*], 63)—signifies for him, first and foremost, the incurable unknowability of things and thus the profound incoherence of scientific rationality. Analytic reason comes at last only to the discovery of the irresolvable contradictions lying at the heart of all logical terms; its inherent tendency is to unravel itself.

> Objective and subjective things [the man of science] thus ascertains to be alike inscrutable in their substance and genesis. In all directions his investigations eventually bring him face to face with an insoluble enigma; and he ever more clearly perceives it to be an insoluble enigma. (*FP*, 49)

The inscrutability of its objects would seem to render the exercise of scientific reason a hopelessly futile enterprise. Spencer, though giving voice to so striking a vein of modernist disenchantment, remains none the less determined not to succumb to philosophical despondency or, as he says, to commit "that error of entire and contemptuous negation, fallen into by most who take up an attitude of independent criticism" (*FP*, 8). He adopts therefore a series of compensatory maneuvers, of which the most flagrant is the assertion that radical relativity implies the "absolute" existence of things (which remain unknowable even so, however). At the same time, he describes his philosophy of relativity not as a denial of knowledge but as merely a warning against intellectual hubris and as an antidote to the pathological spiral of skepticism engendered by an anarchy of theories. "The conviction . . . that human intelligence is incapable of absolute knowledge, is one that has been slowly gaining ground. Each new ontological theory, propounded in lieu of previous ones shown to be untenable, has been followed by a new criticism leading to a new scepticism" (*FP*, 50), he says, in terms often echoed by commentators of the time.[11] To extricate oneself from this futile recurrence by renouncing "absolute" knowledge is not to disavow reasoned investigation, he insists; it is merely "to submit ourselves to the established limits of our intelligence, and not perversely to rebel against them" (*FP*, 80). Yet in *The Principles of Psychology* he takes the desperate expedient of nearly disavowing reason altogether in favor of "simpler intellectual processes." Reason in the sense of rigorous logical analysis is greatly overvalued by philosophers, who unjustifiably hold it in "awe," he declares here. Having played a great historical role in overturning the despotism of prejudice and tradition, reason "tends to play the despot in their stead." "By extinguishing other superstitions, Reason makes itself the final object of superstition." The direct perception of reality "immeasurably transcends reasoning in certainty," he declares, forecasting a line of argument that will be strongly developed by Mach, Durkheim, and others around the turn of the century (*Principles of Psychology*, 2:316, 314, 315, 493).

Spencer's most significant compensatory maneuver of all, however, is simply to deny that the theoretical propositions of relativity carry any practical consequences whatever for the conduct of scientific research—and to proceed to act accordingly. The abolition of the absolute, to which he gives primacy in all his philosophy, is purely hypothetical, he reassuringly tells his readership. Though our ideas of such entities as force and matter have only an unknown and unknowable relation to "absolute truth," he says, still "we may unreservedly surrender ourselves to them as relatively true, and may proceed to evolve a series of deductions having a like relative truth" (*FP*, 184). He enacts the unreserved surrender in his imposing scholarly works of later years, where he elaborates the various fields of the Synthetic Philosophy of evolution without allowing his theorizing to be in any way compromised by the "upheaval in logic" that Schiller, Dewey, and Nietzsche saw to be one of the cardinal and inescapable effects of the evolutionary hypothesis itself.

A working Victorian scientist might thus have felt able safely to ignore Bain's and Spencer's relativistic commentaries on logical method (though a prescient physicist could not have failed to recognize the challenge posed to contemporary theory by Spencer's deconstructive analyses of the concepts of force, matter, the ether, and other physical entities). Such a scientist might also have shrugged off the critique of scientific logic that is developed in Newman's *Grammar of Assent* [hereafter GA] (1870), on the grounds that it was avowedly written, like Mansel's *Limits of Religious Thought*, to advance the cause of dogmatic religion against the influence of progressive modernity. Turning away from Newman's book may have been the safest course, for it articulates a relativistic critique of reason likely to give any scholar committed to rigorous rational inquiry bad dreams; and it is a text that retains its compelling effect even today. I myself feel this effect as I write about what Newman's inveterate adversary, Leslie Stephen, called "that most remarkable book" (*Agnostic's Apology*, 204). It is a text that places the charter of logical inquiry—what the antirelativist Paul Carus in 1913 called nostalgically "the old scientific ideal of objectivity" (*The Principle of Relativity*, 44)—in grave difficulty.

Newman's treatise professes to be an inquiry into the nature of assent or belief, defined as a state of "absolute and unconditional" intellectual acceptance (GA, 47). Stressing "the completeness and absoluteness of assent in its very nature" (35), he takes for granted that such a state is attainable psychologically, and he defends its philosophical legitimacy by asserting dogmatically that a belief may in fact be "absolutely true" (162). With this insistent invocation of the category of the "absolute," he sets himself apparently unequivocally in opposition to the relativistic and evolutionary trends of thought of his day. "Truth cannot change," he proclaims; "what is once truth is always truth; and the human mind is made for truth." Similarly, "the general, fundamental, cardinal truths are immutable," in particular "the primary truths of science," which Newman takes to be fully established for all time (181, 194). It is thus possible to reach absolute and justified certitude about propositions through the exercise of reason, Newman assures us (186). But the analysis of logical reasoning he provides in *Grammar of Assent* makes it hard to see how this could be. As though seized by the dangerous imp of the modern despite all his convictions, he shifts his inquiry from the ontological and the religious plane to what we might call a pragmatist plane *avant la lettre*, with startling consequences. The reader is transported by this shift into an intellectual domain where the only truths are relativistic ones and where certitude based on the infallibility of logic is merely an idealized hypothesis, or a delusion. Reasoned analysis can lead in principle to truth—but *cannot actually be put in practice* in real circumstances, where intellectual problems require for their solution processes of interpretation in which logic is inseparable from and powerfully governed by other, nonlogical mental influences. To imagine that any human conviction has been produced by "mere logical inferences" (241), Newman declares, is purely a fantasy, as is the notion that significant differences of belief can be adjudicated by appeals to logic—by what William James was to call "any emaciated faculty of syllogistic 'proof' " (James, *Meaning*, 305). According to Newman, the realm where formal logic breaks down turns out to lie not in some remote zone of the stellar system, or even in other societies, but in the everyday world of normal intellectual inquiry.

The analytic method Newman employs in *Grammar of Assent* is a forecast of the one employed by Einstein in 1905, and the reorganization of the field of speculation is equally profound in the two cases. Einstein begins "The Electrodynamics of Moving Bodies" by insisting on displacing the concept of "time" in physics from the plane of idealized a priori assumptions and absolute categories to that of actual practical measurement and apprehension. It is scientifically nonsensical, he declares, to speak of time as having any existence in nature apart from human means of perceiving or constructing it by the use of clocks. To analyze rigorously how clocks work is to discover, as he shows, that simultaneity at a distance, and thus "absolute time" itself, cease to have clear meaning. We are tipped by Einstein's paper into an ironic world consisting on its experiential level of only relativistic measurements, where one observer's time inevitably will disagree with another's, will flow at a different rate, simply because they occupy different points of view. Newman's *Grammar* performs much the same operation, taking the supposed absoluteness of logic as its subject.

How, Newman asks, does an inquirer come to a logical conclusion about any subject of investigation? To analyze closely the ratiocinative process not as an idealized abstraction but as it actually works, taking into account the almost indecently mundane and personal factors that actually constitute it (chiefly, the factors of "human nature" that govern knowing in the first place), is to realize, says Newman, that logical computations are inflected inescapably by individual circumstances, that they can only occur in a real-world setting where they are affected by a host of influences alien to logic itself. The process of logical proof, supposedly guaranteed by a set of flawlessly precise technical protocols, is in fact subject necessarily to factors like emotional and ideological predispositions just as the process of physical measurement, supposedly guaranteed by perfected technical protocols of its own (such as the incomparably precise interferometer Michelson built to detect ether drift), is inflected by the observer's state of motion; both processes by their very nature *yield different results at different points in space and time*. The relativity of logical reason in such an account is as much an element of the structure of nature as is the relativity of Einsteinian space and time. The difference is that while relativistic effects are undetectably small for ordinary physical phenomena, those in the field of logical phenomena, as Newman emphasizes, are likely to be very large. For any real-life purposes whatever, and for any scientific purposes in particular, the "absolute" truths so strongly affirmed by Newman are displaced by this analysis of reasoning into the zone of what in Spencer's vocabulary is called "the Unknowable," for they are, after all, wholly alien to human thinking. Given the way human beings must actually perform the quest for knowledge, there is no way to go beyond sets of relativistic interpretations among which no degree of expertise in the researchers can possibly ensure agreement. Faced with contradictory theories such as the Lorentz and the Einstein (the mechanical and the relativistic) interpretations of the discrepancies of length measurements associated with motion, one decides between them, Newman would insist, on the basis of esthetic and affective factors having nothing to do with pure logical proof, though the prestige of logic is such that scientific researchers will always portray conclusions as logically arrived at.

Newman argues, then, as Schiller would argue four decades later, that Cartesian reasoning is an imaginary operation only, that human beings are by the nature of their relationship to reality prohibited from employing it. He singles out Locke as especially deluded in constructing a "theoretical and unreal" model of thinking consisting of pure logical inferences, instead of "interrogating human nature, as an existing thing, as it is found in the world" (GA, 139). ("'Thinking,' as the epistemologists understand it," writes Nietzsche in the mid-1880s [*Will to Power*, 2:7], "never takes place at all: it is an absolutely gratuitous fabrication.") Ratiocination of the kind employed in scientific or philosophical inquiry is relativistic in two fundamental senses, Newman explains.

First, logic deals with propositions only with respect to their relations with other propositions. In our activities of classification and abstraction, "we regard things," says Newman, invoking the mantra of the Victorian relativity movement in all its different fields of expression, "not as they are in themselves, but mainly as they stand in relation to each other" (GA, 44). For Newman, whose nominal standard of reference is that of "absolute truth," the relativity of rational operations dooms them to unreliability. Inference being purely relational, it is fundamentally "conditional and uncertain" (65), he thus declares, and can always be plunged by skeptical analysis into the quicksand of an infinite regress. We are compelled, he says, to fall back perpetually upon previous syllogisms to prove our assumptions, "and then, still farther back, we are thrown upon others again, to prove the new assumptions of that second order of syllogisms. Where is this process to stop?" As for the first principles needed to ground a logical sequence, these "are called self-evident by their respective advocates because they are evident in no other way" (216; see also 90, 91). Logic in itself, as a way of knowing reality, is consequently a tissue of fatal defects. "As to Logic," says Newman, "its chain of conclusions hangs loose at both ends; both the point at which the proof should start, and the points at which it should arrive, are beyond its reach; it comes short both of first principles and of concrete issues" (227).

Furthermore, our vainglorious use of logical procedures in science and philosophy is compromised fundamentally by prejudices that vitiate all attempts at rational stringency of argument. Newman thus claims to demonstrate "how little syllogisms have to do with the formation of opinion; how little depends upon the inferential proofs, and how much upon those pre-existing beliefs and views, in which men either already agree with each other or hopelessly differ, before they begin to dispute, and which are hidden deep in our nature, or . . . in our personal peculiarities" (GA, 221–22). Even the most severely rigorous logical reasoning, Newman asserts, involves "subtle assumptions . . . traceable to the sentiments of the age, country, religion, social habits and ideas" of the reasoners (217). ("The material which a scientist *actually* has at his disposal, his laws, his experimental results, his mathematical techniques, his epistemological prejudices . . . is . . . *never fully separated from the historical background*"—Feyerabend, *Against Method,* 51.) It is impossible to resolve controversies purely by logical appeal to evidence, says Newman, because one's predispositions determine what counts as evidence, and what "reasonableness" means, to begin with. When different writers argue in favor of conflicting theories, therefore, "the conclusions vary with the particular writer, for each writes from his own point of view and with his own principles, and these admit of no common measure" (GA, 287). Inevitably, "what to one intellect is a proof is not so to another" (233). A proof, in effect, is necessarily a matter of opinion and is not subject to "any syllogistic compulsion" (245). In developing this argument, Newman lays the basis for Schiller's later insistence that syllogistic rules have no effect on controversies in science, since "the contending parties do not use or acknowledge the same premisses, and, therefore, do not draw the same conclusions" (Schiller, *FL*, 196).[12] Together they form the basis for Thomas S. Kuhn's proposition, a crux of postmodern theory and a main site of philosophical controversy in the late twentieth century, that there exists in science "[no] set of rules which will tell us how rational agreement can be reached or what would settle the issue . . . when statements seem to conflict" since "the competition between paradigms is not the sort of battle that can be resolved by proofs" (Kuhn, *Structure of Scientific Revolutions*, 148; and see 151).

Scientific knowledge being a cultural phenomenon, fundamental discoveries are bound in most cases to be made simultaneously by different researchers. It is worth noting in passing, therefore, that the theory of the incommensurability of scientific paradigms was also proposed by another writer, John Tyndall, in the same watershed year in which *Grammar of Assent* and Bain's *Logic* were published (1870)—and moreover that Tyndall expressly identifies the principle of the variable logical force of evidence upon different observers with "the law of relativity." In the course of a discussion of the Victorian dispute about the alleged spontaneous generation of reflective particles in the air, Tyndall claims (in an essay proselytizing on behalf of the theory of evolution—another sign of the inseparability of the two great Victorian theories) that the differing views of scientific authorities on this point "may perhaps be, to some extent, accounted for by that doctrine of Relativity which plays so important a part in philosophy. This doctrine affirms that the impressions made upon us by any circumstance . . . depend upon our previous state." "In our scientific judgments the law of relativity may also play an important part," Tyndall muses. "An amount of evidence which satisfies the one entirely fails to satisfy the other," one finding it "perfectly conclusive," the other finding it worthless (Tyndall, *Fragments of Science*, 446–47).

Tyndall's cautious qualifications cushion the impact of this radical line of speculation, and, not surprisingly for a writer arguing that a science of unbending "mechanical laws" should be forcefully advanced until "the malady of doubt is completely extirpated" (*Fragments*, 437), he does not pursue the point about the indeterminacy of logical proof or even seem to grasp its potentially destabilizing effect on the philosophy of scientific research. Newman does grasp it fully, and recklessly pursues it. "Logic then does not really prove," he says again and again (GA, 217). But the incoherence of logical reasoning is of minor consequence after all, says Newman, since human beings appraise arguments not on the basis of rational principles, though they may believe that they do, but on that of an active intuitive faculty Newman names the Illative Sense, the capacity for "right judgment in ratiocination" (269). Ostensibly a factual or descriptive element only, the Illative Sense appears in Newman's discourse, in fact, as an ideological operator charged with a vital mission. It provides the vehicle by which he is able to salvage "the certainty of knowledge" (270) in the specific area of religious dogma from the threat of scientific critique or of philosophical skepticism. *Grammar of Assent* thus stands as a vivid reminder that the Victorian relativity movement, championed by writers like Feuerbach, Spencer, Stallo, Nietzsche, and Clifford as (among other things) a mode of revolt against the influence of organized religion, was itself, like all other arguments, subject to relativity and to extreme indeterminacies of application. But it would be a misreading of *Grammar of Assent*—and would wrench it from its important affiliation with strains of modernist freethought—to give it too narrow a sectarian or apologetic implication.

The illative sense belongs fundamentally to the "humanist" current of radical nineteenth-century argument. It asserts the capacity of human beings to fashion their own world by active exertion and by the force of imagination—a principle given a very broad extension subsequently, as we shall see, by scientific theorists like Karl Pearson, Sir Arthur Eddington, and others. "Man begins with nothing realized. . . . It is his gift to be the creator of his own sufficiency; and to be emphatically self-made," says Newman memorably (GA, 274). Squaring such a proto-Nietzschean, existentialist, implicitly atheistic principle with Christian dogma would seem to represent a formidable challenge, but it interlocks powerfully with the radical "humanism" of a philosopher like Schiller and with the oft-proclaimed creed of twentieth-century anthropology that there is no such thing as human nature, only the manifold diversities of human history.[13] Human intelligence is a prodigiously creative and thus above all a centrifugal, diversifying agency, these movements of thought proclaim. In keeping with this perspective, absolutist formal reasoning increasingly takes on in *Grammar of Assent* the sinister aspect of a structure of essentially inhuman, mechanical *compulsion* designed for the suppression of human motives and creative values. This is the polemical implication of Newman's statement, for example, that "certitude is not a passive impression made upon the mind from without, by argumentative compulsion, but . . . an active recognition of propositions as true" (271). Or, similarly, "it is the mind that reasons, and that controls its own reasonings, not any technical apparatus of words and propositions" (276). In practical matters, skill is "a sort of instinct or inspiration, not an obedience to external rules of criticism or of science," and ratiocination similarly is not "commensurate with logical science" (280). We rightly use logic as far as seems profitable to us, says Newman, but we are not subject to the tyranny of "logical formulas," "syllogistic compulsion," or "scientific necessity independent of ourselves" (283, 245, 251). "The Illative Sense, that is, the reasoning faculty . . . is a rule to itself, and appeals to no judgment beyond its own" (283). Newman in most respects is J. S. Mill's ideological opposite, but we can fairly class *Grammar of Assent* alongside *On Liberty* as one of the most uncompromising texts of the age seeking to unmask and to challenge the insidious system of "compulsion" built into modern culture—built into it particularly, in Newman's analysis, in the form of the ideology of compulsory obedience to logical rationality. In one of the most penetrating comments on Newman, indeed, Leslie Stephen specifically aligns him with Mill—noting that Newman's "dangerous" skepticism at times goes beyond Mill's own (Stephen, *Agnostic's Apology*, 179).[14]

Newman casts off the chains of intellectual compulsion with unparalleled brashness: "My ideas are all assumptions, and I am ever moving in a circle," he declares (GA, 272), with something of the same audacity Paul Feyerabend (a professed disciple of *On Liberty* and an agitator for "the rejection of all universal standards and of all rigid traditions") conveys in his notorious motto, "*anything goes*" (*Against Method*, 12, 19). Still more audacious was Newman's famously scandalous statement, in an 1843 sermon, about the apparent conflict between the biblical assertion that the sun revolves about the earth and the modern scientific view, which has it the other way around. " 'How can we determine which of these opposite statements is the very truth," asks Newman, "till we know what motion is?" If the properties of matter and thus the idea of motion "are merely relative to us," then, he says, in an uncannily modern-sounding formula, "both [propositions] are true," at least "for certain practical purposes in the system in which they are respectively found" (Newman, "Theory of Developments," 347, 348). Such a contemporary as J. B. Stallo would presumably have saluted this statement, a very early version of Eddington's conundrum about magnetic fields or of many similarly paradoxical parables of relativistic theory, as a concise and lucid formulation of the scientific principle of physical relativity.[15] But Newman's remark seemed so brazenly defiant of the law of noncontradiction that J. A. Froude, upon reading it, abandoned his faith in Newman and in Newman's creed of "the surrender of reason" on the spot, remarking on the insidious similarity of "deepest credulity and deepest scepticism" and expressing his view that "a doctrine so monstrous" would, if it became widely credited, unleash an outbreak of "suicide and madness" among Christian believers (Froude, *Nemesis of Faith*, 157–59). Froude's histrionic reaction prefigures Henry Adams's turning "green with horror" at reading Poincaré's analysis of the "endless displacement" of scientific logic (Adams, *Education*, 455) and, as we have seen, a long series of later commentators on the perils of the "abyss" of relativity.

Newman does not merely state his frightening and "monstrous" creed abstractly; he illustrates the alleged inefficacy of logic with detailed case studies of learned investigations into literary and historical problems. One involves the establishment of a famous textual crux in Shakespeare, the Hostess's description of Falstaff on his deathbed (*Henry V*, II.iii.16). In the 1623 folio, this line reads, senselessly, "and a table of green fields"; a possibly contemporary annotation to the 1632 edition corrects the line to "on a table of green frieze"; another anonymous annotation of uncertain date gives " 'a talked of green fields"; and Theobald's wholly conjectural eighteenth-century reading, now generally accepted, is " 'a babbled of green fields." Newman shows that one's decision on which variant to prefer depends less on logical evaluation of the available evidence than on the weight one chooses to ascribe to this or that essentially arbitrary principle of textual authenticity. Which is more authoritative in the science of textual editing, perceived poetic quality or specific textual authority? Would it be inadmissible to say that Theobald's invention has acquired authenticity simply by virtue of having been generally accepted for many years (GA, 220)? Such questions, says Newman, "open upon us a long vista of sceptical interrogations" that ultimately even throw into doubt the meaningfulness of the concept of a single historical author of the Shakespearean corpus: "perhaps, after all, Shakespeare is really but a collection of many Theobalds, who have each of them a right to his own share of him" (221). The entity named "Shakespeare" may best be thought of, says Newman, as a sheer phenomenon of textual dissemination and of editorial interpretation! Newman pursues at some length his amazing investigation of what we have come to call "the author-function," shrewdly analyzing the cultural mechanisms by which "the writing subject cancels out the signs of his particular individuality" (Foucault, "What Is an Author?" 143), throwing into question "the solid and fundamental unit of the author and the work" (Foucault, 141), and seeming to step straight through the looking glass into the ambit of poststructuralist theory. The scholarly controversy over this ever-fluctuating line of Shakespeare's, says Newman, will in any case always be inconclusive, and will forever exemplify "how impotent is logic, or any reasonings which can be thrown into language, to deal with . . . first principles"; it shows "how little depends upon the inferential proofs, and how much upon those pre-existing beliefs and views, in which men either already agree with each other or hopelessly differ, before they begin to dispute" (GA, 218, 221–22). "An inference," he concludes, "can never reach so far as to ascertain a fact" (222).

The same lesson emerges from the scholarly literature surrounding key controversies in classical studies, such as the question of the factuality of the Trojan War or of the nature of the Homeric poems. Citing such authorities as Niebuhr, Clinton, Sir George Lewis, Grote, and Mure (GA, 285), Newman shows that the most rigorous logical analysis of the evidence by the most eminent specialists in these cases—which is to say, in all cases of significant scientific disagreement—leads inescapably to divergent conclusions, since each scholar "writes from his own point of view and with his own principles, and these admit of no common measure." "Hence the categorical contradictions between one writer and another, which abound" (288).

Given Newman's earlier insistence on the reality of "absolute truth" and his insistence here on the fatal deficiencies of logical analysis, one would expect these case studies to figure in *Grammar of Assent* as parables of hopeless futility. Leslie Stephen interprets them in just this way, and replies to what he takes to be Newman's unrelieved epistemological pessimism with a series of declarations of the attainability of "universal objective truth." What counts in evaluating a man's beliefs, says Stephen, is "whether the man's mind is rational, or whether he deals with the evidence in accordance with logical rules," which "express the conditions which secure a conformity between opinion and fact" (*Agnostic's Apology*, 213, 217). With these heavy-handed dogmatic pronouncements, Stephen simply dismisses Newman's argument (which is that different scientists, each scrupulously "rational," may follow logical rules with equal fidelity and yet reach incompatible conclusions)—but also makes him more of a skeptic than he admits to being. Newman's lesson is not that the unattainability of determinate answers to such questions as the genuine reading of Shakespeare or the factuality of the Trojan War, no matter how much evidence we might uncover or how much professional expertise we might muster, renders such investigations futile or devalues the knowledge they generate. We assuredly cannot know the truth on points such as he has cited. "Ingenuity and labor can produce nothing but hypotheses and conjectures," he declares, quoting the historian Sir George Lewis, "which may be supported by analogies, but can never rest upon the solid foundation of proof" (GA, 289). Nonetheless, he insists on the indispensability of logical thinking even as he demonstrates its use in his own exceptionally rigorous inquiry. Thus he confronts the reader of *Grammar of Assent* with the coexistence of two propositions: "in no class of concrete reasonings, whether in experimental science, historical research, or theology, is there any ultimate test of truth and error in our inferences besides the trustworthiness of the Illative Sense that gives them its sanction" *and* "it is absurd to break up the whole structure of our knowledge, which is the glory of the human intellect, because the intellect is not infallible in its conclusions" (281, 187). Considered as a decisive instrument of proof, logical reasoning can only fail; considered as an instrument for the production of knowledge, it possesses sublime virtue.

Newman does not in *Grammar of Assent* spell out what we could call an ethics of belief adapted to the relativistic world of evidence and inference that he discloses with such fullness. Indeed, he ultimately takes refuge from his own compelling arguments with the help, precisely, of a *deus ex machina*. Absolute objective truth, categorically ruled out of the intellectual arena by all the argument that comes before, turns out to be attainable after all—thanks to divine intervention. "It does not prove that there is no objective truth, because not all men are in possession of it," says Newman toward the end of his book. "But this it does suggest to us . . . that we need the interposition of a Power, greater than human teaching and human argument, to make our beliefs true and our minds one" (GA, 293). Evidently this fantastic expedient, which nullifies at a stroke all of Newman's previous insistence on the free creative action of human intelligence, is at least potentially operative in all fields of knowledge, "whether in experimental science, historical research, or theology," though the persistence of conflicting theories in all of these fields suggests inescapably that the superhuman Power has declined, in fact, ever to intervene there. Certainly it has declined to visit the vexed sphere of Shakespearean textual editing, where there seems to be no room at all for the "absolute and unconditional" certitude of Assent, and where Newman's earnestly professed creed of "the certainty of knowledge" (270) seems by his own demonstration wholly ungrounded in experience. His invocation of the supernatural remedy occurs in fact altogether arbitrarily and unpersuasively in *Grammar of Assent*, and carries on the rhetorical level none of the passionate conviction with which he praises throughout the book "the keen and subtle operation of the Illative Sense" (298).

By calling *in extremis* upon the providential regime as a surrogate for human reason, in any case, Newman escapes having to offer detailed epistemological prescriptions to those who might otherwise feel plunged into an abyss of indeterminacy by his previously expounded relativistic doctrines. This lapse in his text is filled by W. K. Clifford, who seems to have read *Grammar of Assent* with deep understanding. Clifford the anticlerical freethinker sets forth at all events a relativistic epistemology strikingly close to Newman's and, again, strikingly premonitory of the differently inflected "humanist" pragmatisms of Schiller and Nietzsche. "The end of all knowledge is action," he writes in 1879 (Clifford, *Lectures*, 2:280), echoing Newman's crucial pragmatic assumption, "Life is for action. If we insist on proofs for every thing, we shall never come to action" (GA, 91). (Engels invokes exactly this doctrine in the next year, 1880.)[16] Clifford, as we saw in the last chapter, stresses no less forcefully than Newman does the principle that "human knowledge is never absolutely and theoretically certain" (Clifford, *Lectures*, 2:316). We have no grounds for supposing that the world is wholly reasonable or "that the order of events is always capable of being explained," he declares (*Lectures*, 1:169–70). From this principle, he draws the ethical conclusion that the intellect, while not declining to adopt on a permanent basis such ideas "as lead to action" (in other words, refusing to lapse into a paralysis of negative capability), "must maintain [toward all others] an attitude of absolute receptivity; admitting all, being modified by all, but permanently biassed by none" (*Lectures*, 1:116). This may seem an impossibly exacting or even paradoxical ideal, but it is the response to crude denunciations of relativity, in the critique that it directs at logic, as synonymous with nihilism.

Balfour's Conundrum

"Doubt itself is a positive state, and implies a definite habit of mind," says Newman in 1870, "and thereby necessarily involves a system of principles and doctrines all its own" (GA, 294). A notable text from the end of the decade, Arthur James Balfour's *Defence of Philosophic Doubt* (1879), offers what amounts to an extended gloss on this statement. Balfour's book rehearses a version of Newman's argument, but with the redeeming complement of the Illative Sense and of the radical "humanism" it professes sharply pared away. Without referring to Newman, Balfour picks up his central themes and focuses them on an uncompromising denial of the logical soundness and coherence of modern science itself, striving at the same time to justify the conception of doubt as a "positive state" with a rigorous intellectual program of its own.

The object of Balfour's exercise in demystification is "the theory of the *logical deduction of scientific doctrine from empirical data*" (*Defence*, 254). The failure of this theory to prove itself by the production of securely verified truths seems to Balfour an obvious fact. Just as Newman invokes the disagreements that persist indelibly among expert scholars interpreting the same bodies of evidence, so Balfour emphasizes the historical instability of all systems of knowledge:

> The multitude of beliefs which, in obedience to a mechanic and inevitable law, sway for a time the minds and actions of men, are then for ever swept away to the forgotten past, giving place to

> others, as firmly trusted in, as false, and as transitory as themselves, form a spectacle which is not only somewhat melancholy in itself, but which is apt to suggest uncomfortable reflections as to the permanent character of the convictions we ourselves happen to be attached to. (*Defence*, 261–62)

Even in the field of modern science, with its highly refined and rigorous methods of investigation, there is no immunity, Balfour declares, from the joint laws of difference, of proliferation, and of evolutionary flux. But whereas writers like Mill, Bain, and Clifford, and later the pragmatists, found in the incompleteness, provisionality, and changeability of our convictions the marks of human intellectual vitality, Balfour, a disconsolate absolutist at heart (and finally a spokesman, like Newman, for religious verities), finds only the sign of the woeful failure of knowledge.

Logical analysis is incurably fallacious in its own right, according to Balfour, even were it possible to exercise it in its pure form, in isolation from the conditions of its actual employment in scientific practice. But it is not, and thus is hopelessly subject to the law of the Relativity of Human Knowledge, which decrees that knowledge is a function of individual and cultural-historical circumstances, and thus is as variable as they are. "A man's beliefs are very much the results of antecedents and surroundings with which they have no proper logical connection," declares Balfour, expressing what by 1879 was already a well-established commonplace. "It must always have been known that there were causes of belief which were not reasons" (*Defence*, 260). Just as Newman and Tyndall already had, Balfour denies the decisiveness of any appeal to objective evidence in controversial matters. "Even when a man attempts to form opinions only according to evidence, *what he shall regard as evidence* is settled for him by causes over which he has no . . . control"—a fact that, if we take it seriously, causes us to incur, he cautiously says, "a sort of skeptical uneasiness" with regard to our own scientifically verified conclusions (261). We readily admit and in fact insist that the beliefs of others—of medieval divines, for example—reflect the inescapable imperatives of their "time and circumstances," but from sheer arrogance we refuse to apply this principle to ourselves, Balfour declares (268). The nonlogical cultural-historical supplement is a parasite infesting our entire system of thought.

The instability that is the curse of knowledge systems is inescapable, says Balfour, again invoking a principle raised by others before him, since scientific rationality cannot help anchoring itself in preliminary assumptions or first principles lying outside the scope of logical demonstration. "My ideas are all assumptions, and I am ever moving in a circle," says

Newman, proclaiming in an affirmative mode the irrationality of reason. But Balfour makes of this principle the basis of a sweeping impeachment of the credibility of science. "The only beliefs of which, according to received scientific theories, we may say with certainty that they *can* have no reason, but *must* have non-rational causes, are those on which the certitude of all other beliefs finally rests" (*Defence,* 264). Scientific logic suffers from another fatal flaw in the attempted explanation of any phenomenon of nature: the impossibility of determining the effects of multitudes of persisting factors on the phenomenon in question, and thus the impossibility of ruling out wholly unknown causes in the analysis of any event. This means, says Balfour, that "we are for ever debarred from a theoretical knowledge of any absolute law of Nature: from a knowledge, I mean, of *all* the phenomena required to produce a given result"; therefore, "all historical inference is thrown into confusion" (281, 284). In this predicament, we can only cast off the conviction "that Science rests on a solid and rational foundation" (277–78).

The main lines of this powerful argument are by 1879 not original, though *A Defence of Philosophic Doubt* stands out for the relentlessness of its condemnation of scientific logic. Nor does Balfour base his argument on the relativistic principles that a growing network of contemporary writers was elaborating, not just into a protest against the cult of infallibility of orthodox science, but into an alternative scientific outlook with a paradoxical logical method of its own. Yet Balfour's book, in its critique of scientific absolutism, played a notable role in the nineteenth-century reconstruction of the canons of reasonableness that enabled forms of full-blown relativistic theory and science to emerge. In this process, the calling into question of "the theory of the *logical deduction of scientific doctrine from empirical data*" was a crucial step, for this theory formed the essential structure of the whole ideology of absolute values that it was the mandate of the relativity movement to challenge. It is significant, for example, that Einstein (though he by no means denied absolute values in science) often declared as one of the key points of his scientific philosophy, essentially quoting Balfour (and contradicting Mill), that concepts and theories "are all—when viewed logically—the free creations of thought which can not inductively be gained from sense-experiences," and that there is a "gulf—logically unbridgeable—which separates the world of sensory experiences from the world of concepts and propositions" (Einstein, "Remarks on Bertrand Russell," 287). However scientific logic may operate, it does not, he declared, construct unshakeable chains of reasoning leading from observed data to true theoretical formulations.

Balfour is perhaps most distinctive in addressing the anticipated criticism that his assault on the logical foundations of science is an act of dangerous intellectual anarchism that will tend to "give free scope to the simultaneous existence of any number of creeds," to "a chaos of conflicting creeds" (*Defence*, 315, 316). For one thing, he offers at the end of the book a gesture of encouragement to the cause of religious belief, though a considerably more guarded one than Newman's or even than Spencer's. "Religion is . . . no worse off than Science in the matter of proof," he says (319); both are logically indefensible.[17] As for his conclusions regarding science itself, they are essentially twofold. If the complaint be made that the critique of scientific method in *A Defence of Philosophic Doubt* is too uniformly destructive, says Balfour, "I reply that speculation seems sadly in want of destructive criticism just at the present time" (293). He offers, in other words, the same self-justification as does J. B. Stallo, the first edition of whose *Concepts and Theories of Modern Physics* two years later (1881) brought upon him "the charge of being a mere destructionist" (Stallo, *Concepts and Theories*, 15). The role of radical intellectual critique according to both authors is to reveal the pervasive irrationality of contemporary thought, its idolatrous worship of myths and of metaphysical entities, even, or especially, in its most progressive-seeming branches.

Not only is it wholesome and necessary to challenge the contemporary cult of infallible science as a social element, says Balfour, it is imperative to cleanse science itself, as much as possible, of a profusion of logically unsound conceptions that have monopolized it heretofore. In carrying out this intention, he follows Spencer closely, just as Stallo also will do, in condemning as specious and irrational a series of fundamental concepts of physics: the ether, force, matter, causality. Does not this program of unremitting "destructive criticism" amount to sheer nihilism? Balfour replies with the remarkable argument that his own systematic assault on scientific belief is just as stricken with impotence as scientific reasoning itself is, and is bound to fall short of its own goal! The word "skepticism" is ambiguous, he says: "it may mean either the intellectual recognition of the want of evidence, or it may mean this together with its consequent unbelief" (*Defence*, 296). This latter state of mind is, however, beyond the power of rational argument to produce. "Scepticism of the far-reaching character required by the reasoning of this Essay can be produced by no rigour of demonstration"—and properly so, for Balfour trains his destructive criticism also upon the maxim, "occasionally uttered as if it were a moral law . . . that we are in duty bound to make the strength of our beliefs vary exactly with the strength of the evidence on which they rest" (297).

This coercive principle, with its aura of moral obligation, flies in the face of the dynamics of real belief, says Balfour. People will believe in their knowledge no matter how it may be disproved by rational argument, for "general unbelief can hardly be regarded as a possible frame of mind" (298). "Practically we need not or cannot regulate our beliefs in strict accordance with the results of rational criticism"; the only true mode of skepticism is therefore the kind "which does not destroy belief" (314, 297). "Rational criticism" is without efficacy, since, just as Newman insisted, belief does not rest on rational foundations to begin with.

Scientific reason systematically defeats itself, then, and so, too, does Balfour's own program of destructive criticism, which *knowingly adopts a methodology that cannot attain its ends*. Balfour does not elaborate this striking point, but one can make out here, in embryonic form, the image of that figure portrayed by Albert Camus as the defining myth of the modern intellectual: Sisyphus, the man condemned to an eternity of futile striving. As we shall see, James Frazer invokes the same myth to define the futility of his own quest for truth through the labyrinths of scholarship he explores in *The Golden Bough;* and Newman himself invokes it in a notable passage in *Grammar of Assent:* "to seek . . . with the certainty of not finding what we seek, cannot in any serious matter, be pleasurable, any more than the labour of Sisyphus or the Danaides" (GA, 171). This in effect is the implicit moral of the movement of thought stemming from Spencer's declaration that scientific and philosophical inquiry was bound to fall short of discovering the real nature of things, of overcoming the distorting effects of personal bias in research, or of mastering the "incalculable complexity" of social influences so as to be able to predict even with minimal accuracy the consequences of social actions (Spencer, *Study of Sociology*, 74, 15, 271). "The sceptical philosopher" does not disavow the use of reason, says Balfour, but, unlike "the man of science," employs it "in the full consciousness . . . that the system he is dealing with is, as a whole, incapable of any rational defence" (*Defence*, 315). This claim of unsparingly paradoxical lucidity is the credo of a genuinely modern rationality, fated as it is, says Balfour, to exercise itself in an intellectual universe in which the possibility of "philosophic certitude" (317) is defunct.

The relativistic/evolutionary critique of logic, and the implicit or explicit advocacy of modes of rationality emancipated from the "superstition" of the infallibility of logical reasoning, had matured therefore into a distinct philosophical tradition (the one summed up and systematized in Schiller's *Formal Logic*) by the last two decades of the nineteenth century[18]—a tradition dense enough that the lines of transmission of ideas within it are impossible to follow with any assurance. It remained heretical, disreputable, unstable, and many regarded it with abhorrence: hence the increasingly militant counterdiscourse we have frequently noted.[19] (Sometimes the two opposed strains of thought coexist bewilderingly within a single text, as we shall see, for example, in a discussion of Karl Pearson's *Grammar of Science* [1882]). Yet the radical critique of scientific reason became increasingly incorporated into the discourse of scientific inquiry itself, preparing the way, as it seems in retrospect, for the advent of the sublime illogic of special relativity. I will close this chapter by sketching the appearance of this theme in what may seem a heterogeneous set of authors: the anti-Darwinian evolutionary theorist Samuel Butler, the anthropologist Lucien Lévy-Bruhl, and, once again, the mathematician Henri Poincaré.

Science and Noncontradiction

One of the most vivid texts of the movement we are tracing is Butler's fragmentary manuscript of a sequel to his 1877 treatise *Life and Habit*, posthumously published in his *Collected Essays* [hereafter *CE*] under the title "Life and Habit, Volume 2." This work of around 1885 offers a concentrated iteration of the critique of logical reason instituted by writers like Spencer, Bain, Newman, and Balfour (though it expresses no indebtedness to them)—not, however, with the goal of debunking science, but with that of cleansing it of its ideology of "absolute truth" and establishing for it a sounder footing in relativistic "common sense" and "convenience." "We must of course be careful not to aim at absolute convenience," says Butler, however; "this would be as complete a chimera as absolute truth" (*CE*, 1:163–64). As much as for Stallo, the term "absolute" encapsulates for Butler all that needs to be reconstructed in modern intellectual life.

"With every one but men of science," he declares, "the fact that we have no certain absolute knowledge *va sans dire*" (*CE*, 1:116). This is so because direct, unproblematic cognition of objects is denied to our minds. It is impossible, says Butler, "[to] see one single little thing not as through a glass darkly, but face to face." Echoing Spencer, he thus proclaims as his key philosophical doctrine the principle that "the ideas we connect with exterior objects are just as arbitrary as words, and have just as little inherent essential connection with, and resemblance to, the realities that underlie them"; human awareness can claim no "foothold," no "*point d'appui*," in unmediated or presymbolic reality, he says (1:202–3, 111–12).[20] This being so, conceptual construction on every level of abstraction is necessarily an arbitrary process that originates in acts of intellectual "violence or chicane," which is to say, in the willful forcing of human desires upon external reality (1:113).[21] We construe the world necessarily as suits our own "convenience," according to the fundamental relativistic and pragmatic principle that "we have no concern or interest in it except in its relation to ourselves" (1:158). But since in reasoning about things "we may look at . . . everything from two different points of view," as unified or as distinct from another thing (1:116), all cognition is characterized by an ineradicable irrationality. Every intellectual formation, Butler says, is found "to involve absurdity or contradiction in terms, or some other deadly logical sin" and "every proposition, nay every idea, carries within itself the seeds of its own undoing" (1:112). It is therefore folly to imagine a perfectly (as opposed to a provisionally or pragmatically) rational, logically coherent scheme of knowledge: this is Butler's most insistent theme.

> People say that whatever happens there must be no contradiction in terms. If a thing is one, it cannot at one and the same time be more than one. If it be more than one it cannot at one and the same time be one only. Whereas in point of fact it is only in virtue of faith in incessant onmipresent contradictions in terms that we can speak, think, or act at all. (*CE*, 1:124–25)

The cultural politics of this sustained polemic against the ideal of noncontradiction are not made explicit in this text, but clearly they belong to the broad movement of Victorian freethought aimed at dismantling rigid principles of orthodoxy and all "fixed immutable arrangements of ideas" (*CE*, 1:164) masquerading as natural verities vouched for by logical demonstration. The idolatry of logical coherence and of that fictive thing "absolute truth" is at bottom, Butler suggests, ideological, an unreasoning idolatry of things as they are in all fields of human existence. Significantly, "Life and Habit, Volume 2" ends with a precocious statement of moral relativism, set here, as it is so often in radical Victorian writing, under the scientific sign of evolution. "The permanence of organic forms affects the question of the fixity of the basis on which ethics and all questions of right and wrong rest," Butler declares. "No one who holds that body and instincts vary will suppose the moral nature of any action to be more constant than the organism which is to execute it." The conclusion to be drawn from this proposition is a distinctly relativistic one: it is "absurd," says Butler, "to judge [an action] in respect of any other environment than its own" (1:165–66).[22]

Similar themes are developed in a different scientific idiom in the work of Lucien Lévy-Bruhl, a brilliant disciple of Durkheim's who devoted himself to the Durkheimian project of creating "the positive science of social phenomena" and proclaimed in *Ethics and Moral Science* (1903) an Enlightenment credo of scientific reason seemingly at the antipodes of the skeptical arguments of writers like Newman and Balfour. "There is a social objective reality, as there is a physical objective reality," he says, defining his and Durkheim's fundamental creed, "and . . . man, if he is logical, should . . . endeavour to learn its laws in order to be as much master of them as possible" (*Ethics*, 6, 19). From this proclamation of rigorous scientific logic as the sole guide to "objective reality," Lévy-Bruhl develops an uncompromisingly relativistic approach to the study of human cultures, and to that of ethical systems in particular, echoing Spencer and Clifford in denying the postulate of "an unchanging 'human nature,' always similar to itself."[23] The invariable consequence of this postulate, he says, is the moral absolutism that presents a particular set of ethical precepts "as universal, and consequently as compulsory, with equal force on all reasoning, free, human beings without distinction of time or place" (*Ethics*, 54, 58). Human conscience, declares Lévy-Bruhl, is "a relative reality," appropriate only within the cultural frame of reference to which it belongs; thus the study of human society is transformed totally by the principle of "the relativity of . . . traditional beliefs" and, more comprehensively, by "the idea of universal relativity" (*Ethics*, 108, 144). (Two years later, in what Stephen Toulmin takes to be a historical "coincidence," Einstein declares that physical measurements are only valid within the "coordinate system" from which they are taken.)

As this argument unfolds in *Ethics and Moral Science*, a significant tension is generated within it. Without seeming to notice the possible instability that this move introduces into his endorsement of scientific method, Lévy-Bruhl extends "the idea of universal relativity" from the field of moral value systems to include that of logical systems as well. People in other cultures "have not the same linguistic and logical habits" we do, he declares; in the mental worlds of many peoples, "underneath the logic of signs is a logic of images . . . and underneath the logic of images, a logic of feelings, doubtless as old as the species itself." The close study of primitive societies allows us to glimpse the processes of thought that lead them "to conclusions, that is to practices, disconcerting or inexplicable to our logic, but as necessary in their eyes as the conclusions of our syllogism are for us" (*Ethics*, 63, 64). It is rash to take for granted that Western syllogistic logic is applicable, after all, to the analysis of bewilderingly complex cultural systems in foreign societies, Lévy-Bruhl goes so far as to say: "*a priori* there is nothing to certify that the complexity covers a logical order, nor that it can be brought back to a few guiding principles" (68). Other systems of thought may be so foreign to our own as to be wholly unintelligible to us and to escape scientific analysis altogether. (Clifford: we have no grounds for supposing "that the order of events is always capable of being explained.") In simple human societies, we thus glimpse again that hypothetical zone, situated by Mill in the unknown reaches of the stellar system, where logic ceases to work. Could it be, as Lévy-Bruhl suggests, that "the idea of universal relativity" applies to our logical practices and to scientific knowledge fully as much as to our ethical practices? The deeply subversive hint percolates through *Ethics and Moral Science*. Just as any society's ethical system is perpetually in flux, "always provisional," though it imagines itself as "universal [and] absolute," so, too, says Lévy-Bruhl, "positive science" must abandon the ideal of acquiring certain, permanent results. Enlightened self-reflexive modern inquiry "repudiates neither the relative nor the provisional," he aphoristically declares, for "it knows that it can scarcely attain anything else" (*Ethics*, 115, 144, 120). For a moment, at least, he sounds distinctly like Balfour in his polemic on the foundationlessness of the scientific method.

This problematic becomes still more acute in Lévy-Bruhl's later and more celebrated work, *Les fonctions mentales dans les sociétés inférieures* [hereafter *FM*] (1910), ill-translated into English as *How Natives Think*. In accordance with his earlier critique of the idea of "human nature," he claims in this book to have discovered that primitive peoples conceptualize the world differently from Europeans. "Different mentalities will correspond with different social types," he declares (*FM*, 27), again echoing Spencer.[24] In primitive communities "as much as in our own, perhaps even more so, the whole mental life of the individual is profoundly *socialized*," which is to say, determined by the categories of thought contained in "collective representations" handed down by tribal tradition. Even bare sense perception falls under the sway of profound cultural relativity, says Lévy-Bruhl; "the external world [that primitives] perceive differs from that which we apprehend," since "primitives see with eyes like ours, but they do not perceive with the same minds" (*FM*, 106, 24, 43, 44). This is the doctrine of "the Relativity of Knowledge" given an extreme evolutionary and sociological inflection.

According to Lévy-Bruhl, the distinguishing element in primitive cognition is its imposition of "mystic interpretation" upon facts of nature, a process that associates these facts in ways unknown to European thinking. Its central element is what he calls "*the law of participation*," in obedience to which "objects, beings, phenomena can be, though in a way incomprehensible to us, both themselves and something other than themselves" (*FM*, 76). This "prelogical" consciousness, says Lévy-Bruhl, is likely to be "wholly indifferent" to the logical principle of noncontradiction, and its distinctive formations, such as the complex primitive notion of multiple souls, or the relations among the individual, his totemic group, and his totemic species, are commonly "irreducible to logical intelligibility" (78,

87). They are pervaded with "contradictions which rational thought cannot possibly tolerate" (104).

> The ubiquity or multipresence of existing beings, the identity of one with many, of the same and another, of the individual and the species—in short, everything that would scandalize and reduce to despair thought which is subject to the law of contradiction, is implicitly admitted by this prelogical mentality. (*FM*, 363)

Lévy-Bruhl is carried by the impetus of his argument to reflect in notable ways upon our own logical practices. The system of rationality does not present itself to the mind naturally by its own self-evidence, he says, but, as cultural formations or "collective representations," "the claims of logical thought" are coercively implanted in the individual and are endowed in this way with absolute, unquestionable necessity. Logical laws in this analysis thus have an ideological character in that they seem inescapably natural, whereas in fact they are emanations and sustaining mechanisms of a certain set of sociocultural conditions. They have as well, for Lévy-Bruhl, the unmistakable character of a system of tyrannical power. "Logical discipline is . . . imposed upon [each individual's] mental operations with irresistible force," he says (*FM*, 107). The law of this disciplinary system is the rigid exclusion and suppression of other, "prelogical" modes of apprehension pervaded with emotional and "mystical" associations. "The laws of logic absolutely exclude, in our own thought, everything that is directly contrary to itself" (106), says Lévy-Bruhl, unconsciously quoting Feuerbach's analysis of religious faith. Intolerance is inscribed profoundly and indelibly in the basic structure of our thinking, in other words. Not to put too fine a point on it, the Western logical mentality is devoted, Lévy-Bruhl tells us, to the methodical elimination of systems of thought native to foreign social orders. In his account of it, this process bears more than a slight resemblance to the liquidation of indigenous cultures and of their associated mentalities in many areas of the world by the powerful agencies of Victorian colonialism, then at their apogee. "The preconnections of collective representations are gradually dissolved and decomposed," and the law of participation is forcibly subdued by experience and logic, he says (109). By very notable contrast, in the primitive thought-world, he declares, logical and prelogical modes *tolerate one another and harmoniously coexist;* the law of participation evidently is not marked by the same need for hegemony as is the law of noncontradiction. One notes the congruence of this idealized model of the primitive mentality and its "feeling of mystic symbiosis" (366) with Schiller's myth of contemporary science as a realm where "Freedom of Thought" prevails and "a plurality of theories . . . are treated with tolerance if they promise to promote the growth of science." For authors elaborating the relativity paradigm in this period, this conception of an ideal state of tolerance and coexistence of contraries, a state freed from the politics of absolutism, obviously forms the most compelling of themes. Yet it is a sign of the complex predicament of advanced thinkers at the turn of the century that Lévy-Bruhl's manifesto on behalf of scientific and relativistic anthropology should link itself to such a strain of primitivism and should be preyed upon in this way, semi-covertly but insistently, by what could seem like a reactionary reversion to Romantic mystification.

Not only does logical thinking, with its rigid insistence on noncontradiction, take on the aspect of a cultural and psychological dictatorship; it also is shadowed with inevitable failure, according to this writer. The failure immediately at hand is that of anthropological science itself, which is doomed by Lévy-Bruhl's relativistic principles to inescapable self-contradiction. Following a long survey of primitive initiation practices, he makes the point explicitly: "the attempt to make such practices 'intelligible' is often likely to prove contradictory. If it attains its end it is a failure. In fact, that which is 'intelligible' to logical thought is very unlikely to coincide with the idea of the prelogical mind" (*FM*, 351). Lévy-Bruhl quotes approvingly a 1904 essay on Maori medical lore by Elsdon Best: "We hear of many singular theories about Maori beliefs and Maori thought, but the truth is that we do not understand either, and, what is more, we never shall. We shall never know the inwardness of the native mind" (*FM*, 70). These dire diagnoses are in the direct line of Spencer's analyses of the aporias of scientific method in anthropology long before. "Even if [a researcher] . . . attempts to see things from the savage's point of view, he most likely fails entirely; and if he succeeds at all, it is but partially," said Spencer in 1872–73. "Yet only by seeing things as the savage sees them can his ideas be understood" (Spencer, *Study of Sociology*, 115).[25]

Not only in anthropology, but across the whole field of scientific inquiry, logical analysis inevitably falls prey to a principle of uncertainty, Lévy-Bruhl concludes. This is so because even among modern Europeans, the evolutionary triumph of logical rationality over mystical collective representations is only partial. The primitive prelogical mentality is never utterly extinguished, and thus the human mind persistently craves not only logical order in its cognitions, but also that which logic forbids, the affective gratifications that come from the sense of "participation" in the object of thought. Indeed, "the need of participation assuredly remains something more imperious and more intense, even among peoples like ourselves, than the thirst for knowledge and the desire for conformity with the claims of reason. It lies deeper within us and its source is more remote"

(*FM*, 385). Consequently, "the satisfaction which is derived from the most finished sciences . . . is always incomplete" (383). Lévy-Bruhl, sounding ever more like Balfour or like Spencer before him, sums up in one maxim the core principle of modern science once it has crossed the threshold of relativity: "actual knowledge in conformity with the claims of reason is always unachieved" (385). The intellectual faculties of modern people are profoundly self-divided, logical science contradicts itself, and knowledge obtained by rational investigation is "always unachieved."

In the literature of physics in this era, one rarely hears the accent of unappeasable frustration that thus emerges in Lévy-Bruhl; but his theme of the incapacities of reason, which is that of the whole Victorian relativity movement, stamps this literature in a decisive fashion nonetheless. It stamps, for example, the relativistic positivism of Ernst Mach, who argues that the claim of science to intellectual rigor is belied by its production of a vast scaffolding of "metaphysical" concepts—force, substance, causality, absolute space and time—without rational warrant and sustained only by "fetishism" (Mach, *Popular Lectures*, 254). In constructing a logically ordered universe that inevitably allegorizes cultural assumptions, scientific reason, according to Mach, "idealizes its objects" (*Knowledge and Error* [hereafter *KE*], 100) and misrepresents the logically uncontrollable character of reality. He describes this process in by now familiar terms as a kind of ideological violence inflicted upon human thought. "Thought does not of itself run on . . . smooth logical paths," he says. "Given the clusters of different cases and all kinds of difficulty with mutually crossing and contradictory considerations, abstraction must be imposed almost by compulsion" (*KE*, 100). (Nietzsche: "The whole apparatus of knowledge is an abstracting and simplifying apparatus—not directed at knowledge, but at the *appropriation* of things"; "that is to say, the will to *logical truth* cannot be consummated before a fundamental falsification of all phenomena has been assumed"—*Will to Power*, 2:24, 28].) The way to rid physical science of its reliance on "surmises and parables" (*KE*, 181) is to create a physics limited strictly to the description and ordering of particular sense perceptions, and the only way to give such description a genuinely scientific character is to start from the principle articulated, as Mach says, by Stallo: that things exist "only through their relations with other things, so that all our conceptual knowledge of things must be relative" (*KE*, 102–3). In this new relativistic science, the metaphysics of "a hypothetical constant reality . . . is no longer admissible"; it is replaced by rigorous attention to the process of scientific observation itself. Indeed, achieving "a full grasp of all conditions of a finding"—that is, of everything affecting the relativity of data—"alone is of practical or theoretical interest" (6).

The most searching fulfillment of Mach's dicta in the literature of physics at the turn of the century is found in Henri Poincaré's great *summa* of modernist philosophy of science, *Science and Hypothesis* [hereafter *SH*] (1902), discussed in chapter 1. Poincaré moves from his cardinal principle that "the aim of science is not things themselves, as the dogmatists in their simplicity imagine, but the relations between things," and that "outside those relations there is no reality knowable" (*SH*, xxiv), to his own skeptical interrogation of the validity of logical reasoning in science and to his own analysis of why this mode of inquiry leads in physics to a world not of definitive certainties but of idealizations, surmises, and parables. By now, we are fully familiar with the basis of the argument. What Poincaré calls "for the sake of abbreviation, *the law of relativity*" (76) undermines those conceptual entities that are indispensable to logic, things definable by essential characteristics rather than by virtue of their "participation" in (all) other things. These fixed entities are surmises and parables in pure form; they lack the properties of natural things. Thus a severe doubt appears as to whether the relativity postulate, the first principle of reasonable thinking, is compatible with strict scientific logic. In unfolding this argument at length, Poincaré makes clear his determination not to succumb to epistemological pessimism, but his strategy for avoiding that fate is finally to embrace the principle that henceforth, knowledge and certain built-in failures of knowledge must be understood as inseparable coefficients of one another, and that rationality must make common cause with modes of unreason if it is not to lapse into futility.

Science and Hypothesis opens on a note of crisis. "The very possibility of mathematical science seems an insoluble contradiction," declares Poincaré in his startling first sentence. If such a science is inductive, he asks, where does it get its "perfect rigour"? If it is deductive in accord with formal logic, "how is it that mathematics is not reduced to a gigantic tautology?" (*SH*, 1). I will not seek to trace the many forms in which Poincaré expresses his fundamental insight into the paradoxical, unrationalizable character of scientific inquiry; I will only highlight a couple of exemplary passages. One would be the remarkable demonstration early in *Science and Hypothesis* of the inherent illogic of "the physical continuum" as determined by experiment. In the archetypal laboratory situation stipulated by Poincaré, magnitudes A and B are experimentally indistinguishable, as are B and C; but A is clearly less than C. "Thus the rough results of the experiments may be expressed by the following relations," says Poincaré: "$A = B$, $B = C$, $A < C$, which may be regarded as the formula of the physical continuum. But here is an intolerable disagreement with the law of contradiction. . . . Although we may use the most delicate methods, the rough results of our experiments will always present the characters of the physical continuum with the contradiction which is inherent in it" (*SH*, 22–23). *Empirical observation and the law of noncontradiction are hopelessly incompatible*. Lévy-Bruhl's primitives, it turns out, are not the only peoples whose system of thought defies with impunity the fundamental law of logic! In Poincaré's account here, the mathematical continuum, "the concept of a continuum formed of an indefinite number of terms," is invented as the main expedient of modern science for salvaging an appearance of conformity with the law against logical contradiction. "The idea of the physical continuum, drawn from the rough data of the senses[,] . . . leads to a series of contradictions from each of which in turn we must be freed. In this way we are forced to imagine a more and more complicated system of symbols" (27). Mathematics—itself vouched for, he says elsewhere, only by "indemonstrable axiom[s]" (12) and intuitions—is imported into physical science to mask the fundamental illogic of the latter.

Poincaré subjects the law of noncontradiction to another and, if possible, even a more drastic affront. If, as he repeatedly asserts, all we can know of things in nature are the measurable formal relations pertaining among them, things in themselves being forever closed to our knowledge, then scientific theories, particularly those involving mechanical explanations of phenomena, can never possess more than hypothetical validity. They are at best potentially useful conventions of scientific work, like the "surmises and parables" Mach grudgingly accepted as temporary expedients in developing areas of inquiry. From this doctrine proceed two of the dominant themes of *Science and Hypothesis*: the vacuousness of the concept of cause and effect in science (an argument made by a series of previous writers and prominently by Mach) and, as was discussed in chapter 1, the in principle infinite number of theoretical explanations that can be constructed for any physical phenomenon. Just as the axioms of geometry are purely conventional, which means that "one geometry cannot be more true than another; it can only be more convenient," and just as the Newtonian principle of the equality of action and reaction "*can no longer be regarded as an experimental law but only as a definition*," and just as in science at large "principles are conventions and definitions in disguise" and thus "unverifiable," so, too, no explanation in science can ever claim a monopoly over all others. Logically contradictory theories may be "all simultaneously true" (*SH*, 50, 100, 138, 104, 162). Nor can we invoke at this critical philosophical moment the ultimate axiom of scientific rationality, which states that the simpler a true scientific law is, the more profound its correspondence with reality. This great principle is itself, Poincaré says, just a convention, and a peculiarly arbitrary and dubious one at that, as impossible to square with experience as is the law of noncontradiction. It conflicts irreparably with that central implication of relativity, the infinite relatedness of things. "For if all things are interdependent, the relations in which so many different objects intervene can no longer be simple," says Poincaré (147). To grasp the importance of this principle is to see that scientific inquiry is destined by "the law of relativity" to endure a perpetual oscillation of uncertainty.

> No doubt, if our means of investigation became more and more penetrating, we should discover the simple beneath the complex, and then the complex from the simple, and then again the simple beneath the complex, and so on, without ever being able to predict what the last term will be. (*SH*, 148–49)

Poincaré thus refutes as sweepingly as Balfour did, or Newman or Nietzsche, "the theory of the *logical deduction of scientific doctrine from empirical data*" (Balfour). Still, purely for practical purposes, science properly operates, he says, "as if a simple law were, other things being equal, more probable than a complex law" (*SH*, 130). Such a formula bases scientific logic squarely upon the "sceptical '*as if*' " that John Tyndall decried as "one of the parasites of science, ever at hand, and ready to plant itself . . . on the weak points of our philosophy" (Tyndall, *Fragments of Science*, 437). Science admits the parasite in this case and in others, Poincaré concedes, only for the indubitably circular reason that it is necessary to its existence. "One thing alone is certain. If this permission [to derive general principles on the basis of experimental facts] were refused to us, science could not exist; or at least would be reduced to a kind of inventory . . . of isolated facts" (*SH*, 129–30). Better a sharply destabilized system of scientific logic that has been deprived of coercive authority and definitive results than no science at all: in a relativistic post-Christian intellectual world where infallibility no longer makes sense, this, Poincaré concludes, is as close as one can come to a fundamental verity.[26]

Arendt's Lessing

The modern elaboration of the theory that nothing is one thing just by itself is thus historically inseparable from a determined critique of absolutist binary logic—not one that is alien to scientific thinking, however, but one almost identical to the most advanced forms of such thinking in different fields of inquiry. What needs to be stressed, in any case, is that the Victorian and post-Victorian decline of faith in what Spencer called the superstition of logical reason and William James called "rectilinear arguments and ancient ideals of rigor and finality" (James, *Meaning*, 205) came about not in the form of a gradual deterioration of cultural self-confidence or cultural integrity but in the form of a militant campaign aiming to impart to science a deeper self-knowledge and to preserve it from ideological perversion. Nor does it show much if any commonality with that debased (and, as I have claimed, essentially imaginary) form of relativism said to be guilty of "the displacement of the idea that facts and evidence matter by the idea that everything boils down to subjective interests and perspectives," in Larry Laudan's parodic formula (*Science and Relativism*, x). The guiding principles of the movement we have sketched were summed up in very different terms by Eddington. "We have been aiming at a false ideal of a complete description of the world," he says in 1928. "It seems more likely that we must be content to admit a mixture of the knowable and unknowable" (*Nature of the Physical World*, 228). "Uncertain knowledge" is not illegitimate, he insists in 1939; "assurance of truth" is a futile criterion for science as for other fields (*Philosophy of Physical Science*, 1, 2). Logic, which "presupposes an analysis into fixed things and terms with fixed meanings," is incongruous with natural reality and "at best . . . can only be approximately applicable to experience," says the physicist P. W. Bridgman in 1936. "Complete logical rigor" may after all be an inappropriate goal, he declares, for "any rational intellectual process" (*Nature of Physical Theory*, 44, 45, 108). Such propositions have evidently a crucial role to play in the epistemological discourse of modern physics, and their usefulness in this context is arguably not a function of their cultural history. However, underlying this whole critical development, as we have seen, is the recurring, insistent intuition that the assurance of truth on logical grounds is not just an intellectual fallacy but at bottom the expression of an ideology of violence.

This intuition expresses itself plainly in the essay on Lessing by Hannah Arendt, "On Humanity in Dark Times" [hereafter HDT], which I have cited before as a touchstone of this book. Arendt eulogizes Lessing as a great champion of "the fundamental relativity" of human thought and value (HDT, 27) and, among other things, as an antagonist of the use of logical means to establish "truth." Lessing, she says, regarded "freedom of movement" as the supreme need of human beings (9). She quotes him approvingly on the illogical character of much of his own polemical writing, which always was driven, she says, by nondoctrinaire, ever-shifting computations of the relation of ideas to the needs of the contemporary world and of the cause of freedom. "May my ideas," he wrote, "always be somewhat disjunct, or even appear to contradict one another, if only they are ideas in which readers will find material that stirs them to think for themselves" (8).

> [Lessing] not only wanted no one to coerce him, but he also wanted to coerce no one, either by force or by proofs. He regarded the tyranny of those who attempt to dominate thinking by reasoning and sophistries, by compelling argumentation, as more dangerous to freedom than orthodoxy. (HDT, 8)

Arendt thus positions Lessing, but chiefly positions herself, in the modern intellectual lineage—Mill, Stallo, Feyerabend—in which the cause of human emancipation, the theme of relativity, and the critique of binary logic form an indissoluble nexus. We are so obsessed with history and with ideology, says this scholar of totalitarianism, that "we are astonished that Lessing's partisanship for the world could go so far that he could even sacrifice to it the axiom of noncontradiction, the claim to self-consistency, which we assume is mandatory to all who write and speak" (8). The axiom of noncontradiction, she says, bearing testimony to the persistence of the tradition I have been seeking to reconstruct, is prone ultimately to serve as an instrument of "the tyranny of those who attempt to dominate thinking by reasoning and sophistries."[27]

Poincaré, exhausted by his great dialectical wrestle with the paradoxes of scientific inquiry, looks forward, as Mach and others had done, to the day when science will resign itself to the purely mathematical study of formal relations, "when physicists will no longer concern themselves with questions which are inaccessible to positive methods," and when the development of theoretical constructions in science will be left "to the metaphysicians" (*SH,* 223). His unconditional embrace of the postulate of relativity—the postulate that we can have no knowledge of objects, only of relations—leads him to long for a science from which imaginative creation has been evacuated as far as is consistent with maintaining the activity of science. Others, exhilarated by the overturning of the "divine necessity" of logic (James, *Pragmatism,* 34), take the same premise to very different conclusions.

CHAPTER FOUR

Karl Pearson and the Human Form Divine

Nature is no great mother who has borne us. She is our creation.
—Oscar Wilde, "The Decay of Lying"

In Protagorean relativity, if nothing is one thing just by itself, then "a man is the measure of all things" (Plato, *Theaetetus*, 16). As expounded by Hamilton, Spencer, Bain, Stallo, and others as the nineteenth century progressed, the theory of the Relativity of Human Knowledge came to focus ever more provocatively on this claim, differently formulated by different authors, that human awareness was in some sense the foundation of reality, to which it inevitably imparted human characteristics. Mechanistic science, possessing no vocabulary for entertaining such ideas, pronounced them nonsensical and wholly at odds with the concept of scientific objectivity. For both sides in this momentous debate, which can be seen in retrospect to have prepared the intellectual universe for the advent of quantum thinking, what was evidently at stake was not merely a set of rational values but certain deep-running themes of modern cultural politics.

So intimately is this principle of the human-centered character of reality bound up with every statement of nineteenth-century relativity that we have necessarily touched upon it at numerous points already. I isolate it somewhat artificially for separate treatment in this chapter simply for convenience of exposition.

The Dehumanization of Science

In a 1917 apologia for psychoanalysis, Freud portrays the history of Western science as a series of parallel revolutions, each inflicting chastisement upon the retrograde (in psychoanalytic terms, infantile) influence of human "narcissism." The quest for scientific truth and the repression of anthropocentrism—of human longings for belief in the privileged status of our species, of human values or a human point of view—become in this narrative essentially indistinguishable, like the mysterious identity of inertial and gravitational mass in physics. Thus, Freud argues, the belief that humanity occupies a central and somehow determining position in the scheme of nature received its initial refutation by Copernican astronomy; Darwinian biology taught the same doctrine, overturning the human sense of singularity and of superiority to the animals; and psychoanalysis now teaches the similarly humbling but therapeutic lesson that the human ego is unable to establish control over instinctive drives or even to comprehend fully its own subjection to the unconscious (Freud, "A Difficulty in the Path of Psycho-Analysis," 140–43). It appears from this analysis that the paramount principle of science is the insistence on the subordination of the specifically or ideally human to the processes of external nature. To think as a scientist, according to Freud's schema, is to adopt as one's mission "the destruction of . . . narcissistic illusion" (140). This program implies as its cardinal rule the need to purge scientific inquiry of all "subjectivity," all the distorting medium of human sensibility, in the name of attaining "objective" truth. This definition of scientific method as (in Freud's repeated term) the "wounding" of human self-love fits closely with his analysis in texts such as *Civilization and Its Discontents* of the cruel mutilations of libido civilization necessarily exacts upon mankind. Given the sadistic impulse he may be said to identify as the crucial imperative in science, it is no wonder that Freud worries in this essay that many regard his own new science with "aversion and resistances" (144).

Freud's identification of science with anti-anthropocentrism is of course an orthodox one close to the philosophical thinking of most practicing scientists down to the present day; his historical schema of the progress of this principle in science is itself canonical.[1] "Has there not been since the time of Copernicus an unbroken progress in the self-belittling of man and his *will* for belittling himself?" asks Nietzsche in 1887 (*Genealogy of Morals*, 201), raising the issue of the possibly morbid character of the official ideology of science. But this ideology has served various functions since its emergence in the late Renaissance, and around the turn of the twentieth century and in the following several decades it was proclaimed with a frequency and a rhetorical urgency that seem to signal some noteworthy cultural perturbation. One scientist after another exalts the specifically antihumanistic conception of science in this period, branding any form of anthropomorphism or anthropocentrism as the fatal intellectual fallacy, in fact as the specious negative against which scientific method should define itself—as precisely the antithesis and the nullification of science. Writing in 1901, Durkheim, for instance, pronounces anti-anthropocentrism to be the doctrine on which genuinely scientific sociology rests, though he warns that the disillusioned modern outlook has yet to establish itself securely in this field. "There still subsist numerous vestiges of that anthropocentric postulate which . . . blocks the path to science," he declares (*Rules of Sociological Method*, 46).[2]

In the self-laudatory figure repeatedly invoked at the turn of the century, modern scientific knowledge, the sole standard of truth, is routinely contrasted to the thinking of what was complacently called "primitive man" specifically on the grounds of anthropomorphism. Spencer had stressed this theme in 1862: primitive thought is marked by the anthopomorphic character of deities, and scientific advances are marked correspondingly, he says in proto-Freudian language, by the "laceration" of human feelings they inevitably produce (*First Principles*, 85). John Tyndall made a similar argument in 1871. Savages, he asserted, explain natural phenomena by reference to a population of anthropomorphic gods and spirits. "But observation tends to chasten the emotions and to check those . . . efforts of the intellect which have emotion for their base." Thanks to this wholesome chastening tendency, and to the ascendancy of the scientific concept of mechanical law, "the idea of direct personal volition mixing itself with the economy of nature is retreating more and more" (*Fragments*, 357). Frazer reiterates this narrative at great length in *The Golden Bough* (1890–1915), not only proclaiming the supremacy of scientific method—a "golden key" to truth, he declares (*Golden Bough*, 825)—and portraying modern science specifically as the eradication of superstitious anthropomorphic thinking, but attempting a vast evolutionary reconstruction of that thinking. Myth may be defined, Frazer declares, in a formula stressing the delusiveness of this mode of thought, as "a description of physical phenomena in imagery borrowed from human life" (*Golden Bough*, 770). By analyzing the dynamic of primitive anthropomorphism, Frazer means to drive a scientific stake at last into the heart of this immemorial source of folly, ignorance, and cruelty. His conception of science as the method "which strips nature of personality" (*Golden Bough*, 633) is entirely consonant with Freud's in this respect, and his attempt to install the caste of scientists such as himself in the vacated cultural office of wizards and shamans forms the plainest motive of the whole work.

The declaration of a war of extermination on anthropocentrism may constitute the permanent charter of true science, as the above-cited writers take for granted, but the urgency of this theme in turn-of-the-century discourse indicates its distinct period-specific character, too, as one major component of the wave of new thought and expression known as modernism. Some of the cultural dynamics that it expresses may be glimpsed in the portrayal of the new order of sensibility given by José Ortega y Gasset in his important 1925 essay "The Dehumanization of Art" [hereafter DA].

Serious modern artists, in revolt against the reign of self-aggrandizing and narcissistic Victorian sentimentalities, use their art, Ortega declares, to dramatize "an explicit act of dehumanization" (DA, 22)—and take reactions of "aversion and resistances" from the general public as evidence of their success, as Freud implicitly did. Modernist artists, says Ortega, constitute themselves as an elite avant-garde driven by "a real loathing of . . . forms of living beings"; their works symbolize the obliteration, the "shattering" of the "human aspect" of reality (40, 21). "For the modern artist, aesthetic pleasure derives from . . . a triumph over human matter. That is why he has to drive home the victory by presenting in each case the strangled victim," that is, imagery of the human form destroyed by imaginative violence (23). In characteristic modernist metaphors, "the weapon of poetry turns against natural things and wounds or murders them" (35). The "elimination of the human . . . elements" in art by such means constitutes, Ortega says, an esthetic "purification" (12). He describes this symbolic act not as a figurative genocide akin to the exterminationist fantasies of Conrad's Professor but as an attempt to view reality in its "objective purity," as "purified in an exemplary objectification" and a "conversion of the subjective attitude into the objective" (28, 29, 30). Modernist discourse on science, where scientific objectivity is made synonymous with the suppression of the human point of view, converges unmistakably at this point with modernist literary interpretation. In particular, "The Dehumanization of Art" brings vividly to the fore the sadistic imagery of wounding and laceration for which exponents of the modern theory of scientific objectivity have had a notable propensity. One can hardly contemplate such materials without remarking on the affinity between all this complex of discourse and the rhetoric of national purification by violence not only professed but put into practice by various totalitarian regimes a few years subsequently. We will discuss the "purification" theme more fully in chapter 5, in connection with *The Golden Bough*. To highlight these stylistic effects (centered on the theme of the salutary "elimination of the human" by a cadre of enlightened leaders) in ostensibly apolitical texts is not to suggest complicity on the part of any of the writers I have named in dire political events of the 1920s and 1930s in Europe; it is only to seek to practice a mode of intellectual history keyed to the primacy of figurative expression, and thus to an awareness that ideological phenomena may propagate themselves in multitudes of surprising forms along the circuitry of "thought style."[3]

The antinarcissistic motive celebrated by Freud and Ortega is open to the suspicion that it serves as the vehicle for a narcissistic ideology of its own. Just as the "dehumanization of art" functioned, according to Ortega, to consolidate the elite cultural status of a vanguard caste of artists and art consumers defined by their emancipation from Victorian sentiment and by their contempt for the general public, so, too, for example, the imagining of science as the intellectual agency for rebuking human narcissism functioned to promote the enormously influential cultural mythology of scientists as a wizardlike caste of European professional men freed from ordinary mental and moral failings and thus uniquely able to command authority in post-Victorian civilization. Karl Pearson's 1892 *Grammar of Science*, the focal point of this chapter, with its invocation of "the scientific man" as the "ideal citizen" by virtue of his ability to form "a judgment free of personal bias," makes this argument plainly (*Grammar*, 6). Such a text tells us that the scientific repression of "narcissism," or rather, the conception of science as the instrument of this repression, was inseparable from an extravagant fostering of the cultural power of contemporary scientists themselves. Freud's immodest self-alignment with Copernicus and Darwin makes this maneuver almost too apparent.

The cultural structure of this development is too plain to be mistaken: the vogue of the theory of science as specifically the repression of human vanity can only be seen as the signal of the usurpation by the newly ascendant institution of science of the moral ideology of the previously ascendant one, puritanical Victorian Christianity. Usually, cultural transvaluations of this kind—recodings or displacements of established systems of symbolism—need to be relatively surreptitious, but in this case, the process was quite direct, reflecting perhaps the swift obsolescence that overtook religious ideology in the last decades of the century. The path to truth can be found only through a painful discipline of self-abnegation: this principle, which still held compelling force for late-Victorian minds despite the withering of the religious convictions on which it was based, was harnessed to the service of the very institution most implicated in the decline of nineteenth-century religion. A whole moral system was transported intact from the discourse of Christian faith to that of professional science.[4]

So flagrant a shift could hardly have escaped notice. It was diagnosed, with deep ambivalence, by Nietzsche under the heading of "asceticism." Modern scientific thought, he says, represents the "*latest and noblest form*" of the ascetic impulse, particularly that of Christian religion (*Genealogy*, 192). The decisive event of nineteenth-century culture, says Nietzsche, is precisely that of "the Christian morality itself . . . the confessor-subtlety of the Christian conscience translated and sublimated into the scientific conscience into intellectual cleanness at any price" (*Genealogy*, 208).[5]

All the foregoing serves to confirm the argument made by Lorraine Daston and Peter Galison in "The Image of Objectivity," to which I referred in the introduction. In their focus on the epistemological anxieties experienced by turn-of-the-century atlas makers, Daston and Galison, describing the fusion of scientific objectivity and a code of moral compulsion, emphasize the intensity of the ethic of "heroic self-discipline" and of "*self-surveillance*" in the science of the period (Daston and Galison, "Image," 83, 103) rather than the construction of late-Victorian science as a system of harsh antihumanistic values. When one looks from a slightly different angle at the rhetoric of scientific objectivity in this period, however, what stands out is not its moderate-seeming insistence upon intellectual self-discipline, but rather, the polemical excess and even the semipathological character of all the exhortations to wounding, lacerating, shattering, defeating, murdering, strangling, and exterminating that mark the late-Victorian campaign on behalf of the dehumanization of science. The violent vindictive tone so characteristic of this literature will be a mystery—considering that the battle against anthropocentric science had supposedly been won for good in the sixteenth century—if we fail to see it as in some degree linked to the emergence in the same period of a dissident movement antithetical to the received definition of scientific method as a suppressing of anthropocentric fallacies. "The subject of science is the human universe," declares, for example, the philosophical mathematician W. K. Clifford at the early date of 1872; "that is to say, everything that is, or has been, or may be related to man" (*Lectures*, 1:141). Clifford's relativistic formula implies a radically different mission for science than that of exploring the various domains of nature as though they were self-existent things independent of human thought. The proponents of the intellectual movement based on this principle direct their critique not against the delusive narcissism of human sentiments, but against that of would-be "objective" science claiming more or less exclusive knowledge of what are called, in the passion-laden phrase much bandied about (always with aggressive polemical intent) by late-Victorian writers, "absolute truths." They argue that scientific rationality and all the acquisitions of scientific knowledge are, to the extent that they are genuinely scientific, necessarily anthropocentric and anthropomorphic, and that the great superstition to enthrall modern thought has been in fact that of the possibility of knowledge purified of "imagery borrowed from human life." In striving to dehumanize itself, they argue, thought becomes fatally incoherent and sacrifices its intellectual legitimacy. They question the implicitly theological imperative underlying the fierce assault on anthropocentrism and suggest that the credo of science as the persecutor of human narcissism originates not in the disinterested love of truth but in a will to tyrannical domination.

The logic linking this attempt at the rehumanization of knowledge to the relativity movement is obvious: once it is recognized that knowledge of the external world is in some fundamental fashion "relative to us," then the very presumption of a natural world existing otherwise than as a function of humanity or as a projection of human consciousness comes into doubt. Not that this movement is consistent in its arguments: on the contrary, its texts are volatile, marked chronically by self-division and by equivocation.[6] Karl Pearson identifies this instability as the defining mark of speculative discourse in an age of turbulent change in fundamental ideas. "The same individual mind, unconscious of its own want of logical consistency, will often exhibit our age in microcosm," he lucidly remarks, in a formula that turns out to have great pertinence to his own work (*Grammar*, 3). But in all the forms that the revisionist Protagorean project assumes, it holds up to severe critical scrutiny the widespread, almost compulsive determination of the philosophical authorities to shield from view the human image stamped indelibly upon all the intellectual productions of our species.

Pearson and the Demon

Having discussed Ludwig Feuerbach in chapter 2, I will not dwell on him here, but any account of the late-Victorian controversy over anthropocentrism in science, and any attempt to gauge its bearing on our own intellectual situation at the turn of the twenty-first century, must begin with his *Essence of Christianity* [hereafter *EC*] (1841). Feuerbach argues in this work that the fundamental human impulse is that of projecting the human image and human values onto all of reality. Christian religion, insofar as it is authentic and legitimate, bases itself, he says, on an act of anthropomorphic imagination: the creation of a divinity in human shape as an idealized projection of mankind. Since human feeling constitutes for man "the absolute, the divine," he "unconsciously and involuntarily creates God in his own image" (*EC*, 10, 118). Religion, the ultimate expression of human narcissism, thus originates in and symbolizes nothing less than "the deification of man": "the beginning, middle and end of religion," according to Feuerbach's famous motto, "is MAN" (*EC*, 51 n., 184).

But the unsubdued creative power of human imagination represents a standing threat to the dominance of religious authorities, who strive, as their most vital order of business, to deny, disable, and suppress it. Theology thus misrepresents the human character of the divine, portraying God as an independent being, and so "[attaching] religion to an external object" to which it is the religious duty of believers to submit utterly (*EC*, 11). Feuerbach sees theology as an agency of the same banishment of anthropomorphism that Freud and others ascribe honorifically to modern science, in other words. For Feuerbach, however, this banishment is the reverse of a rolling back of superstition. Severing the linkage of identity between God and human faculties and thus dispossessing mankind of its supreme imaginative creation, theology produces a profound psychological and spiritual fracture, "the disuniting of man from himself"; indeed, it "sets God before [man] as the antithesis of himself," in a relationship of estrangement and hostility (33). "Religion alienates our own nature from us, and represents it as not ours" (236). All the fanatical obsession of Christian thinkers with the category of absolute truth, all their morbid hostility to and vindictive repression of human nature, all the conspiracies against freedom and all the sanguinary abominations to which religious history testifies—all these things flow, according to Feuerbach, from the theological denial of the truth that God is constructed wholly of "imagery borrowed from human life." Thus it is that religious belief has become inseparable not only from authoritarian structures of power but from that epicurean form of sadistic fanaticism that "delights in the blood of heretics and unbelievers" (52). In *The Essence of Christianity*, the rejection of relativistic anthropomorphism is a purely diabolical moment.

In its stress on the world-transforming potential of human imagination and on the extreme physical and ideological violence necessary to keep it in check, *The Essence of Christianity* differs from later and more pessimistic narratives of the growth of systems of authoritarian domination in modern society like Spencer's in *The Study of Sociology*, Mill's in *On Liberty*, Nietzsche's in *The Will to Power*, or Dostoyevsky's in "The Grand Inquisitor."[7] Such works stress, instead, the feebleness and the malevolent mediocrity of the modern citizenry, who readily collaborate in the persecution of dissidents (these writers argue) out of devotion to any regime that promises to relieve them of the intolerable burden of freedom. A version of the Feuerbachian paradigm reappears, however, in Karl Pearson's *Grammar of Science* [hereafter GS], published in 1892, fifty-one years after *The Essence of Christianity*.

Pearson, an eminent statistician as well as a self-described "freethinker," a socialist and a feminist (and subsequently, as we shall see, a proponent of eugenics), exhibits in his work a striking example of the nineteenth-century convergence of relativistic theory, evolutionist science, and emancipationist critique. He was acutely sensitive to the ideological properties of science, and, in particular, to what he took to be the sinister valences of the movement of "dehumanization" that was implied by the late-Victorian "moralization of objectivity." His project as a theoretician was to reveal the concealed workings of this system and to institute in its place a remoralized, as it were a rehumanized, science—a science philosophically expressive of an ideal of human freedom rather than of the "serfdom" he declared to be weighing heavily on modern people.[8] His link to Feuerbach lies in his insight that religious orthodoxy moves, at this important juncture of intellectual and social history, exactly in concert with the scientific orthodoxy that glorifies science as a system for persecuting human narcissism and for suppressing the claim of human beings to shape and reshape reality by "direct personal volition." Pearson never refers to Feuerbach but seems to have his analysis of theology at least unconsciously in mind as he focuses upon the systematic perversion of another department of knowledge, that of modern physics. The vocabulary of the polemic has shifted, in keeping with the times: what Feuerbach calls "theology," Pearson in effect names "materialism."

The Grammar of Science begins on a note far removed from Feuerbachian skepticism, professing in its early pages what looks like a conventional late-nineteenth-century celebration of scientific knowledge, particularly in the burgeoning field of physics. Pearson delivers here an unqualified homily in praise of the rationalistic ethic of science. Legitimate science appeals strictly to reason and disavows imagination, Pearson declares, in a wholly anti-Feuerbachian spirit, seemingly oblivious to the critiques of scientific rationality various writers had elaborated by then. (Notable among these writers was of course Newman, to whose *Grammar of Assent* Pearson's own title makes what seems at first to be a wildly improbable allusion.) Pearson makes clear, furthermore, that the idolizing of contemporary science has a fundamental ethical dimension, particularly in connection with the sphere of politics. Scientific reasoning forms an ideal training for citizenship in a period of rampant controversy and uncertainty, since it depends, he says in the conventional way, on what he terms "self-elimination" (*GS*, 6), the principled setting aside of personal bias—of that element of narcissism or anthropocentrism that Freud also tells us is the enemy of rational understanding. Evincing his acknowledged indebtedness to Einstein's own early philosophical guide, Ernst Mach, Pearson declares here that science takes as its sole enterprise "the cautious and laborious classification of facts"; it is a stringent, ascetic discipline, dehumanized in exactly Ortega's sense, treading unswervingly "the hard and stony path of classifying facts and reasoning upon them" (20, 17). Its extraordinary results are obtainable only through "the long and patient toiling of many groups of workers, perhaps through several centuries" (18). Thus Pearson praises science in terms that give powerful expression to the emotionally enthralling moral discourse of Victorian popular culture, the discourse that flows, for example, through the glorification by novelists like Dickens and George Eliot of the ethic of renunciation and self-sacrifice, of the suppression of egoism and frivolity, of the spiritual value of work. In short, Pearson eloquently exemplifies "the moralization of objectivity" just as it is described by Daston and Galison. The superiority of science over philosophical speculation, he says, is manifest in its power of producing "unanimity of judgment," whereas philosophy, subject as it is to "the influence of individual bias," is all too likely to lead only to "an endless number of competing and contradictory systems" (19)—in other words, to that state of unimpaired intellectual ferment that Mill held up as an ideal, or, alternately, to that "cancerous and dangerous proliferation of significations" that Foucault identified as the abiding dread of the Western intellect. By abolishing individuality, science represses the alarming instability of free speculation and helps to ensure a society where all think alike. The theory of science as an engine for the mortification of human narcissism could hardly find a fuller expression than this opening movement of Pearson's treatise. Nor could the epistemological dictatorship of science find a stronger endorsement. "The scientific method," declares Pearson, "is the sole gateway to the whole region of knowledge" (24).

The political and ideological reverberations of this imperial philosophy of science are easy to recognize, but even so, they are not left to the reader's imagination. Expressly treating his vision of science as a template of civic values, Pearson argues that the wholesome repression of anthropocentric thinking in science corresponds to a political code based on the sublimation of selfish interests in "the interests of society at large" and glorifying the willing acceptance of "long and patient toiling [by] many groups of workers, perhaps through several centuries" (*GS*, 5, 18). Scientific method in this interpretation allegorizes an ideal sociopolitical regime in which the supreme duty of workers is to accept their lot without complaint and without disrupting the prevailing unanimity of thought. The institution of positivistic science, like Methodism in Elie Halévy's interpretation of British history, is portrayed in Pearson's preamble as a bastion against disturbance of the political status quo.

This paean to scientific method has an acutely equivocal relationship to what follows in *The Grammar of Science*. We may take it as an exercise in rhetorical misdirection, or, more plausibly perhaps, as the sign of the persistence of an earlier absolutist phase of Pearson's thinking—one that would return disturbingly to the surface in his subsequent career as a eugenicist. The latter view would be supported by the evidence of Pearson's early rhapsodic novel *The New Werther* (1880), which centers, as George Levine observes, on an epistemological "quest for the absolute" (Levine, "Two Ways to Be a Solipsist," 19) and invokes an extreme ideal of renunciation both as the means of discovering "truth" and—as Pearson says, foreshadowing Ortega—as "a means of purifying mankind" (quoted in Levine, "Two Ways," 20). Similar themes are set out in a less overwrought idiom in Pearson's 1883 essay "The Ethic of Freethought," where the contemporary freethinker is declared to be the missionary of "new truth" and where, Pearson says, "this very word truth itself denotes some absolute, fixed, unchangeable law"; an "ordered outcome of cause in effect"; "not a finite changeable thing, it is absolute, infinite, independent of all conceptions of time or change" (Pearson, *Ethic*, 22, 26). By 1892, by contrast, he had become committed to a Stalloesque relentless critique of the concept of the "absolute" in any context whatever, and in particular had made the rejection of causality one of the keystones of his scientific philosophy.

With what seems to be calculated rhetorical effect, but may be just a symptom of the extreme instability that (as Pearson says at the outset) marks speculative thinking at this historical moment, *The Grammar of Science* moves without warning into another discursive register altogether—and flies, in fact, squarely in the face of Pearson's opening credo. The human creative imagination, a few pages previously cast out in disgrace from the domain of respectable science, now is identified as the source "of all great scientific discoveries," and the laws of science, no less than works of art, are said to be "the product of the creative imagination" (*GS*, 30, 36). And now, in place of the austere discipline of scientific analysis plodding laboriously along "the hard and stony path" of positivistic fact, science, like art, proves to be largely driven by the *jouissance* proper to the experience of beauty: "both afford material for the gratification of the aesthetic judgment," he says (36). The philosophy of science seems here to have come under the subversive influence of the aesthetic movement, just at its zenith in 1892. In thus displacing renunciation and "the moralization of objectivity" by personal gratification in the scheme of scientific values, Pearson does not go so far as to assert that artistic beauty could be a criterion of the truth of a scientific theory, as others have since, but his argument points unmistakably in this direction, with potentially disruptive philosophical consequences.

Taking as his point of departure the doctrine known in the nineteenth century as the Relativity of Human Knowledge, Pearson takes a bold step further: "the universe is largely the construction of each individual mind," he declares; it is therefore "a variable quantity," undergoing alteration as the result of the imaginative work performed on it by each succeeding generation of scientists (GS, 15). In proposing this radically relativistic interpretation of the principle of evolution (and thus suggesting once again the essential identity of the two great nineteenth-century theories), *The Grammar of Science* may in fact be the text that marks the entry of "constructivism" into the vocabulary of philosophical discourse. Underlining the considerable leverage he means for this term to exert in his argument, Pearson credits its coinage to Lloyd Morgan, in *Animal Life and Intelligence* (1890–91). "At the bidding of certain stimuli from without," Morgan had said, "we construct that mental product which we call the object of sense. It is of these mental constructions—'*constructs*' I will call them for convenience—that I have now to speak" (Morgan, *Animal Life*, 312). "External objects," a phrase regularly placed within quotation marks by Karl Pearson, are thus not self-existent entities confronting human perception and forcibly dictating human awareness. Rather, in accordance with what Pearson, naming his chief principle of scientific analysis, calls "the pure relativity of all phenomena" (GS, 394 n.), external objects are constructions of human faculties, as Morgan had insisted they were. Scientific laws of nature, by the same token, are necessarily manufactures of human thought, without a basis in any world of external reality. "It may seem at first sight strange to the reader," says Pearson in a considerable understatement, "that the laws of science should thus be associated with the creative imagination in man rather than with the physical world outside him" (36), but such is the case, he insists. "Law in the scientific sense is . . . essentially a product of the human mind and has no meaning apart from man. It owes its existence to the creative power of his intellect. There is more meaning in the statement that man gives laws to Nature than in its converse that Nature gives laws to man" (87). As he had memorably put it in "The Ethic of Freethought," "only what man *thinks*, can possibly *be*. . . . *It is the mind of man which rules the universe*" (*Ethic*, 31). The human being, as W. K. Clifford had sacrilegiously phrased it several years previously, with a genuflection to Protagoras the Sophist, is nothing less than "the Lord of Nature and the Measure of all things" (Clifford, *Seeing and Thinking*, 156). Hence, says Pearson, Newton's statement of the formula of mutual acceleration "was not so much the discovery as the *creation* of the law of gravitation" (GS, 86).

For Pearson, in other words, once it falls under the sign of "pure relativity" scientific thought becomes of necessity anthropocentric and anthropomorphic—in Freud's sense, narcissistic. From being an insignificant creature on an insignificant planet ruled by iron systems of natural law, the human being is scientifically reconceived in *The Grammar of Science* as nothing less than the creator and ruler of the universe—a ductile universe that varies with the passage of human generations—and the author of its laws. Science in Pearson's hyperbolic conception, we may say, becomes the vehicle of "the deification of man," as pretheological religion is in Feuerbach's. Yet the so-called laws of nature that scientists, driven by the craving for esthetic pleasure, labor to articulate are in fact not so much inventions as simply the principles of human cognition itself. "Science is in reality a classification and analysis of the contents of the mind," says Pearson; "the logic man finds in the universe is but the reflection of his own reasoning faculty" (*GS*, 52, 91)—the latter formula, with its metaphor of the reflecting mirror, bringing the Narcissus myth distinctly to mind. In the compulsive and misguided search for absolute truths, the modern intellect trains its analytical instruments on the phenomena of the physical universe, only to find there, according to Pearson, the lineaments of what Blake called "the human form divine." Science no longer imagines deities in human form within nature, but it rests upon an analogous mental operation, for it imagines particles, waves, and forces behaving in accordance with human laws of logic. Pearson's doctrine, to paraphrase Feuerbach, is that "the beginning, middle, and end of science is MAN."

This radical proposition is central to the historical significance of *The Grammar of Science*, particularly with respect to relativity physics and its environing realm of discourse (the advent of special-relativity theory is distinctly foreseen by Karl Pearson, as we shall notice in a moment). Much of this book still seems challenging in the stringency of the skepticism it brings to bear on a broad range of received scientific concepts, and the doctrine it preaches can pass muster as sophisticated postmodern theory.[9] When Henry Adams read it, he found it disturbing and "revolutionary" in its invocation of themes of "supersensual chaos," which is to say, the chaos of the unknowable realm of things-in-themselves. But it is important to insist that much of Pearson's argument belonged to what was already by 1892 a well-established line of thought. Adams, writing in 1905, speaks of him a little slightingly as echoing ideas formulated by Stallo twenty years earlier, and as providing evidence of how the latter's heresies had achieved a degree of acceptability in the intervening years (Adams, *Education*, 449–50).[10] Pearson's arguments had by no means outlived their power to scandalize intellectual traditionalists—they have not outlived that power to this day—but they drew on ideas fully elaborated and in wide circulation by the

1880s, indeed by the 1850s and 1860s, ideas that were woven (sometimes only in invisible thread) into late-Victorian speculative discourse and that enjoyed, as Adams reminds us, a kind of subversive orthodoxy. It was this same complex of ideas that served as the point of origin of the special theory of relativity in 1905.

Karl Pearson, then, elaborates in *The Grammar of Science* a set of themes enunciated by Spencer, Stallo, Mach, Clifford, and others; his and Lloyd Morgan's special contribution to them lies in their bold insistence on the principle that if human thought is "relationing," in Spencer's term, this process must be understood as an actively creative intervention in the structure of the world, and that science is above all a work of imagination. Morgan stresses "the terms 'construct' and 'construction,'" he says, to underline his refusal "to reduce the human mind . . . to the condition of a mere passive recipient instead of a vital and active agent in the construction of man's world" (Morgan, *Animal Life*, 320, 332). This Blakean and Nietzschean creativism seemed a dangerous, unscientific doctrine even to extreme relativists of Morgan's day, perhaps because its overtones of political radicalism were too distinctly audible for comfort. As Mach declared in 1893, the year following the appearance of *The Grammar of Science*, "science does not create facts from facts, but simply *orders* known facts" (*Popular Scientific Lectures*, 211). But Pearson ultimately could not be satisfied with any such notion, his close affinities with Mach and Spencer and his ardent devotion to the cause of scientific rationality notwithstanding.

The reign of false doctrines of what Pearson calls, synonymously, "materialism" and "metaphysics" has occurred in accordance with exactly the dispossessing, traumatically alienating event diagnosed by Feuerbach in the domain of religious imagination—and, in the meantime, by Marx in the field of capitalist production, to account for the fetishistic character of commodities (Marx, *Capital*, 71–72), and by Nietzsche, to account for the "irrefutable" appearance of the laws of reason (Nietzsche, *Will to Power*, 2:29).[11] The materialist philosopher of science, says Pearson, "unconsciously severs himself from the products of his own reason, projects them into phenomena, only to refind them again and wonder what reason put them there"; this same materialist philosopher constructs a universe of objects related by natural laws "and then—as his habit is—forgetful of his own creative facility, has converted [them] into a dominant reality behind his perceptions and external to himself" (*GS*, 91, 167). Pearson, like Mach, claims that science has no business explaining the phenomena it studies, and so does not attempt to explain the appearance of this syndrome of perverse misunderstanding, which shrouds scientific rationality in metaphysics and in the process inculcates an ethical-political philosophy of which the cardinal principle is that of inescapable subjection to a "dominant reality." As far as Pearson's text explicitly states, scientists fall into metaphysics simply in a fit of forgetfulness. He does not explore further this mechanism of amnesiac self-alienation, but its logic would not be hard to interpret. It undoubtedly lies in the whole apologetic tradition, tinged with religious urgency, being reaffirmed by Freud and so many others at this period, to the effect that the progress of scientific knowledge can only come about by means of—in fact, is identical to—the repression of the vice of human narcissism. Science in this official view, the one founded on "the moralization of objectivity," is that intellectual enterprise dedicated to teaching man his subordinate place in the universe, where he is to toil uncomplainingly and eat his bread in humility. The indoctrinated subject of such science is defined by that characteristic attitude analyzed by Spencer as "the awe of embodied power" (Spencer, *Study of Sociology*, 175), professing as this subject does a creed that equates moral value with the idolizing of "a dominant reality . . . external to himself." Abjuring human creativity, imagining one's own perceptions as external objects, and revering "absolute truth" in science as in other domains of human life—these are the epistemological corollaries of the moral commandment of poverty and self-abnegation: so runs the implicit logic of *The Grammar of Science*.

To follow this logic is thus to move with Pearson's help toward understanding the anti-anthropocentric polemics of turn-of-the-century science as a residue of a cultural dynamic deeply associated with the "Victorian" moment in Europe, and as playing to some degree a dissonant, unstable, even regressive role with regard to the modernist enterprise at large. A definition of science as first and foremost an agency for "the destruction of . . . narcissistic illusion" is symmetrical with Feuerbach's analysis of theological authoritarianism or with, say, Mill's assertion that the defining feature of the Victorian milieu was its domination by what he called "the engines of moral repression" (*On Liberty*, 13) and identified with tyrannical perversions like those of Mansel, Comte, Governor Eyre, and others. The tyrannical frame of mind was in fact, as we have seen, diagnosed as the cultural axis of the age by one Victorian observer after another and was symbolized, for instance, in the grim, repressive, sadomasochistic Evangelicals who populate the novels of Dickens, Thackeray, the Brontës, George Eliot, Trollope, Samuel Butler, and others. Pearson and other writers of the relativistic party make it possible to recognize the discourse of anti-anthropocentrism, which advertises itself as the voice of enlightened intellectual modernism, as a renewed manifestation of this same malign spirit. Thinkers around 1900, we may suppose, turned to the affirmation of a scandalous anthropocentric conception of science as a means of seeking to extricate themselves from the broad sociopsychological regime of which the official epistemology appeared to be a symptom, just as *fin-de-siècle* aesthetes cultivated scandalous attitudes of their own for a similar purpose. Strictly from the political point of view, a scheme in which the central proposition is that "man gives laws to nature" rather than being helplessly subject to laws established immutably and imposed upon him from without, and in which the universe is conceived as ductile to human influence rather than rigidly fixed, must have seemed laden with subliminal ideological messages which to a socialist and feminist like Pearson, an agitator against all forms of contemporary "serfdom," could only have been seductive.

Having established this broad polemical framework, he turns to the central project of *The Grammar of Science*, a systematically deconstructive survey of the field of modern physics that bears a strong family affinity to Stallo's in *Concepts and Theories* and to Schiller's survey of the doctrines of logic in *Formal Logic*. The principle from which the whole enterprise of *The Grammar of Science* proceeds is what Pearson defines as "the *relativity* of natural law" (GS, 83) and illustrates by quoting at length the famous passage from Maxwell's *Theory of Heat* describing the imaginary "demon" who, thanks to his preternaturally keen perception, is able selectively to control the movement of molecules of gas within a closed system by opening and closing a microscopic shutter. The demon in this way produces inequalities of temperature within the system, contradicting the second law of thermodynamics. "Such a conception," says Pearson,

> enables us to grasp how relative what we term nature is to the faculty which perceives it. Scientific law does not, any more than sense-impression, lie in a universe outside and unconditioned by ourselves. Clerk-Maxwell's demon would perceive nature as something totally different from our nature, and to a less extent this is in great probability true for the animal world, and even for man in different stages of growth and civilisation. (GS, 84)

Thus "law in the scientific sense is . . . essentially a product of the human mind and has no meaning apart from man," says Pearson, repeating (not for the last time) his key principle. "It owes its existence to the creative power of his intellect" (GS, 87). Insofar as we speak of certain natural laws as universal, says Pearson (who here again follows the lead of Lloyd Morgan), we implicitly take as our point of reference "the [world] . . . of normal civilised man" (84–85)—a tripartite category each component term of which is obviously fraught with grave ambiguity. Pearson gives great theoretical weight to this theme. "The 'universality' of natural law, the 'absolute validity' of the scientific method, depends on the resemblance between the perceptive and reflective faculties of one human mind and those of a second," he says (101). For members of other species, other societies, other eras of history, perhaps for another gender than "man," sense-impressions and the interpretations made of them necessarily vary. For all such observers, as for those individuals in our own society who see the world so differently from the conventional point of view that they are labeled "madmen," Pearson says unjudgmentally, "our laws of nature are without meaning" (101). He does not recommend abrogating the regime of scientific knowledge in the name of epistemological relativism, and even suggests that natural selection may dictate the mode of scientific reason most advantageous for different populations of beings (103–4, 184); but the overriding lesson of his analysis is that nature is defined not by intrinsic properties of its own but by human consciousness and that, given the variability of consciousness, claims to "absolute" knowledge are indefensible. (He does not point explicitly to another aspect of this demonstration that is of a piece with the large subtextual argument of *The Grammar of Science*, namely, the imagining of Maxwell's mischievous demon as a figure of the subversive individual discovering his or her power to bend the supposedly permanent realities of nature—the subordinate position of women in society or the persistence of a social substratum of extreme poverty, for example[12]—to his or her will.)

From this disquisition on "the *relativity* of natural law," Pearson proceeds to the deconstructive analysis of a series of fundamental "metaphysical" entities introduced with pernicious effect into science, he argues, in consequence of its determined denial of its human origin. Among these superstitious "fetishes" of a "crude metaphysical materialism" (*GS*, 332)—the intellectual vermin spawned by the will to impose a scheme of "absolute" reality upon a world in which nothing exists but relativity—are the concepts of force, mass, matter, causation, the ether, the electron, the atom. I will refer just to a high point or two of this material, seeking to illuminate the implicit dynamic of the argument.

Consider the long critique in *The Grammar of Science* of the concept of causality, which was for Mill, we recall, identical with logic itself, and which Spencer considered the anchoring principle of all scientific thinking. We are likely to think of the law of cause and effect, which we have made into "the dominating factor in phenomena," as an "absolute truth" and "as a necessity existing in 'things in themselves,'" says Pearson (*GS*, 152), hinting plainly in all this evocative language of domination, absolutism, and necessity at issues concerning the relations of contemporary freethought to systems of external power. Pearson declares that replacing the concept of causation with that of statistical correlation and contingency represents an "epoch-making" change, one that "conceptualises the universe under a new category" (165, 166). His demonstration here rests on two principles. The first states that instead of a crudely binary conception in which things are classed either as causally connected or as independent, "the newer, and I think truer, view of the universe is that all existences are associated in a higher or lower degree" (166). As was stressed in chapter 1, this postulate of the infinitely nuanced interconnectedness of all things is nothing but the fundamental axiom of relativity thinking itself: nothing is one thing just by itself. In expounding it, Pearson hints at the moral and political economy of such a precursor text as *Bleak House:* "many things pass in the universe for absolutely independent, which a finer power of analysis or observation would demonstrate to be associated" (163), he says. The second main principle is Pearson's denial, which he conveys with a polemical emphasis that can only be called impassioned, of the possibility of "*absolute* sameness" in natural phenomena (153). Events may resemble one another, and certain events may exhibit a high degree of correlation with others; still, "everything in the universe occurs but once, there is no absolute sameness of repetition" (157). We treat things and events as identical for certain practical or philosophical purposes, but in reality, "existences are individual" and all things are marked by "individuality and change," says Pearson, who goes so far as to venture the thesis that even "the absolute sameness of the molecule is only a statistical sameness" (166, 156). If then "we realise individuality at the basis of all existence," he says, "from this standpoint the universe appears as a universe of variation rather than as a universe controlled by the law of causation in its narrowest sense" (156, 174). This law has been inscribed upon reality, according to Pearson, just as the tyrannical law of theology was inscribed upon it in Feuerbach's analysis. The averaged curve superimposed on a blurry contingency table, says Pearson, "is the 'causality' which man extracts from his experience and thrusts back into nature as if it had actual existence there" (169). Science in the modern mode will seek to measure degrees of variable contingency, not to identify one thing as the "cause" of another, and above all not to imagine the world as subject to a rigid "principle of absolute causation" (158).

The received idea of causality "gives no measure of the variation in experience, and it has trammelled the human mind, because it has led to a conceptual limit dominating actual experience" (*GS*, 170), Pearson says, in his usual pointedly suggestive language. In such a formulation as this, the continuity of his plea for a sophisticated science of contingency values with morally and politically emancipatory freethought is easy to detect. The juxtaposition of *The Grammar of Science* with the nineteenth-century touchstone of freethought, *On Liberty*, is instructive here. Needless to say, Pearson's insurgency against a doctrine taken by establishmentarian science to represent "absolute truth" vividly exemplifies the sort of heretical challenge to "the tyranny of the prevailing opinion and feeling" (*On Liberty*, 4) that Mill idealizes as the very definition of intellectual liberty. More particularly, in insisting so strongly on "individuality" as the fundamental characteristic (along with interrelatedness) of physical events, Pearson echoes the doctrine that for Mill is central to political values: that "the individual is sovereign," that the preservation of "individual liberty" is the highest function of society, and that "individuality should assert itself" against the deadening effect of systems of thought that compel conformity to a single dominant pattern (*On Liberty*, 10, 8, 56).

Pearson's critique of the "trammelling" of physics by the fetishizing of causality as "a conceptual limit dominating actual experience," and his recommendation of the alternative concept of statistical correlation, thus translate immediately into Mill's polemic against the dictatorship of received ideas. We would have only a truncated understanding of *The Grammar of Science* if we were to shy away from this sort of interpretive translation. A reader freed by Pearson's teaching from the mind-forg'd manacles of scientific ideology (which treats scientific laws as facts of nature rather than as products of human imagination) will see that the theory of causality, which depends on imposing the principle of "*absolute* sameness" on the world of nature, operates symbolically as a consecration of the established order; on the textual level at least, it abolishes the power of human "creative facility," which for Pearson is basic to nature itself, to construct new systems of experience; it equates righteousness with an attitude of submission to fixed, "objective" and "natural" outside forces—the "dominant reality" of conventional scientific metaphysics. It expresses allegorically the aggravated tendency of modern people "to prescribe general rules of conduct, and endeavor to make every one conform to the approved standard" (Mill, *On Liberty*, 70), just as Maxwell's demon personifies the human capacity to subvert such a standard by acts of emancipatory sabotage. The subtext of Pearson's disquisition on causality in physics is on one level the implication that the current social order has no natural warrant of rightness and permanence; rather, it is open to reconstruction by human initiative. As he definitively puts it in an 1887 essay, "the right to re-shape itself is the one birthright of humanity" (Pearson, *Ethic*, 320). "It is [man's] gift to be the creator of his own sufficiency . . . and to be emphatically self-made," Newman had said (GA, 274).

Some of the same implications arise from Pearson's relativistic critique of the Newtonian category of "absolute, true, and mathematical time" in physics (GS, 216). His basic doctrine here is precisely the one that undergirds Einstein's special-relativity paper thirteen years later: that time has no existence in the natural world except as "a relative order of sense-impressions," and that there is therefore, as he says, "no such thing as *absolute* time" (GS, 213). Pearson follows in this section the essential lines of Stallo's earlier demolition of absolute time (time, said Stallo, is "measured by the recurrence of certain relative positions of objects or points in space, and . . . the periods of this recurrence are variable, depending upon variable physical conditions"—*Concepts and Theories*, 219), but elaborates the point more fully than Stallo did. He imagines a sequence of clocks, each one presumably truer and more authoritative than the one preceding it: one's pocket watch; the standard clock at the Greenwich Observatory; the revolutions of the earth on its axis as measured by the passage of circumpolar stars about the horizon; astronomical periods defined by the motion of the planets. But how can we ever know that any timekeeper is in fact regular, and how can we hope to measure absolute time "if there be nothing in the sphere of perception which we are certain flows at a constant rate" (GS, 216)? What if all the bodies in the solar system had *together* gained or lost time over the course of millennia? Nor can we measure time absolutely by a physical standard based, say, on the distance traversed by light in one second: if a measurement a thousand years hence were to differ from the standard, how could we decide whether time had changed its flow or the speed of light had changed? "No more than the astronomer can the physicist provide us with an *absolute* measure of time," Pearson concludes. "So soon as we grasp this we appear to lose our hold on time" (215). Absolute time is a metaphysical figment, he declares.

As for space, "it is peculiar to ourselves," declares Pearson—it is purely anthropocentric. We create it conceptually as a mode of grasping the relations of objects, and in itself it has no existence (GS, 184). Only a thoroughgoing relativistic analysis is appropriate to such a thing. "If space be a mode of distinguishing things, we must have at least two things to distinguish before we can talk about position in space. Position of a point is therefore relative, relative to something else" (233). We can recognize here, in this statement of the doctrine of sophisticated modern physical science, the formula running almost verbatim through nineteenth-century relativistic discourse from Hamilton, Mansel, Bain, and Spencer onward. Pearsonian space, we may say in Saussurean phraseology, is for Pearson wholly a system of differences, "*without positive terms*."

Pearson is led by this argument to speculations on the interconnectedness of space and time so audacious that he confines them to a remarkable endnote first appearing in the 1900 second edition of *The Grammar of Science* (GS, 394) in which he proposes a "conception . . . of time as a problem in relative motion." If "a colleague of Clerk-Maxwell's demon" were traveling away from the earth with a velocity greater than that of light, he says, "clearly all natural processes and all history would for him be reversed." Or, "by motion to or from the earth, our demon could go forward or backward in history, or with one speed—that of light—live in an eternal *now*." Thus "irreversibility of natural processes is a purely *relative* conception." There is still a substantial conceptual distance to be traversed between this conceit of Pearson's (which he credits to a suggestion from L. N. G. Filon) and Einsteinian special relativity, but Pearson's denial of absolute time and his refusal to declare that an earthbound observer's frame of reference has any theoretical priority over the demon's implies the most amazing aspect of special relativity, the relativity of simultaneity. (Since Einstein postulates the speed of light to be an unsurpassable natural limit, the notion of historical reversibility is excluded for him; Eddington suggests, indeed, that he postulates it precisely to exclude this wild possibility.)[13] Pearson here plainly states, in any event, the principle of the inseparability of space and time—which is to say, of the dependency of the passage of time upon the relative motion of different frames of reference—that forms the foundational principle of twentieth-century scientific relativity. The reading of his text here proposed makes clear, as well, the origination of this principle in themes of subversive ideological critique.

Pearson mounts a parallel demonstration of the relativity of motion, tracking the variable motion of a man's waistcoat button relative to various other objects: the staircase, the sun, the constellation of Lyra (GS, 235–36). I will not rehearse this passage in detail, if only because it duplicates closely Spencer's demonstration in *First Principles*, thirty years earlier, of the relativity of motion, using the illustrative imagery of a captain walking the length of his ship (FP, 41–43). "We can only say how fast one thing is moving *relatively* to another," says Pearson, "since all things whatsoever are in motion, and no one can be taken as the standard thing, which is definitely 'at rest'" (GS, 236). Any object in the universe, says Pearson, constitutes a "frame of reference" (235) for the motion of any other object. It is here that Pearson puts forward as an inescapable scientific conclusion the notorious paradox formulated by Newman in 1843, as we saw in the preceding chapter. "Is it correct to say that the earth actually goes round the sun, or that the sun goes round the earth?" asks Pearson. "Either or neither; both are conceptions which describe phases of our perception"; we prefer the heliocentric model simply as a convenience of expression, he declares (236). Pearson never refers to Newman by name in *The Grammar of Science*, any more than he refers to Feuerbach, but the covert allusion to *Grammar of Assent* in his title appears to link this treatise on physical science closely to Newman's critique of the cult of logic and scientific proof, as though the two contrary-seeming endeavors were at last two manifestations of a single trend of thought. That this trend expresses itself as bewilderingly either radical or reactionary, socialistic or theological—depending, as it were, on each author's relative state of motion with respect to it in intellectual space—is a reminder that the law of "the pure relativity of all phenomena" applies to the textual realm fully as stringently as to that of physics. I need hardly underline the seeming paradox of a text that could claim descent from Cardinal Newman while clearly situating itself at the same time within the fiercely anticlerical heritage of authors like Feuerbach, Stallo, and Spencer. Such are the complicated paths and often subliminal-seeming nexuses by which relativity thinking finds its way toward Einstein in the nineteenth century—and which intellectual history must be careful not to abridge in the way that causality, according to Karl Pearson, abridges the complexities of correlation in the physical world.

Other writings by Pearson provide signs that the ideological valences traceable in this one work were not inadvertent (that is, not mere artifacts of critical interpretation), and that in promoting the credo that "*it is the mind of man which rules the universe*" he sought consciously to perform an act of "demonic" sabotage of his own: an untwisting of the symbolic equation by which the reign of absolute values in the privileged discourse of science meshes with and lends credence to intolerant regimes of authority in society. One could cite, for example, an 1887 essay of Pearson's in which his polemic against absolute values for time, space, and motion in physics seems to trope itself directly into the language of cultural relativism:

> There is no absolute code of morality, no absolute philosophy nor absolute religion; each phase of society has had its own special morality, its peculiar religion, and its own form of sex-relationship. Its morality and its religion have been stamped as immorality and superstition by later generations. . . . One thing only is fixed, the direction and rate of change of human society at a particular epoch.[14] (Pearson, *Ethic*, 428)

Read in conjunction with the relativity physics of *The Grammar of Science*, this 1887 text ought all by itself to make clear the harm to historical scholarship that is wrought by the dogmatic suppression of "the tired old canard that Relativity Theory can be meaningfully related to cultural relativism" (Dutton, "Knowledge Replacement Therapy," 213) and by assertions that the historical contemporaneity of the two was merely a "coincidence," and that it is "wildly implausible" that relativity theory was related to "something in the ambient culture." What counts as "meaningful" is of course a *de gustibus* matter, as Newman so strongly insisted. Still, the convergence of texts in Pearson's case should show once again how heavily the burden of disproof falls upon those determined to maintain in the face of evidence the ideological *cordon sanitaire* isolating scientific discovery from its cultural correlatives. In seeking to define the correlation in a given case such as Pearson's, we need not assume that one complex of references (the physical, say) is the vehicle of this figurative construct and the other (the social and the moral, say) is the tenor; rather, each is a function and a reflection of the other in a neighboring register of discourse. Each revolves around the other, as Newman and Pearson said of the sun and the earth.

If we follow the skein of implication that gradually unwinds from Pearson's deconstructive exercises in *The Grammar of Science*, and if we have attuned ourselves to his subtly satirical literary idiom, we will come at last to view the physical world fantasized by Newtonian mechanics—an outlandish world of massive bodies, occult forces, and implacable, unintelligible laws of causality, void of human consciousness or volition—as, among other things, the characteristic self-expression of a cultural and intellectual system based on the repressive rule of a caste of law-giving authorities wed to an absolutist and "metaphysical" scheme of thinking. This, Pearson makes clear, is exactly the state of affairs his theory of science is meant ultimately to address.

What he refers to as the "maze of metaphysics which at present obstructs the entry of physics" (*GS*, 221) signifies for this author, in other words, not an innocent intellectual miscalculation after all, but an imprisoning structure of mystification, the function of which is "to reduce the mind to a mere passive recipient" and to lock human critical and imaginative faculties helplessly within a system of external absolutes. Physicists' bewilderingly circular, illogical, self-contradicting theories of such entities as force and matter, which Pearson (collapsing together science and literary analysis) documents at length with reference to writings of Maxwell, Kelvin, Tait, and other eminent scientists, embody for him the same pernicious mode of discourse allegorized by Dickens in *Bleak House* as the "monstrous maze" of legal mumbo-jumbo (an impenetrable labyrinth of "numerous difficulties, contingencies, masterly fictions, and forms of procedure"): webs of language designed in both cases to produce a condition of stupefied acquiescence in the established order (*Bleak House*, 482, 760). When Pearson declares that the unintelligible notion of causality in science "has trammelled the human mind, because it has led to a conceptual limit dominating actual experience," this is what he means. His extravagant stress on human creativity and on the doctrine of "the pure relativity of all phenomena" springs from an expressly emancipatory intention, in other words—one that does not confine itself strictly to the field of Newtonian mechanics. As Henry Adams put it in his comments on what he called Pearson's "revolutionary" book, what Pearson seeks to dismantle is nothing less than the whole order of thinking that had "created a universe the essence of which was abstract Truth; the Absolute; God!" (*Education*, 456). Adams understood with perfect clarity the Feuerbachian impulse that drives Pearson's critique of physics and his insistence, at the risk of being charged with "narcissism," on reinstating the idea of "direct personal volition mixing itself with the economy of natural law."

Anthropocentrism in Clifford, James, and Schiller

Karl Pearson's radical anthropocentrism may well, as he warns, strike many readers (for whom materialist metaphysics and common sense are likely to be identical) as overstated and paradoxical or even as willfully fraudulent (see above, n. 6). But the line of thought elaborated in *The Grammar of Science* found sympathetic echoes in writings of other avant-garde thinkers around the turn of the century.

W. K. Clifford, whom Pearson cites often and whose *Common Sense of the Natural Sciences* (1885) was posthumously edited by Pearson, offers early versions of many of Pearson's themes,[15] and goes far toward elaborating a philosophy of science based on the principle of the mental character of the natural world. He proposes in an 1873 essay, as we saw, that each of our physical sensations consists ambiguously of two elements, "a message that comes to us somehow" plus a mental "supplement," "something that we imagine and add to the message" (*Lectures*, 1:308, 309). The solid objects of nature are thus at least in part imaginary, Clifford insists—or even wholly imaginary, a sheer figment of human thought: "suppose it is all supplement, and there is no message at all!" (*Lectures*, 1:338). If this is so, he says, "the universe which I perceive is made up of my feelings . . . in fact it is really *me*"; in this case, "*all* the sciences will become pure sciences, all knowledge will be *a priori* knowledge; and we may construct the universe by sitting down and thinking about it" (*Lectures*, 1:341, 338).[16] So in Clifford we meet again the radically anthropocentric image of the human mind as the architect of the universe: "this world which I perceive *is* my perceptions and nothing more," he categorically declares (*Lectures*, 1:343). Nor is it a surprise, by now, to find that this doctrine is correlated in Clifford's writings with the political and moral orientation that marks other proponents of nineteenth-century epistemological relativity. Clifford was a disciple of Spencer and the radical reformer Mazzini, and was driven, says his biographer Sir Frederick Pollock, by "a conception of freedom as the one aim and ideal of man," a conception that "included republicanism as opposed to the compulsory aspect of government and traditional authority in general" (Pollock, "Biographical," 43–44). This last phrase sounds as though it had been borrowed from Spencer's or Einstein's autobiographical statements of their own character, or from Mill's analysis of the Victorian scene in *On Liberty*.

Clifford does finally declare his faith in the existence of a nonimaginary, nonrelativistic substrate of external reality lying behind perceptual phenomena and in some sense knowable by means of its representation in human consciousness. He introduces the concept of an element he calls "mind-stuff" or feeling that inheres, he says, in every molecule of reality, organic and inorganic, and with which human thought is somehow continuous both in an evolutionary and a phenomenological sense. Mind-stuff, he declares, disobeying Spencer's banishment of all absolutes whatever from philosophical discourse, "is *Ding-an-sich*, an absolute, whose existence is not relative to anything else" (*Lectures*, 2:69). I will not seek to trace the convolutions of Clifford's reasoning on this theme except to suggest that this allegory of human mind somehow projected outward in fragments as "absolute" externalized realities in nature seems to rehearse exactly that act of self-alienation (and self-mystification) analyzed by Feuerbach in the field of theology. Clifford's resulting model of the relation of thought to world is a kind of metaphysical poem (in both the seventeenth- and late-nineteenth-century senses of "metaphysical"), a sequence of ratios whose appearance of mathematical definiteness seems to hold out the prospect of computing scientifically the properties of those philosophical phantasms, things in themselves:

> *As* the physical configuration of my cerebral image of the object
> *is to* the physical configuration of the object
> *so is* my perception of the object (the object regarded as complex of my feelings)
> *to* the thing-in-itself. (*Lectures*, 2:72)

Levels and categories multiply here like orders of celestial beings in medieval theology. In reading this passage, Karl Pearson must have felt that Clifford, in his attempt to restore absolutes and things-in-themselves to the philosophy of science and so to avoid the extreme human-centeredness that Pearson himself espoused, had gone hopelessly astray in the fantastic maze of metaphysics.[17]

The same ambivalence and inconsistency come out in the writings of

William James. James professes at times an anthropocentric creed of human creativity nearly as extreme as Pearson's own. He quotes approvingly, for example, Schiller's *Personal Idealism* to the effect that "the world," being ultimately nothing but "what we make of it," is "*plastic*" (James, *Pragmatism*, 117; see also 119, 123). Once again we see the human intellect promoted to the divine role of creator of the world itself—not the role of an absolutist creator who dictates the shape of things once for all and demands loyalty to a single truth ever after, but one perpetually engaged in a work in progress. Like Poincaré and Dewey, James stresses strongly the plurality of potentially true theories, the necessary incompleteness of even the best theory, and, like Pearson, *the mutability of the universe itself* under the impact of human consciousness. "Why should anywhere the world be absolutely fixed and finished?" (James, *Meaning*, 222). The implications of this doctrine for discussions of the contemporary social order are again plain to see. Yet James shows in the waverings of his argument the potential costs in intellectual rigor associated with systematic rejection of "the pretence of finality in truth" (*Pragmatism*, 31) and with the disavowal of a pretense of finality even in one's own philosophy. He boldly asserts at times that "reality is an accumulation of our own intellectual inventions" (*Meaning*, 209), that humanity builds reality for its own purposes. At other times, scathed by the bitterness of "absolutist" criticisms of his views, he shows a willingness to praise the notion of "existent objects," to declare himself a faithful philosophical realist, to deny that pragmatism or, as he terms it, "relativism" is in any way opposed to the conception of "absolute truth," and to uphold a blurry Cliffordesque version of the correspondence theory of "truth" that elsewhere he expressly denies (*Meaning*, 172, 283, 309). The doctrine that truth consists in "agreement" with a "determinate" external reality, "some mirrored matter," is impossible to square with his own relativistic and anthropocentric claims that "truth may vary with the standpoint of the man who holds it" (*Meaning*, 284, 272, 301). James goes so far in watering down the pragmatist critique of dogmatism as even to declare the new philosophy compatible with religion and with "the hypothesis of God" (*Pragmatism*, 143–44)—seeming here to mean the God of the theologians, not of Feuerbach.

By contrast, in the steadfast "humanism" of James's pragmatist colleague F. C. S. Schiller, the anthropocentric and anthropomorphic character of reality is insisted upon unequivocally. His philosophy rests, says Schiller, on "the great principle that man is the measure of his experience, and so an ineradicable factor in any world he experiences" (*Studies*, 13). Pragmatism in his definition of it is very closely identifiable with the doctrine "that reality for us is relative to our faculties," and thus, he says, this school of thought "may be taken to point to the ultimate reality of human activity and freedom," and consequently "to the plasticity and incompleteness of reality" (34, 19). It is merely a fantasy, and one with menacing implications, to speak of knowledge decontaminated of its inherent human bias, he argues at length. "A man's personal life must contribute largely to his data, and his idiosyncrasy must colour and pervade whatever he experiences" (18); there is and can be no scientific method rigorous enough to reach knowledge unenveloped in "imagery borrowed from human life." In order to be free of stultifying metaphysics, truth must establish itself by means of procedures of testing and verification, says Schiller; and all such procedures are determined inescapably by "human ends and values" (7). "Human interest, then, is vital to the existence of truth" (5). "All actual truth is human, all actual knowing is pervaded through and through by the purposes, interests, emotions, and volitions of a human personality" (171). To think of another species of known truth, of a knowledge of some external order of things undetermined by the needs, desires, interests, values, purposes, historical circumstances of the knower, as we are bidden to do by the "absolutist theory of knowledge," is to think of a category unintelligible and wholly nonexistent. "*Purpose is logically vital*" to meaning (175, 112).

If it is nonsensical to invoke things-in-themselves in philosophical discourse, Schiller reasons, then one must acknowledge as a primary reality "a real freedom of human choice," the immediate consequences of which are "a real plasticity in reality at large" and the inescapability of "*alternative* constructions of reality" (*Studies*, 125). In preaching this Protagorean and Poincaresque gospel, Schiller invites his readers to take comfort from the philosophical vertigo they may be experiencing by looking forward to a day when "Absolutism . . . has ceased to oppress us and to be a menace to the liberty of thought" (139). In writing this phrase and setting radically anthropocentric philosophy as the antithesis of a dark, threatening cloud of absolutist oppression hanging in 1904 over European life, he had surely in mind a danger that extended beyond such immediate philosophical opponents as F. H. Bradley. One clear implication of his repudiation of the view "that Truth is determined by Fact, by which it is 'dictated' " (123) is that the very meaning of the dictatorship of Fact and of the supposedly obligatory allegiance to this mode of Truth is that such a regime presupposes the evacuation of human value and will indeed insist upon it as a condition of its own survival. Schiller sees his championing of Protagorean relativism as a way of forestalling that evil day. What he proposes is thus, we might say, a new and different phase of "the moralization of objectivity," one based not on obeisance to "fact" but on relativity and freedom and on the radical capacity of human beings to alter the world they inhabit. By Schiller's account, the mechanistic and authoritarian theory of knowledge unravels unavoidably, since the relation between knower and known is dynamic and variable and since, as he says, in the prophetic formulation I have cited before, "*mere knowing always alters reality*" (439). Human spectatorship is invariably active, invasive, disruptive, and cognition always entails an impact on the world; this is the founding principle of anthropocentric scientific philosophy in its modern form.

1905 and After

I will conclude this chapter by briefly noting how the nineteenth-century theme of "the deification of man" imprinted itself on the subsequent discourse of relativity physics, beginning with Einstein's special-relativity paper of 1905, "On the Electrodynamics of Moving Bodies" [hereafter EMB]. It is true that the paragraphs of equations that largely make up this paper would seem initially to afford no purchase at all for "imagery borrowed from human life," and it is no less true that orthodox commentators on Einstein's thought would reject out of hand any account of his discoveries that would identify them with philosophies of the construction of the objective universe by human thought. Yet the anthropocentrism of Einstein's scientific imagination in 1905 can be seen as its defining, radical characteristic. Not to take account of the varied textual manifestations of this attitude is to rend Einstein away from some of his significant moorings in the history of thought.

His epochal assertion, which follows the introductory section of the paper, that "we must not ascribe *absolute* meaning to the concept of simultaneity" (EMB, 145), does not embody an original discovery of his own. As we have seen, the nonexistence of absolute time (of absolute simultaneity, in other words) had been repeatedly announced over the four preceding decades, with varying degrees of specificity, by a series of writers including Spencer, Stallo, Pearson, Mach, Poincaré, and others (among them, of course, H. A. Lorentz, whose transformation equations relating space and time Einstein said he had not seen prior to duplicating them in "Electrodynamics"). By 1905, the denial of absolute time, along with the denial of absolute space, absolute rest and motion, and absolute causation, not to mention any "absolute code of morality," had become a commonplace of radical theory—one that carried at its every iteration powerful overtones of ideological significance, as I have been seeking to show. Einstein exactly rehearses the conclusions of many of his predecessors,[18] conclusions typically arrived at on the philosophical grounds of a generalized relativistic denial of absolute values, but which Einstein attains, as far as the text of his paper tells us, by means of mathematical deduction alone. The mathematics does not operate in a philosophical vacuum, however. It is dictated by one massive intervention on Einstein's part: the proposition of the absolute value of the speed of light and the use of light signals to measure time. Underlying this structure of reasoning, tacitly but irresistibly, is the anthropocentric imperative.

The "absolute" time Einstein shatters is a time uniform for all points in the universe; more fundamentally, it is time conceived as a self-existing thing in the structure of the natural world, independent of human experience and human cognition. Special relativity testifies that by the turn of the century, the growth of a habit of criticism keyed to the exposure of metaphysical fetishes masquerading as rational statements caused such a concept inevitably to strike a modernist intellect as an absurdity, a thing glaringly incapable of comprehensible definition. The primary doctrine of special-relativity theory, though it is never stated in these words, is that *time can only be understood scientifically as a human construct*, and can therefore only be given precise significance by means of an intensive investigation into the physical operations humanly used to measure it. Hence Einstein's famous analysis of the phrase "the train arrives here at 7 o'clock," with which the argument of the 1905 paper begins. It is crucial to note that the inquiry here does not take the form of a question regarding the supposed intrinsic characteristics of this or that constituent element of external nature: rather, it takes the form of a question about what that exemplary manufacture of human imagination, a linguistic utterance, is understood conventionally to *mean*. Relativity physics here at its canonical point of origin is an *explication de texte*, a parsing of linguistic symbols. (That it is a drastically selective one that ignores, say, the rich symbolism of locomotives, the dramatic circumstances and emotional condition of the narrative protagonist waiting for the train to arrive, and other such factors does not alter the point.)

This exercise in what we might call, not to be facetious about it, "the grammar of science" reminds one immediately of the antimetaphysical exercises of a text that has regularly figured in the margins of the present study: Derrida's *Of Grammatology*.[19] Einstein instructs us that it is vain to look for reference to a transcendent signified in the sentence about the train, which means, he asserts, absolutely nothing more than "the pointing of the small hand of my clock to 7 and the arrival of the train are simultaneous events" (EMB, 141). There is therefore no meaning in the idea of "time," scientifically speaking, except in reference to that supremely ingenious construction of the human intellect, the clock. (Even the moons of Jupiter or the decay of a radioactive ion are of course ingenious human constructions in this sense when used scientifically as clocks). The time of science exists only when it is read off the clock by an individual human being, one who is a competent interpreter and understands that the imagery of the clock face is a figurative signifier making figurative reference to the daily revolution of the earth on its axis. *The mental activity of this clock-reader is what creates time*: there is no escaping the centrality of this dictum to special relativity. (Again, we must not be misled by the drastic selectivity that allows one to elide, for scientific purposes, all reference to such factors as individual psychology and subjectivity, and to speak of measuring devices as though they functioned uncannily without human participation.)[20]

The surprising emergence of a fictionalized Einstein himself in the first person singular, timing the arrival of the train and wondering what it means (EMB, 141), thus is not an insignificant stylistic detail of this paper, any more than is its profusion of references to different "observers," imaginary scientists stationed in one coordinate frame of reference or another relative to particular physical events, either anchored in "the system at rest" or taking the position of "the observer [who] co-moves with the . . . measuring rod" (144)—the latter, the physicist's equivalent of the "participant observer" in post-Malinowskian anthropology. Time in Einstein's conception, to reiterate the main point, is not discovered but humanly *made*; it is a phenomenon not of physical observables in their own right, but of *meaning*. To study rigorously the implications of these primordial principles, the 1905 paper seeks to show, is to become aware of a string of fantastic-seeming phenomema, shrinking objects and dilating periods of time, that hitherto had escaped scientific observation. "When space and time are relegated to their proper source—the observer—the world of nature which remains appears strangely unfamiliar," says Eddington (*Space, Time and Gravitation*, v). This of course is the fundamental difference between Einstein's account of why objects contract with motion and the account given earlier by FitzGerald and Lorentz. The latter theorists attributed contraction to hypothetical mechanical effects of speed such as the compression of molecules, but for Einstein, no further explanation is needed or possible than the radically relativistic kinematic one that reality in its phenomenal character depends on an observer's point of view.

No less humanly manufactured is the special theory of relativity itself, as Einstein stresses dramatically in one place after another. How, he asks, can we speak of a "time" common to two points *A* and *B*? "The latter can . . . be determined by establishing *by definition* that the 'time' needed for . . . light to travel from *A* to *B* is equal to the 'time' it needs to travel from *B* to *A*" (EMB, 142). Time in the sense of simultaneous moments in distant locations, Einstein's italics emphasize strikingly, is an arbitrary conception for which no empirical evidence exists. Human beings create simultaneity, he says in effect, because it assuages what Mach (*Knowledge and Error,* 102) called "the intellectual [and also perhaps spiritual?] needs of humanity" to do so. In order to open that conception up to critical analysis, Einstein proposes other definitions fully as arbitrary and imperious as that of absolute time—or "time," as he puts it. The literary effect of this mode of argument is made breathtakingly vivid on the first page of the paper, where Einstein sets out his theoretical presuppositions.

He refers here to certain theoretical issues in contemporary physics that, he says, lead to "the conjecture" that there is no absolute rest in nature and that natural laws have the same form in all inertial physical systems. "We shall raise this conjecture (whose content will be called 'the principle of relativity' hereafter) to the status of a postulate," he declares, "and shall introduce, in addition, the postulate, only seemingly incompatible with the former one, that in empty space light is always propagated with a definite velocity *V* which is independent of the state of motion of the emitting body." Had he announced his intent to postulate that 2 plus 2 equals 7, the effect would not have been greater. How could the same ray of light have the same speed relative to two observers, one "at rest," the other, say, moving at high speed toward the light source? It is by founding all the reasoning that follows on this wildly irrational-seeming postulate—a postulate is a fundamental axiom assumed to be true and requiring no proof—that Einstein is able to devise the system of translation among systems of symbolism (that is, the mathematical symbolism defining each co-ordinate system) that vindicates and in a real sense *creates* the relativistic universe long foretold by nineteenth-century writers.[21] Eddington commented wryly about the status of the velocity of light in special relativity that "in asserting its absoluteness scientists mean that they have assigned the same number to it in every measure-system; but that is a private arrangement of their own" (Eddington, *Nature of the Physical World,* 56). We would not perhaps want to call the creation of special relativity an act of "narcissism," but Einstein's amazing postulate, set forth rhetorically with the sublime assurance of one who is declaring his capacity to reinvent the universe and to confound scientific rationality in the process, may represent the most compelling instance in the history of Western thought of what Tyndall called "direct personal volition mixing itself with the economy of nature." As far as the text of Einstein's 1905 paper is concerned, he simply revises the economy of nature by fiat, by an act of personal creation. He confirms this reading of special relativity, one might add, in various passages of his later writings on the history of science, where he suggests persistently that the creation of new scientific theories entails nothing less than the reconstruction of the natural world itself. In seeking to develop the concept of energy, he says, for example, contemporary researchers are attempting "to create a new substance"; in the development of the concept of the field, similarly, "a new reality was created" (Einstein and Infeld, *Evolution*, 51, 158). Lorentz achieved his great advances "by divesting the ether of its mechanical, and matter of its electro-magnetic properties" (Einstein, *World*, 126). The recurring hint that fundamental reality is subject to the transforming interventions of human thought is a calculated one that Einstein stresses as he does because it is so vividly implicit in his own scientific work and, as I am suggesting in this chapter, in relativity thinking itself.

From their varying interpretations of "the principle of relativity," Karl Pearson and other authors surveyed in this chapter derive a twofold lesson. First, they insist that the anthropocentric component of human knowledge is primary and inescapable, and that the scientific analysis of knowledge needs to develop refined theoretical instruments for understanding its modes of operation. Denunciations of narcissism in the name of an ideal of a dehumanized scientific world are inadequate to the task of modern inquiry, they assert. Moreover, they put forward versions of human-centered theory as an affirmative principle with conscious remedial and therapeutic intent. They suggest that the rise of the antihumanistic dogma of modern science is associated with broad pathologies in European culture: with a deep, persistently reenacted experience of alienation, with a characteristic sense of helplessness, insignificance, and bewilderment, with moral and esthetic dysfunction, and, in a vicious circularity, with vulnerability to the dangerous intellectual and political mystique of the "absolute." In idealizing the "laceration" science supposedly inflicts on human feelings, they suggest, we signify our subjection to a sadomasochistic syndrome that forms the psychological correlative of subjection to authoritarian social regimes. One source or symptom of this complexly disabling trauma, they suggest, is the identification of scientific inquiry with an ideology of value-free rigor and discipline and with a dogma of absoluteness according to which there can only be a single truth, all other versions of things being evil imposters deserving pitiless elimination. So ingrained is this ideology in the mentality of modern science that even W. K. Clifford, with his idealistic devotion to intellectual freedom and with a revulsion from institutional violence as keenly honed as Ivan Karamazov's, betrays a trace of it in praising (in the course of an appeal to the "sacredness" of nationalistic loyalty) "the purifying and organising work of Science," the mission of which, he says, is to extirpate all nonscientific thinking "without mercy and without resentment" (*Lectures*, 2:49, 50–51). One can imagine these writers asking, for example, to what extent did and does admiration for the sheer scientific achievement of the building of an atomic bomb (the achievement habitually cited to refute those who seem to doubt the almost miraculous efficacy, and thus the proven "truth," of modern scientific method), allied to the deeply ingrained conception of science as the engine for the repression of prideful error, account for the failure of the West to stand appalled at the nuclear bombing of the civilian populations of Hiroshima and Nagasaki? Is humankind in the age of science so profoundly demoralized by what Feuerbach called "the disuniting of man from himself" that it has come to require displays of terroristic power to reclaim its own self-respect? Surely the quasigenocidal use of such a weapon seems like a supreme, unsurpassable form of "dehumanization," of a drive to obliterate the human form divine, not by the laceration of feelings but by recourse to a fantastically heightened technology of violence.

If so, we can only conclude that the long-germinating cultural and intellectual movement that came to fruition in relativity physics (and thus, subsequently, in the Manhattan Project) underwent a radical deformation in its process of transition from intellectual critique to practical science. But one need not look so far afield for an example of the brutalization of relativistic science. Karl Pearson himself offers one in a lecture published in 1901 as *National Life from the Standpoint of Science* [hereafter *NL*].

In the years following *The Grammar of Science*, Pearson the archrelativist pursued evolutionary theory and his specialization in statistics into a hardening infatuation with eugenics, the ideological implications of which he sets forth in *National Life*. England's initial reverses in the Boer War point disturbingly to the biological decline of national "stocks," he declares. The science of heredity tells us that the rebuilding of national strength (measurable directly, it seems, in terms of the capacity for military violence) requires a politics based on pitiless competitive struggle. "History," he declares, "shows me one way, and one way only, in which a high state of civilization has been produced, namely, the struggle of race with race, and the survival of the physically and mentally fitter race" (*NL*, 19). It is thus the scientifically prescribed duty of European nations not just to subjugate but effectively to extirpate altogether "inferior races" such as Africans and American Indians. "The only healthy alternative" to the evil of racial coexistence is that the white man "should . . . completely drive out the inferior race," he says (21). Like Freud worrying about the "aversion and resistances" anti-anthropocentric scientific thought arouses in the popular mind, Pearson here acknowledges that the dictates of eugenics may initially seem "harsh, cold, possibly immoral" (14); but such unscientific squeamishness needs to be overcome, he insists, in the name of national welfare. "From the scientific standpoint," one can consider, say, "the romantic sympathy for the Red Indian generated by the novels of Cooper and the poems of Longfellow and then—see how little it weighs in the balance!" (13, 23). "The scientific view of a nation" is that it achieves progress "chiefly by way of war with inferior races" (43–44) and by the cultivation of intense nationalistic sentiments, or what Pearson calls "patriotism" (51). Science and violence have never been more explicitly fused into one. At the same time, it is the duty of enlightened leaders in the area of domestic policy to ensure that "social sympathy" not be carried so far "that the intellectually and physically weaker stocks multiply at the same rate as the better stocks" (54). This campaign of racial improvement is to be led by "a band of men trained to observe and reason" and unswayed by sentimental pity for the victims of natural selection: that is, "men of science" (37).

In a classic instance of Feuerbachian self-alienation, Pearson here abandons with a vengeance his previous doctrine that the human mind creates the laws of nature in favor of a new doctrine of relentless evolutionary determinism against which it is futile to struggle. With this philosophical shift, he undergoes his gruesome transformation from high-minded socialist and feminist into a late-Victorian Dr. Strangelove and sets forth this early prototype of the National Socialist plan to annihilate the Jews and the Slavs and then, the war of conquest once concluded, all Germans with incurable illnesses (Arendt, *Origins*, 391 n., 411). Despite the distinguished scientific credentials of its author, it would be a libel on science to consider *National Life from the Standpoint of Science* as a scientific text, but considered as imaginative literature, it reflects in its exterminationist impulse just that modernist mentality that Ortega associated with "an explicit act of dehumanization" and with "a real loathing of . . . forms of living beings" and that Conrad's terrorist Professor expresses, as we saw, in reveries of eugenic death camps. Set in relation to the ideology of scientific objectivity preached at the beginning of *The Grammar of Science*, this later text lets one see that the goal of eliminating "individual bias" can lead with a powerful logic to idealizing the elimination of living individuals themselves; for ultimately, no other approach to the problem of mind control can fully succeed.

Karl Pearson's Jekyll-and-Hyde intellectual biography, with its crazily tangled threads of emancipationism and state collectivism, idealistic socialism and murderous eugenics, could perhaps never be rendered logically coherent. All one can say for sure is that this episode stands as a parable of the acute ideological instability of scientific philosophy in the late nineteenth century—and more broadly, of that of intellectual formations in general. It is at any rate the sinister logic of eliminationism, so fully spelled out in *National Life*, that the human-centered theory of knowledge takes as its essential mission to overthrow. It sets the principles of the legitimacy and necessity of "*alternative* constructions of reality," of human freedom, and of the almost divine potency of the human creative imagination at the center of the definition of science. We have also seen how consistently it associates itself with the condemnation of authoritarian violence, as though all the rest of its philosophical superstructure arose at last from this originating motive. From the standpoint of this mode of relativistic "humanism," the model of scientific rationality as the instrument of the repression of human narcissism has no place at all.

CHAPTER FIVE

Frazer and Einstein

The emergence of relativity as the distinctive theme of avant-garde speculation in the late nineteenth century signaled a conceptual sea change: a displacement of the mechanical model of the world, hitherto synonymous with scientific thinking, by an intellectual manifold in which the constituent elements of reality were not bodies, forces, and causality, but symbolic processes. Michel Foucault describes the development of the "human sciences" in the period in just such terms, as marking a phase of thought defined by "the primacy of representation" (*Order of Things*, 363). According to this new intellectual regime, which in fact extended well beyond the human sciences, things are defined by their sovereign property of *standing for* other things. Whether this substitutive relation is one properly called symbolism, thus analyzable by a mode of poetics, or an arbitrary one called the sign, thus belonging to semiology, or another, nameless mode of correspondence defined by the unknowability of the referent and thus the utter unintelligibility of the relation, is a dilemma that causes the late-nineteenth-century field of thought to fragment into a number of diverging trends and to give rise to a lot of controversy in which fundamental ideological issues seem to hang in the balance.

In this chapter, I seek to explore this unstable conceptual field by means of a comparative discussion of two of the most ambitious scholarly projects of the age: James Frazer's *Golden Bough* [hereafter *GB*], the three main editions of which appeared in 1890, 1900, and 1911–15; and Einstein's special theory of relativity, published in 1905, as we know, after a ten-year incubation. It may seem an unjustifiable scholarly liberty not only to seek to draw equations between fields as disparate as evolutionary anthropology and mathematical physics, but to treat as coequals in scientific endeavor a writer now often dismissed as a charlatan and an "embarrassment" (Ackerman, *Frazer*, 1) and the sainted figure revered as the greatest thinker of his age. The field of Frazer studies is dominated nowadays by a voluminous literature (produced largely by anthropologists) in which *The Golden Bough* is portrayed as antiquated, reactionary, and conceptually vacuous, even, in René Girard's words, "fanatical and superstitious" (*Violence and the Sacred*, 318).[1] Marilyn Strathern sums up the current view by declaring that Frazerian anthropology has come to seem "not simply erroneous but absurd" (Strathern, "Out of Context," 254). Yet no harm will come from at least temporarily suspending Frazer's banishment from respectable scientific society—quite the contrary. Doing so will help to shed light, for one thing, on those intimate relations between the twin principles of evolution and of relativity that have formed a principal subtheme of the present study all along; it will help to make certain texts of each writer readable in new ways—or readable, period; and it will help to highlight the arrival of the modern relativity movement at a new phase of its history.

From the middle of the nineteenth century onward, the movement of thought centered upon "the principle of Universal Relativity" (Bain) had slowly but surely crystallized in the form of positive, sustained intellectual applications. The doctrine that nothing is one thing just by itself was everywhere implicit in the burgeoning field of evolutionary biology. The great German scientist Hermann von Helmholtz was no systematic relativist, but his researches in another domain, that of the physiology of vision, led him in 1867 to make a magisterial statement of our theme. "If what we call a property, or quality, always implies an action of one thing on another," he declares, "it can never depend upon the nature of one agent alone. It can exist only in relation to . . . the nature of some second object which is acted upon. Hence it is really meaningless to talk as if there were properties of light which belong to it absolutely, independent of all other objects. . . . The notion of such properties is a contradiction in itself" (Helmholtz, "Recent Progress," 187).[2] In another powerful methodological innovation, the economist W. Stanley Jevons decisively identifies himself with contemporary Protagorean theory in announcing in 1871 the overturning of the labor theory of value by the concept of marginal utility. "There is no such thing as absolute utility, utility being purely a relation between a thing and a person," and thus perpetually in flux, Jevons declares; "utility, though a quality of things, is *no inherent quality*. It is better described as *a circumstance of things* arising out of their relation to man's requirements. . . . We can never say absolutely that some objects have utility and others have not" (Jevons, *Theory of Political Economy*, xxxiii, 43). Jevons teaches that rational economic science is incompatible with the tendency of economists "to speak of such a nonentity as *intrinsic value*," when in fact "value in exchange expresses nothing but a ratio, and the term should not be used in any other sense" (77).[3]

These developments in various disciplines represent significant practical expressions of relativistic theory and provide more evidence, if more is needed, of the intense contagiousness of the relativity hypothesis among late-Victorian intellectuals. None of them, however, has room to incorporate a key tenet of Victorian relativity theory: the tenet that the relativity effect is always "twofold" by virtue of the construction of the scientific field not only according to the relations of objects among themselves, but also according to those between objects and the situated observer/analyst/interpreter. If "thinking is relationing," as Spencer taught, and if "*to think* is *to condition*," as Hamilton taught, and if, moreover, the purposes and predispositions of a researcher impinge necessarily on his or her findings, as Newman taught, then the specific relation of the scientist to any body of data—what we might call the marginal utility of any item of knowledge in a given analytic field—cannot be ignored without introducing incalculable falsification into all our analyses. In very different ways, Frazer and Einstein, in engaging with unprecedented fullness the principle of the symbolic nature of the world of science, profoundly explore this intuition that the observer's point of view forms an integral constituent of scientific objectivity.

The Theory of Symbols

As scholars have long observed, the ascendancy of mechanistic theory defined mainstream scientific awareness throughout the nineteenth century.[4] (That it continues to prevail uncriticized in many intellectual domains, not always in forms conducive to lucidity, is another story.) "It seems to me," the illustrious scientist Lord Kelvin opined in 1884, "that the test of 'Do we or not understand a particular subject in physics?' is, 'Can we make a mechanical model of it?' " (Kelvin, *Baltimore Lectures*, 111). The transparent self-evidence of the mechanistic order and its system of deterministic causal explanations was troubled by the emergence of the electromagnetic theory of light, which required the invention of the "luminiferous ether" to fill empty space and provide the logically necessary supportive medium for light waves—but which proved, contrary to all expectation, maddeningly difficult to detect empirically. Yet the ether, for all its elusiveness, was no "vague or fanciful conception on the part of scientific men," declared

John Tyndall in 1865. "Of its reality most of them are as convinced as they are of the existence of the sun and moon. The luminiferous aether has definite mechanical properties" (Tyndall, *Fragments*, 4). Tyndall was no less emphatic than Kelvin in his endorsement of the regnant scientific philosophy. "The tendency of natural science . . . is to bring all physical phenomena under the dominion of mechanical laws," he said. "No matter how subtle a natural phenomenon may be . . . it is in the long run reducible to mechanical laws" (*Fragments*, 258, 410). The future anthropologist James G. Frazer was a student of Kelvin's at the University of Glasgow, and took from his lectures, he later stated, a similar conviction: "a conception of the physical universe as regulated by exact and absolutely unvarying laws of nature expressible in mathematical formulas." This conception, he said, "has been a settled principle of my thought ever since" (quoted in Ackerman, *Frazer*, 14).

In opposition to this orthodoxy, there arose the subversive new doctrine (new though traceable to origins in Humean and Kantian philosophy) that an unbridgeable gulf separates external reality from human thought, and that empirical science therefore deals, after all, not with directly measurable natural objects, but only with images, simulacra, representations, symbolism—nothing that seemed to provide a solid purchase for an epistemological anchor. "Our sensations are for us only *signs* of the objects of the external world," said Helmholtz at the early date of 1853, defining one of his cardinal principles of philosophy, "and correspond to them only in some such way as written characters or articulate words to the things they denote" (Helmholtz, "Recent Progress," 70). Mill in his 1865 book on Hamilton's philosophy declares similarly that "of the ultimate Realities . . . we know the existence, and nothing more" and that human cognition deals exclusively with "the *representations* generated in our minds by the action of the Things themselves" (Mill, *Sir W. H.*, 19). Science describes and analyzes a figurative world of mental imagery, of representations, of signifiers, not of external objects, according to Helmholtz and Mill.

Of all Victorian theorists, the one who developed this trend of thinking most fully was Spencer. He had by no means abandoned a mechanical conception of things, as we noted in chapter 1.[5] But mechanistic ideas lead an ambiguous coexistence in his writings with his insistence on the insurmountable separateness of knowledge from its causes in the external world of nature, and on the impossibility of understanding or even attempting to study the connections between them. The world of nature is wholly and permanently beyond cognition: this is Spencer's primary doctrine. As he declares in *The Principles of Psychology* (1855), "all the sensations produced in us by environing things are but symbols of actions out of ourselves, the natures of which we cannot even conceive" (*Principles of Psychology*, 1:206). A perception of an object, he says in the same work (2:497), is to be understood as "a symbolization in which neither the components of the symbol, nor their relations, nor the laws of variation among these relations, are in the least like the components, their relations, and the laws of variation among these relations, in the thing symbolized." His philosophical writing everywhere insists on the conception of a scientific world evacuated of material things, inhabited only by symbolic imagery. This principle is the core of the doctrine of "The Relativity of All Knowledge" (*First Principles*, 50), in the name of which, as we saw, he mounts his precocious critique of such categories of physical science as absolute space, absolute time, absolute motion, and the luminiferous ether.

The same complex of ideas is central to Karl Pearson's *Grammar of Science* (1892), where the stress on the purely figurative character of knowledge—that is, the figurative character of the real world itself, insofar as it is accessible to scientific study—is radical and unremitting. The two disjunct realms, knowledge and the real world, differ as widely "as language, the symbol, must always differ from the thing it symbolises," says Pearson, collapsing natural science into linguistics as Helmholtz had. "Science is . . . a theory of symbols. . . . It is not an explanation of anything" (*Grammar*, 62, 232). The world of phenomena is undoubtedly the one in which we move, says the investigator of animal intelligence Lloyd Morgan at the same time, "but it is none the less a symbolic world" (*Animal Life*, 317). Sir Arthur Eddington summed up this line of thought in 1928, declaring that ejecting substantial physical entities from scientific thought in favor of a universe constructed purely of mathematical symbolism represented the crucial step in the movement to a fully modernistic and relativistic conception of reality. "It is difficult to school ourselves to treat the physical world as *purely symbolic*," he says, calling at this late date for the same radical program of intellectual reconstruction that Spencer, Mill, Pearson, and others had announced long before (*Nature of the Physical World*, xvii).

In anthropology, the symbolic hypothesis emerged early, in the formative decade of the 1860s. This development effected nothing less than the intellectual and spiritual salvation of the discipline, turning it decisively away from the crudely mechanistic theory of racial determinism with which, as Marvin Harris says, it had become synonymous in preceding decades (Harris, *Rise of Anthropological Theory*, 101). The vehicle of the new paradigm in this field was the theory of evolution, as originally formulated by Spencer (sans the mechanism of natural selection) in the early 1850s and as elaborated in anthropology over the next half century by a long line of scholars including J. F. McLennan, Spencer himself, E. B. Tylor,

William Robertson Smith, Frazer, and Durkheim, among others. This great collective enterprise of evolutionary or, as it was called, "comparative" anthropology has long been expelled from respectable intellectual history as being ideologically disreputable and scientifically valueless. But it deserves a far more respectful portrayal, if only because it served as one of the privileged sites of radical speculative thinking in the Victorian age.

That societies could be ranked higher or lower, more or less advanced, on a supposedly universal evolutionary scale was a principle taken for granted rather than argued by comparatist writers; the point of research in this field was not to prove such an assumption in the way that comparative skull measurements, for example, were intended to give empirical evidence for racial rankings. The point, rather, was to confirm two broad propositions that had, in fact, as finally became clear, a fatally circular logical relation to each other: first, the proposition that human societies passed in the course of their development through a uniform series of evolutionary stages, each defined by its own characteristic institutions; and second, that clues to previous stages of cultural development could be found in the vestiges and "survivals" of primitive institutions that analysis could reveal in subsequent stages of social life. Comparatist theory held that functional social forms as they become obsolete persist in the guise of mere "symbolism"—the symbol being understood in this context as a degraded, enfeebled residue of a social element that once played a robust role in the practical life of the society. "Symbols are facts in decadence," declared McLennan ("Worship of Animals and Plants," 1870, 210). Once placed under the sign of this governing principle, the scientific study of social evolution became very much, in Karl Pearson's phrase, "a theory of symbols." As such, it came of necessity to focus intensively, if not always with full critical awareness, on the act of *interpretation*—specifically, the interpretation of cultural formations of all kinds as traces of forgotten earlier formations, "disguised," as McLennan said, "under a variety of symbolical forms" (*Primitive Marriage*, 7).

The purported findings of this mode of research are embroiled in the logical incoherence of the whole program of comparatist anthropology, as commentators like Margaret T. Hodgen have insisted (*Doctrine of Survivals*, 177). A more enduring legacy of this enormous body of scholarship resulted from the largely unwitting way in which the evolutionary construal of social "symbolism" gave rise to an awareness of the figurative and expressive dimensions of all institutions—an awareness, that is, that every cultural formation, whatever its degree of supposed evolutionary value and however purely instrumental it may appear, can be understood as a structure of symbolic imagery readable in the way poetic texts or, according to Freud, dreams are readable. Once this awareness takes shape, the category of the symbolic ceases to be synonymous with obsolescence and dysfunctionality, and becomes central to understanding the way institutions work to produce and reproduce the mental and ideological order of society.

It is easier to mark the culmination of this trend of thought in anthropology than to mark its point of origin. The former comes after the turn of the century in the work of Durkheim, to which I will return in a moment. The latter might be situated in the work of 1841 that I have treated as one of the *Ur*-texts of radical nineteenth-century speculation: Feuerbach's *The Essence of Christianity* [hereafter *EC*]. I would highlight in the present connection chapter 7, "The Mystery of the Logos and Divine Image." Feuerbach makes the general argument that divine beings have no existence except as projected and idealized images of human characteristics. "Religion," he declares, "is human nature reflected, mirrored in itself" (*EC*, 63); it is purely a construction of allegorical imagery and a revelation of human nature. Christianity expresses this truth in its special glorification of the second person of the Trinity, the Son, a personage, as Feuerbach says, "expressly called the Image of God" (*EC*, 75). The cult of Christ in Feuerbach's analysis thus embodies "the chief and ultimate principle of image-worship"; in fact, Christ is nothing other than the divinized human imagination itself, "the satisfaction of the need for mental images" (*EC*, 76, 75). "To live in images or symbols is the essence of religion," says Feuerbach. Religion by its very nature "sacrifices the thing itself to the image" (*EC*, 182): this supremacy of the image relative to the thing it represents—a supremacy enacted in the form of a sacrifice—is the cardinal principle of what Feuerbach calls "The True or Anthropological Essence of Religion." According to Feuerbach, in other words, Christianity comes into being in order to express the worship human beings instinctively bestow upon symbolic imagery—which is to say, upon their own power of building figurative worlds made of images that stand in a substitutive relation to natural reality. Hence "the sacredness of the image" in all religion, where "the image necessarily takes the place of the thing"; "wherever the image is the essential expression, the organ of religion," he says, "there also it is the essence of religion" (*EC*, 76).

From an early point in the history of the movement of thought we are surveying, Feuerbach's principle of the sacred status of the figurative image relative to the thing itself seems not only to preside over a broad reconceptualizing trend in anthropology, but to give rise to distinct new strands of anthropological research. Among these would be the fixation of Victorian anthropology, from McLennan and Robertson Smith to Freud and Durkheim and afterward, upon the study of totemism and the fetish.

All these writers pronounce totemism to represent the genuine primitive form of religion, and they define it more or less explicitly as the worship of symbolic imagery per se. The theme of the sacrifice of the thing to the image is suggested plainly, for example, in what amounts to the inaugural text of the Victorian study of totemism, McLennan's series of essays in the *Fortnightly* titled "The Worship of Animals and Plants" (1869–70). In these essays, where the theory is put forth that primitive nations universally pass through a very early stage of worship of animals and plants, all of McLennan's argumentation focuses not on the characteristics of sacralized animals and plants themselves (as Mary Douglas's theory of taboo would at a much later date), but on the characteristics of complexes of symbolic imagery of them in such diversified media as sacred graphic representations, sculpture, badges, emblems, insigniae, names, and mythological narratives. McLennan develops in this text a remarkable proto-Freudian discourse for tracking totemic or fetishistic imagery through evolutionary transformations he describes in such terminology as "compounding" and "transference" ("Worship of Animals and Plants, part 1," 211; part 2, 563), and which regularly involve the apparent slippage of the aura of sacredness from living animals and plants to their symbolic representations, as though the derived and secondary form had freed itself from the natural original and seized precedence. "The animal was . . . worshipped under symbols," he says in discussing the totem cult of the bull, for example, "or as represented by images" (part 1, 574). McLennan's analysis goes a long way to confirm Feuerbach's claim that "to live in images or symbols is the essence of religion."

The implicit theme here is the one that runs just below the surface of the whole literature of late-Victorian comparatist anthropology and is crystallized by Durkheim in 1912, in *The Elementary Forms of the Religious Life*. In his own study of totemism, Durkheim strongly emphasizes the way symbolic imagery of the totem tends to proliferate everywhere, in one form after another, in a tribal society, and he concludes specifically, as McLennan had only intimated, that it is the symbolic image itself rather than the totem in its natural form that is chiefly invested with sacredness and worshipped. From an examination of ethnographic evidence, says Durkheim, "we arrive at the remarkable conclusion that *the images of totemic beings are more sacred than the beings themselves*" (*Elementary Forms*, 156). Durkheim's italics seem to mark this phrase as enunciating a new fundamental principle of anthropological science, though, as we have seen, the idea had been in circulation long before he articulated it so strikingly. According to this theory of totemism, we may say, "religion sacrifices the thing itself to the image," exactly as Feuerbach had put it long before. Durkheim draws still broader conclusions, declaring systems of "figurative representations" to be the fundamental constituent, the DNA, of all human society, not just of totemic tribes. "Social life, in all its aspects and in every period of its history, is made possible only by a vast symbolism," he declares (*Elementary Forms*, 264)—a conception that obviously marks a noteworthy step beyond McLennan's and Tylor's view of symbols as "facts in decadence." For Durkheim, symbolism *is* the factual stuff of social life, just as it was the factual stuff of physical science for earlier writers like Spencer, Helmholtz, and Pearson.

Durkheim is silent, I believe, with respect to Feuerbach, but he does often mention his eminent contemporary Sir James Frazer, chiefly to dispute points of theory such as Frazer's claim that magic invariably precedes religion in cultural evolution. Durkheim's dismissive comments on Frazer serve to mask what is in fact a deep and primary intellectual bonding between the two scholars. Of all Victorian texts, *The Golden Bough* is the one in which the reconceptualization of anthropology as a thoroughgoing theory of symbolism, and the consequent identification of it with a radical principle of relativity, is most decisively asserted. This is also the text of this age in which the intellectual vertigo produced by this discovery is most compellingly dramatized.[6]

Frazer: Comparatism and Its Discontents

Frazer's position in the trend of new thought percolating through a widening variety of fields can be indicated by comparing him with his predecessor E. B. Tylor, whom he resembles intellectually on one level to such a degree that one might be tempted to follow Hodgen (*Doctrine of Survivals*, 115–16) in viewing *The Golden Bough* as little more than an elaborate imitation of *Primitive Culture* (1871) and other works of Tylor. For example, the theory of cultural "survivals," so important to Frazer, is laid out by Tylor as the central theme of his research, as is the whole Frazerian theory of the supposed primitive belief that a magical connection runs between things and images that represent them (Tylor, *Researches*, 100, 101). For Tylor, these notions remain firmly attached to what could be called a mechanical and deterministic idea of culture. In accordance with that "fear of interpretation" that Daston and Galison identify as the key syndrome of orthodox late-Victorian science ("Image," 103), Tylor repeatedly decries the intrusion of what he regards as unscientific methods of free interpretation into the anthropology of his day,[7] and he advocates a strongly progressivist Spencerian-Darwinian theory of evolution intended to cause cultural phenomena, as he plainly says, "to come within the range of distinct cause and effect as certainly as the facts of mechanics" (*Primitive Culture*, 1:69, 18). All this methodological scaffolding, designed to legitimize the claim of evolutionary anthropology to genuinely scientific status in the age of Tyndall and Kelvin, drops away, to all intents and purposes, from *The Golden Bough*. The result is the elaboration of scientific discourse of a dramatically new mode.

Frazer does at times pledge allegiance to an epistemology of mechanical cause and effect worthy of his original mentors, Kelvin and Tylor; indeed, he goes out of his way to disavow developments in modern physics that seem "to cut straight at the very root of science by eliminating causality and thereby implicitly denying the possibility of a rational explanation of the universe" (quoted in Ackerman, *Frazer*, 14). Yet except for an occasional perfunctory gesture toward mechanistic explanation, he in fact everywhere subordinates causal evolutionary narrative to his true, all-consuming enterprise, the exegesis of cultural symbolism. This he claims to be able to effect not by reference to external, that is to say, extracultural realities (climate and geography, racial determinants, trade routes), but by conceiving the totality of world cultures as a single complexly articulated, self-contained and self-explicating signifying system, the various elements of which are decipherable *only by reference to other elements of it*. Frazerian science thus disaligns itself irretrievably from mechanistic explanation, and commits itself irretrievably to the ambiguities of the relativity effect, by basing itself on the principle enunciated by his much-admired Spencer[8] and other Victorian theorists: that reality, as far as it is open to scientific investigation, *consists wholly of symbolism*. For Frazer, all social phenomena such as religious rites and mythology and practices like taboo avoidances constitute symbolic expressions of *meanings* that it is the primary task of ethnographic analysis to decipher, and that, embedded as they are in primitive systems of thought deeply unlike our own, are necessarily enigmatic.

> There is an instructive class of [May-tree rituals] in which the tree-spirit is represented simultaneously in vegetable form and in human form, which are set side by side as if for the express purpose of explaining each other. In these cases the human representative of the tree-spirit is sometimes a doll or puppet, sometimes a living person, but whether a puppet or a person, it is placed beside a tree or bough; so that together the person or puppet, and the tree or bough, form a sort of bilingual inscription, the one being, so to speak, a translation of the other. Here, therefore, there is no room left for doubt that the spirit of the tree is actually represented in human form. (*GB*, 144)

In such a characteristic passage, Frazer insistently portrays complexes of imagery as in effect—or, here, quite explicitly—a linguistic system, a network of interdependent signifiers, one in this case actually indexed with built-in keys to correct interpretation. It is this primary intuition of the supremacy of symbolic expression in social phenomena that motivates the many-tentacled research program of *The Golden Bough* and marks its divergence from all modes of explanation that pay tribute to mechanical models.

The research paradigm corresponding to this set of assumptions is virtually synonymous with Frazerian anthropology: "the Comparative Method,"[9] which supposedly allows a scientist to interpret enigmatic symbolic forms or to reconstruct lost ancient ones by means of analogies with other, similar forms. Frazer typically refers to this technique as though it were nothing more than a common-sense practical expedient for filling in gaps in the documentary record. "Direct evidence . . . there is none," he says, for example, in reference to his theory of annual wedding rites at Nemi, "but analogy pleads in favour of the view" (*GB*, 162). That such a method in fact brought into being exactly that regime of radical relativity forecast by Bain, Spencer, Stallo, and other theorists, and that it was subject to all the intellectual hazards of that regime, became ever more apparent.

The object of study is posited by the comparative method not to be metaphysically unknowable (as it is in Spencer's philosophy) but—what amounts to the same thing—to be placed forever beyond the scope of direct inspection. In this predicament, the inaccessible "thing in itself" (here, the homicidal Arician priesthood, which it is "the primary aim" of *The Golden Bough* to "explain" [*GB*, v]) can only be known as a function of its conjectured resemblances to other things (rites, myths, and other social structures of numerous other peoples, ancient and contemporary). Frazerian anthropology thus bases itself on a formidable paradox. A rigorously empirical science in which (as Frazer's critics have never sufficiently acknowledged) the gathering of detailed, authenticated data is fundamental, it starts nonetheless from the proposition that its object is closed to view, incurably ambiguous, and conceivable only as a complex of parallelisms and analogies. It is as though Frazerian science were designed expressly to embody H. L. Mansel's dictum that knowledge "is only possible in the form of a *relation*" (Mansel, *Limits*, 96), or Hamilton's that it "is only possible where there exists a *plurality of terms*" (Hamilton, "Philosophy of the Unconditioned," 37), or Bain's that "there is no escape from the principle of universal relativity" since "there is no possibility of mentioning a thing, so as to be intelligible, without implicating some other thing or things" (Bain, *Logic*, 1:255, 61), or, later, Saussure's that "in linguistics, *to explain a word is to relate it to other words*" (*Course in General Linguistics*, 189). If Frazer imagined that one could with impunity adopt such principles as the basis of a practical research project, he soon learned how wrong he was. He learned that in a thoroughly relativized intellectual environment such as the one constructed in *The Golden Bough*, "our practice and belief," as Mansel shrewdly said, "must be based on principles which do not satisfy all the requirements of the speculative reason" (*Limits*, 27). He discovered also, to his dismay, that the law that "truth itself is nothing more than a relation" (Mansel, *Limits*, 146) fails to satisfy the requirements of textual coherence, as well—that the comparative method, pursued in an uncompromising way, was drastically incompatible with the model of a well-wrought, unified scientific dissertation. Rather than seeking to screen these intractable problems of relativity from view or to explain them away, however, Frazer states them ever more openly and claims a key place in modern and postmodern intellectual history by suggesting that they call for a new conception of scientific explanation itself.

In a noteworthy letter of 1888 to the Plato scholar Henry Jackson, for example, Frazer emphatically denies the possibility of taking up a position of unambiguous objectivity with regard to cultural materials. For one thing, "our present way of looking at the world" is merely a passing fancy; "natural and correct as it seems to us," he says, "[it] will perhaps one day appear as remote, absurd, and unnatural to our descendants as the worst extravagance of savage opinion now appears to us." Nor is there any way for a scientist to stand outside the field of relativity generated by the dependency of every empirical observation and every act of reasoning upon historically contingent fundamental assumptions. On this point, Frazer expresses himself with considerable philosophical sophistication. "You seem to think that man stands for ever on the same spot in the river and sees it speeding past him," he writes to Jackson. "It is not so, he is borne along on the current. There is no *absolute* way of looking at the world" (quoted in Ackerman, *Frazer*, 88). He does not allow epistemologically dangerous reflections such as this into the text of *The Golden Bough*, but he does call the ideology of definite, objective scientific truth into question there in other ways. As we shall see in more detail below, he insistently presents his arguments as tissues of interpretive guesswork couched in "may" and "might" and as based on no links of reasoning less ambiguous than those of analogy. Readers steeped in the creed that "the logic of science is infallible" (Poincaré, *Science and Hypothesis*, xxi) and in "the fear of interpretation" as anathema to the principles of scientific proof could only find Frazerian equivocation gravely unsatisfactory. Lucien Lévy-Bruhl, thinking chiefly of Frazer, thus complained in 1910 that "the 'explanations' of the English school of anthropology, being never anything more than probable, are always affected by a co-efficient of doubt" (Lévy-Bruhl, *How Natives Think*, 23).[10]

Worse still—indeed a mortal failing—in the eyes of the science of mechanical modeling is the alarming fluidity of Frazer's theoretical framework. With reference to the great question of the origin of mythology, for example, he adopts one hypothesis after another: cognitionism (according to which myths represent primitive attempts to explain natural phenomena), ritualism (according to which myth arises from the obsolescence of magic ritual or from the forgetting of its original rationale), euhemerism (according to which myths are based on real historical events). Stanley Edgar Hyman has shown that these different theories come and go in the course of the three main editions of *The Golden Bough*, often appearing contradictorily side by side (Hyman, *Tangled Bank*, 239–43, 245). Hyman and Frazer's thoughtful biographer Robert Ackerman are harsh in their assessment of what they regard as Frazer's "theoretical confusion"; for them, his "seeming indifference to theoretical self-contradiction" (Ackerman, *Frazer*, 233) is a strange intellectual perversion. If Frazer's theoretically polymorphous text was intended to raise a question about the adequacy of the scholarly ideology of unwavering fidelity to a single explanatory mechanism through thick and thin, it certainly failed with these readers. Of course, this ideology is sacrosanct, and it may hardly be possible to call it into question within the framework of reputable scholarly discourse. Yet there are distinct signs that Frazer's long attempt to penetrate the secrets of the Arician priesthood was also an attempt, no doubt a quixotic one, to imagine an intellectual world in which the law of contradiction would have only a qualified, conditional authority—that he was striving to construct for himself a scientific credo, or at least a practical modus operandi, emancipated from the absolutism of immutable theoretical models.

Like Newman and like F. C. S. Schiller, he clearly had an intuitive aversion to the coercive and reductive character of formal logic, and to the potential for delusion built into its insistence on "[an] *absolute* way of looking at the world." In the letter of 1888 cited above, he declares his wariness of general theoretical principles. "I . . . care chiefly for particular cases," he says, "and am apt to regard discussion of general principles as nearly a waste of time" (quoted in Ackerman, *Frazer*, 98). In the prefaces of later editions of *The Golden Bough*, he repeatedly emphasizes the "hypothetical" nature of the explanations proposed in the work (*Magic Art* [1900], xix) and his intense consciousness of the distorting effects inherent in any scholarly method bent on "reducing the vast, nay inconceivable complexity of nature and history to a delusive appearance of theoretical simplicity" (*Magic Art*

[1910], x). In 1921, he protests in the same way against "the vice inherent in all systems which would explain the infinite multiplicity and diversity of phenomena by a single simple principle, as if a single clue, like Ariadne's thread, could guide us to the heart of this labyrinthine universe" (quoted in Hyman, *Tangled Bank*, 276). In such pronouncements, Frazer sounds less like someone afflicted with "theoretical confusion" than like a forerunner of the dissident philosopher of science Paul Feyerabend, who condemns in similar terms any method of inquiry "that contains firm, unchanging, and absolutely binding principles" (Feyerabend, *Against Method*, 14). Given the "fragmentary, obscure, and conflicting" character of the evidence with which an analyst of primitive mythology must work, Frazer says in 1913, even the most penetrating research must despair of reaching fully satisfying or even moderately durable conclusions at the level of theory. "It is the fate of theories to be washed away like children's castles of sand by the rising tide of knowledge" (Frazer, *Balder the Beautiful*, xi). He expresses hope that his work will retain its value "as a repertory of facts" (*Magic Art*, xx), but treats theoretical arguments as heuristic devices without a legitimate claim to permanence or exclusiveness. "Theories," he says, "are transitory" (*Balder*, xi). He hints that research in any field whatever, not just in prehistoric culture, is obliged to manufacture such knowledge as it can out of fragmentary, obscure, and contradictory information.

Declarations that *The Golden Bough* "is a work committed to totalizing explanation, to the discovery of a single system of reference and meaning" (Beer, *Open Fields*, 73) need therefore to be regarded with great caution: one of its most insistent strands of argument is in fact that all such systems are illusory. Particular explanatory theories in *The Golden Bough*, far from claiming totalizing authority, everywhere unravel before the reader's eyes or are made to coexist as best they can with contradictory ones—and ultimately are disavowed as chimerical and unscientific by the author himself. Poincaré argues in 1902, as we saw, that whenever it is possible to construct a mechanical model to explain a given phenomenon, it is possible in principle to construct any number of them, each one equally legitimate. Despite his repeated protests against the failure of theories to do justice to the "multiplicity and diversity" of things, Frazer does not exactly formulate Poincaré's principle, but he does something more daring still: as in his multiplying models of the origin of mythology, he *performs* this principle in concrete terms, as if precisely to free his scientific practice from the code of totalizing explanation, which he knew to be incompatible with the nature of research in a speculative field like his own, constituted wholly of symbolism. "Proliferation of theories is beneficial for science," declares Feyerabend, who, taking Mill's *On Liberty* as his scripture, advocates the need not only to authorize in scientific inquiry "what one might call the freedom of artistic creation," but "*to use it to the full*" (*Against Method*, 24, 38). No better motto could be found for the mode of science invented and then—at the risk of being condemned as "not simply erroneous but absurd"—actually practiced to the full in *The Golden Bough*.

Gillian Beer may be the only commentator on Frazer's work to have recognized how sharply it diverges from its author's occasional Kelvinesque or Tylorian descriptions of it. Frazer's "apparently unsystematic procedure" of following any cultural theme wherever it led by a means of free association, Beer writes, "had methodological and ideological advantages": it "liberated his work from the ruthless developmentalism" of earlier evolutionist writers and "set loose powers at odds with the severely developmental and law-bound world-set within which he believed himself to work" (Beer, *Open Fields*, 94). She seems in these perceptive remarks to corroborate T. S. Eliot's praise of Frazer for his "absence of speculation," which Eliot regarded as "a conscious and deliberate scrupulousness" and saw as basic to his "inevitable and growing influence over the contemporary mind." Frazer, says Eliot, withdraws "in more and more cautious abstention from the attempt to explain" (Eliot, "A Prediction," 29). What looks like a chronic dereliction of logic to Hyman and Ackerman looks to Eliot like the sign of Frazer's principled refusal of dubious intellectual methods and of the rigorously modern character of his work.[11]

The result, in any case, is a species of analysis unknown to the logic of mechanical models. The transformative impact of Frazer's concept of scientific demonstration registers immediately in the opening discussion of the cult of Diana at Nemi, where he introduces the riddle *The Golden Bough* at least nominally represents an attempt to solve. Just as importantly, he offers here a compressed specimen of the investigative method to be employed in what follows—"just as importantly" since the real lesson is that the riddle turns out ultimately to be a function of the comparative method itself, and for that reason to lie permanently beyond definitive solution.

The semilegendary cult at Nemi, Frazer informs us, was presided over into historical times by a series of kingly priests or priestly kings, each of whom came into office by killing the previous incumbent, and was associated, among a host of other ritual forms, with a certain sacred tree from which no branch must be broken. In order to account for the genesis of this strange institution, he says in effect, we must postulate that its every element is symbolic and able in principle to be scientifically deciphered so as to reveal its underlying meaning. Like New Critics analyzing a cryptic poem or a psychoanalyst interpreting a dream, we must postulate, too, that the various elements of this symbolic complex—the sacred tree, ritual themes of hunting and of fertility, fire ceremonies, priestly murder—all fit together in some intricately articulated system. In formulating this concept of cultural practices as integrated, complex symbolic networks, Frazer lays the groundwork of the modernist cultural anthropology pioneered by Malinowski, Mauss, Benedict, and others, which habitually is portrayed as anti-Frazerian.[12] No fair estimate of Frazer's role in intellectual history can fail to stress this point. But his exercise in unpacking bundles of symbolic cultural imagery has barely begun before we are made to discover that it yields few if any simple one-to-one tenor-to-vehicle correspondences; rather, it opens onto *chains* of correspondences that seem to ramify uncontrollably as we train the scholarly gaze upon them. The ruling principle of this mode of investigation, though Frazer never states it in these terms, is that cultural imagery turns out to stand not for referents in the world of external reality, but merely for other forms of imagery, which themselves turn out when we explicate them to refer to still other forms of imagery. To put this state of affairs in the Derridean terms that are exactly appropriate to it, each signified turns out when put under the lens of Frazer's analysis to be just another signifier in its turn. Derrida may not have had Frazer in mind in imagining this formula, but probably there never has been a text that enacts it more vividly than does this one. The suggestion, which Frazer carries in *The Golden Bough* to such mind-boggling length as to announce a sort of intellectual revolution, is that in the realm of human culture it's imagery all the way down, and that "to live in images or symbols" must henceforth be "the essence" not only of religion, as Feuerbach said, but of the modern-style sciences of man. This is the relativistic principle that usurps the role of mechanistic explanation in Frazer and gives him his chief claim to exercise an "inevitable and growing influence over the contemporary mind."

A quick synopsis of Frazer's preamble can suggest how the principle of the comparative interpretation of cultural symbolism translates into practice in *The Golden Bough*. Along with Diana, a minor deity named Virbius seems to have been worshipped at Nemi, he reports. Certain ancient authorities indicate that Virbius was actually Hippolytus, brought back to life by Artemis after suffering his violent death—Artemis being, as Frazer reminds us, "the Greek counterpart of Diana" (*GB*, 5). Hippolytus was established, in disguise as Virbius, as the king of the sacred grove at Nemi. This mythic structure reduplicated itself often in classic mythology, we are told. "The rivalry of Artemis and Phaedra for the affection of Hippolytus reproduces, it is said, under different names, the rivalry of Aphrodite and Proserpine for the love of Adonis, for Phaedra is merely a double of Aphrodite" (8). From the evidence of analogy provided by this complexly imbricated set of erotic myths, Frazer conjectures that Diana in her sacred grove at Nemi was actually the consort of the shadowy Virbius, "who was to her," he says, "what Adonis was to Venus, or Attis to Cybele" (9). He further speculates "that this mythical Virbius was represented in historical times by a line of priests known as Kings of the Wood, who regularly perished by the swords of their successors" (9–10). According to this hypothetical reconstruction, the killer priests are symbolic figures who not only worship but actually *stand for* the mythical Virbius, himself really a disguised figure of Hippolytus, himself an analogue of Adonis and also of Attis, who, as Frazer says later, "was to Phrygia what Adonis was to Syria" (403).

It may be hard to follow this network of equations and mythological ratios in my condensed version of it, but it is not much easier in Frazer's text. The key effect of this opening passage, in either version, is the way the portrayal of the homicidal priesthood turns in Frazer's analysis into a densely layered complex of symbolic counterparts, equivalences, occult identities, substitutive images. Nor is this the end of the chain of references by any means, for Frazer claims to need to pursue this line of exegesis far more widely afield—so much more widely that twelve volumes of scholarly exploration will hardly, in the last analysis, be sufficient for the project. On the contrary: he expressly concludes, as we shall see, that even this prodigious amount of scholarly work necessarily falls far short of its goal, and that any pretense of imposing "theoretical simplicity" upon a reality that by its nature is inexhaustibly "labyrinthine" (being a construct of potentially innumerable equations and analogies) can only be fallacious.

How, he asks here at the outset, did this constellation of myths come to be established at Nemi in the first place? In an especially provocative formula, he says: "clearly they belong to that large class of myths which are made up to explain the origin of a religious ritual and have no other foundation than the resemblance, real or imaginary, which may be traced between it and some foreign ritual" (*GB*, 6). All the Diana/Virbius material associated with the institution of the priesthood at Nemi, in other words, does not belong properly to it in the first place, but represents a symbolic replication of some other, unknown complex of ritual and mythology! The mythological symbolism associated with the Nemi cult, ostensibly its basis and *raison d'être*, thus proves to be a cultural back-formation manufactured *ex post facto* to provide an alibi for a preexisting social structure. And rather than embodying anything like a direct mapping of symbolic imagery onto the thing itself (namely, the priesthood), the process of mythologization that Frazer postulates involves a triangular system of representation in which the image is linked to its referent only by means of a displacement from an unknown third entity, such that the cultural scene at Nemi hallucinates that of the hypothetical "foreign ritual." Objects of religious veneration are constructed fantastically by mimetic substitution in Frazer's model, just as objects of sexual desire are in René Girard's model.[13] Not to aggravate the scholarly seasickness this analysis is sure to arouse, Frazer does not dwell here on the obvious implication that were we to identify the supposed foreign model and examine it in turn, it, too, might decompose similarly into a complex of derivative imagery originating in still other foreign structures. Hoping to avoid so quixotic a project, he proposes instead to turn the investigation in a different direction and to seek to reconstruct by means of the comparative method the aboriginal form of the Nemi cult as it was (or might have been) practiced prior to the grafting onto it of belated mythological references. But notice has been amply served that in the multiplying volumes of scholarly detective work to follow "no other foundation" may ever be discoverable for a theory of the Nemi priesthood than such as can be made out of networks of resemblances, "real or imaginary." Whether these two categories can ever be stably discriminated in the elaboration of "a theory of symbols" like this one is of course the question.

Whatever stability Frazerian analysis can achieve is vested in the notion of meaning. The scientific study of an institution by means of the comparative method is an attempt to determine what it "means." But *meaning* turns out to be a deeply ambiguous term in Frazer's theoretical lexicon—surprisingly so, considering its centrality to the new interpretive model that he and other comparatist anthropologists promote. As we will see below, and as the May-tree passage implies, he avowedly does not mean a conscious set of expressive intentions in the minds of social actors; nor does he identify "meaning" with the discovery of some definite referent for each unit of symbolism. In practice, the discovery of meaning by means of the comparative method entails for Frazer the exhaustive tracing out, along what may be several different synchronic and diachronic axes, of all the multitudinous thematic associations arising from a given symbolic object. Stallo described this general scheme as the paradigm of any sustained scientific inquiry carried out under the auspices of relativity. "The concepts of a given object," he says, in terms that would be echoed closely by Saussure and by Derrida, "are terms or links in numberless series or chains of abstractions varying in kind and diverging in direction with the comparisons instituted between it and other objects"—a formula from which he draws the Spencerian conclusion that "all thoughts of things are fragmentary and symbolic representations of realities whose thorough comprehension in any single mental act, or series of acts, is impossible" (*Concepts and Theories*, 158–59). Stallo's statement very precisely describes the organizing logic of *The Golden Bough*. A glance at the table of contents of the abridged edition is sufficient to show the array of diverging trains of thought that branch out from the original topic of the Arician priesthood: magic in its different forms, "Incarnate Human Gods," the worship of trees, taboo in a score of different aspects, the sacrifice of sacred kings, and so on. There is apparently no legitimate way to truncate this proliferation of reference; adequately to interpret the "meaning" of a symbolic formation such as this one obligates the Frazerian scientist of culture to reconstruct the entire system or family of systems to which it belongs, even if a shelf of scholarly volumes will be required in order to begin to do so. And since comparatist anthropology is committed irrevocably to the idea of the unity of human culture, the field of reference for the explication of any item includes all societies on earth at all periods of history. This implicit operating principle of *The Golden Bough* is codified long afterward in Lévi-Strauss's dictum that any myth must be defined scientifically as consisting of "all its versions," earlier and later, at home and abroad, and that valid structural analysis must take all versions into account (Lévi-Strauss, *Structural Anthropology*, 217). Frazer does not merely give lip service to this almost impossibly demanding requirement, but acts it out in his immense catalogues of ethnographic illustrations from every remote corner of the globe and every period of human history. The "meaning" of a cultural formation, according to such a procedure, can never be reducible to any one point of reference, but rather is diffused and dispersed throughout the ramified system of imagery of which it is a part. This is a conception likely to be productive, again, of severe intellectual malaise for those who think of a meaning as a unitary, definite, localizable entity, and even more so for those whose idea of "explanation" is the construction of a causal-mechanical model.

Frazer does seem to base his interpretation of primitive symbolic representations on the concept of homeopathic or imitative magic. Cultural symbolism according to this theory manifests an original belief in the magical identity or interconnectedness of things and their images, and thus signifies the intent of exercising magical force on this or that object. Ritual fires are lighted in order to stimulate the sun to shine, for example. But this quasi- or pseudomechanical thesis is no sooner formulated than it is subjected by Frazer to major qualification. For one thing, he stresses at the risk of paradox that the primitive peoples said to organize their lives around elaborate systems of symbolic magic do not, in fact, possess any conscious idea of the magical mechanics on which their symbolizing practices are supposedly based. The laws of magic "are certainly not formulated in so many words nor even conceived in the abstract by the savage" (*GB*, 22), he says, disconcertingly. This is a very early formulation of the principle on which cultural anthropology, not to mention psychoanalysis,[14] has depended ever since—the principle that logical systems of thought can be attributed to people who may be unaware of having such thoughts in their heads or to collectivities that can hardly be said to think at all.[15] Thus Frazer acknowledges in a late passage that the worshippers of divine kings have after all no clear conception of the all-important magical bond supposedly connecting the king's physical welfare to that of the community he symbolically personifies: "probably their ideas on the point are vague and fluctuating," he says, "and we should err if we attempted to define the relationship with logical precision" (686). Scientific ethnography requires and licenses one to do exactly this, however. "It is for the philosophic student to trace the train of thought which underlies the magician's practice," Frazer says (13). This is a momentous innovation in cultural science, but it exposes its practitioner to the suspicion that the magical symbolic equivalences that form the logic of his study are not native to a system of primitive thought at all, but are simply manufactures of comparative scholarly analysis. Such suggestions, so recklessly made by Frazer, as though to introduce a dangerous instability into the heart of scientific rationality itself, risk rendering the notion of meaning "vague and fluctuating" indeed.

I will not debate here whether Frazerian comparatist anthropology, trafficking as it does in such equivocal objects of study as symbolic correspondences and unconscious meanings, can properly be called scientific. I will merely emphasize one of this writer's most significant discursive gestures, one that is also acutely characteristic of modern intellectual awareness in general: his insistence on the irremediably unstable character of his own methodology and findings. Ever more distinct in his work is the suggestion, indeed, that the ideal of a scientific method free of relativistic incoherence and grounded in unconditional objectivity may be unrealistic, even ultimately unscientific.

Once the data under investigation in any field are defined as intrinsically symbolic, as allowing only a reflexive access to an absent reality, in other words, and once the task of the scientific investigator becomes as a result inescapably an interpretive one, then scientific work finds it ever harder to free itself from a presumptive character of fictionality—of being "made up" out of analogies like the Nemi mythology itself. No problem could of course be graver in the eyes of official epistemologies wed to the craving for definite, unequivocal truths and thus full of phobic dislike of what Foucault calls "the free circulation, the free manipulation, the free composition, decomposition, and recomposition of fiction" ("What Is an Author?" 159). Frazer is no advocate for anarchic scholarly freedom, but he is no dupe of his own methodology, either, and he never ceases reminding us that the interpretive reconstructions of the comparative method are bound always to fall short of definitive proof, to remain conjectural and provisional at best, affected deeply by that inescapable coefficient of doubt to which Lévy-Bruhl objected. His entire inquiry is, he frankly tells us, an exercise in "inference"—that is, a kind of free interpretive guesswork—where what he calls "demonstration" will never be attained because it is unattainable, and where the best one can hope for is what he calls "a fairly probable explanation" of such a riddle as the real meaning of the priesthood of Nemi (*GB*, 2–3). "A full and satisfactory solution of so profound a problem [as that of the displacement of magic by religion] is hardly to be hoped for," he announces elsewhere. "The most we can do . . . is to hazard a more or less plausible conjecture" (*GB*, 65). At the conclusion of his vast work, he states this overriding dilemma in expressly relativistic terms. "We can never completely replace ourselves at the standpoint of primitive man, see things with his eyes, and feel our hearts beat with the emotions that stirred his," he declares. "All our theories concerning him and his ways must therefore fall far short of certainty; the utmost we can aspire to in such matters is a reasonable degree of probability" (*GB*, 823). The whole project of data-gathering and of interpretation by means of the comparative method will be governed by a kind of Heisenberg Uncertainty Principle of anthropology.

These and similar statements of Frazer's have been read as expressions of "outstanding intellectual humility" (Robert Fraser, *Making of* The Golden Bough, 212), as testimony to their author's uncompromising code of scholarly integrity—and, implicitly, as an apologia for a rationalistic ideal of interlocking facts and theory, which is to say, of conclusively verifiable knowledge, in scientific inquiry.[16] But there is every reason to understand them differently, as positing a necessarily ambiguous and shifting relation between empirical data and the explanatory stories we profess to extract from them, and as acknowledging that once science has recognized itself to be wholly vested in "a theory of symbols," and once the world of material things has thus been replaced for scientific purposes by a world of relations and coded meanings, definite norms of logic cease to apply. Scientific inquiry under these conditions is bound to base itself, Frazer implies, on the sovereignty of analogical interpretation, according to which phenomena remote in time and space may be seen as making reference to and elucidating one another, as though by means of *actio in distans* or as though connected to one another by an occult ether of cultural transmission. This principle of analysis may be what Girard had in mind in calling Frazerian science "fanatical and superstitious." Indeed, the magical character of comparatist study is plainly enough insinuated in Frazer's own text. The "spurious science" of homeopathic magic bases itself, he explains, on what he names "the Law of Similarity," which superstitiously assumes "that things which resemble each other are the same" (*GB*, 12–13); in just the same way, his own modernistic science of cultural analogies founds itself, as he says, on the doctrine of "the essential similarity" of seemingly disparate cultural phenomena "specifically different but generically alike" (2). Comparatist anthropology takes as its declared first principle its own magical Law of Similarity, in other words.

Research in this radically relativized field may yield rich, profound, interesting knowledge, but it will never be able to appeal to any criterion of verification more stringent that that of "probability," or, in other words, the judgment of the illative sense that tells us when a conjecture is "more or less plausible." In some passages, Frazer imparts this lesson in the guise of a philosophical principle, suggesting that logical incoherence and the arrival at merely "tentative and provisional" conclusions (*Magic Art*, 1:xx) are not the marks of scholarly error but the signs under which philosophically knowing research of a modern style inescapably is placed.[17] In other passages, increasingly, he presents the unattainability of definite knowledge as a tragic existential predicament for a scholar "seeking," as he says in one late preface, "to know what can never be known," in defiance of reason, "like Sisyphus perpetually rolling his stone up hill only to see it revolve again into the valley, or like the daughters of Danaus doomed for ever to pour water into broken jars that can hold no water" (*Adonis Attis Osiris*, ix).

Even in the disillusioned intellectual scene of late-twentieth-century postmodernity, anthropological writers seek to shield themselves from such a fate by laying claim to what Frazer would have regarded as vainglorious false ideals of objectivity and certainty. One thinks of Clifford Geertz's declaration that the study of culture is "a positive science like any other" and that "the meanings that symbols . . . embody" are in principle as exactly determinable by empirical study "as the atomic weight of hydrogen or the function of the adrenal glands" (Geertz, *Interpretation of Cultures*, 362–63). For Frazer, who disdains this kind of scientific absolutism as intellectually delusive, empirical facts and the process of interpretation are bound up inseparably together, linked by a twofold relativity; consequently, what Spencer called "the Unknowable" can never be disentangled from knowledge except at a cost of a disabling mystification. "What we call truth is only the hypothesis which is found to work best," Frazer declares (*GB*, 307). Whatever may be its final value as scientific anthropology, his work in accordance with this principle has a powerful claim on the attention of scholars of modern thinking, if only for the insidious deconstructive pressure it exerts on its own code of scientific rationality by alluding so plainly to its own magical or "hypothetical" character. No wonder that later generations of anthropological writers have shown a desperate eagerness to expel Frazer from the canon of reputable scientific literature: branding him "not simply erroneous but absurd" signals an anxiety that the philosophical scandal of Victorian comparatism, if not firmly suppressed, might compromise the claims of scientific rigor put forward by various subsequent anthropologies.

Purificatory Violence

In a preface of 1910, Frazer declares, as though it had taken him the twenty years since publishing the original two-volume edition to fathom his own argument, that the guiding theme of *The Golden Bough* has been that of "men who have masqueraded as gods." Even though his speculations about the origin and significance of the Arician priesthood may prove erroneous, he says here, his work would remain valuable for its demonstration "that human pretenders to divinity have been far commoner and their credulous worshippers far more numerous than had been hitherto suspected" (*Magic Art*, 1:ix). In this statement, he annexes his work directly to the critique of the cult of authority that runs powerfully throughout modern relativity literature. He seems to ally himself, for instance, to the analysis by Spencer, his early guru, of that pathological disposition in modern people that the latter calls the "awe of embodied power" (*Study of Sociology*, 175) and illustrates with reference to the modern history of tyrannical atrocities from Napoleon to Governor Eyre to the Contagious Diseases Acts. Frazer illustrates the same pathology in the idiom of the ethnographic report.

> Thus in the Marquesas or Washington Islands there was a class of men who were deified in their lifetime. . . . Human sacrifices were offered to them to avert their wrath. . . . A missionary has described one of these human gods from personal observation. The god was a very old man who lived in a large house within an enclosure. In the house was a kind of altar, and on the beams of the house and on the trees round it were hung human skeletons, head down. . . . This human god received more sacrifices than all the other gods; often he would sit on a sort of scaffold in front of his house and call for two or three human victims at a time. They were always brought, for the terror he inspired was extreme. He was invoked all over the island, and offerings were sent to him from every side. (*GB*, 110–11)

The Golden Bough is almost devoid of explicit contemporary reference, and Frazer himself is famous for an extreme bookish detachment from modern events.[18] Yet beginning in the second edition of 1900 and in all subsequent editions, he includes in an early section an apocalyptic warning of the possibility of a modern social upheaval on a scale to threaten the survival of European civilization. He does not specify the form that such a calamity might take, but connects it to "the permanent existence of . . . a solid layer of savagery beneath the surface of society." This layer of atavistic tendencies poses, he says, "a standing menace to civilisation"; "we seem to move on a thin crust which may at any moment be rent by the subterranean forces slumbering below" (*GB*, 64). In the third edition, these lines are glossed in a marginal heading: "Latent superstition a danger to civilisation" (*Magic Art*, 1:236). Frazer repeats the dire imagery of this passage in a little-known lecture of 1908. "We appear to be standing on a volcano which may at any moment break out in smoke and fire to spread ruin and devastation among the gardens and palaces of ancient culture wrought so laboriously by the hands of many generations"; "I tremble for civilization," he says, quoting Renan (quoted in Ackerman, *Frazer*, 213). In light of his statement about men masquerading as gods, we can infer that the dreaded upheaval of Frazer's imagination would be linked to the rise of a charismatic political leader who would be worshipped as a god by his nation as the terrifying dictator of the Marquesas was worshipped by his. Frazer cites in this connection, though, not evidence of new forms of state despotism, in the manner of Spencer, but rather, oddly as it must have seemed to his original readers, isolated cases of modern-day superstitious atrocities: a woman in Ireland roasted to death as a witch, a girl murdered in Russia and chopped in pieces to be boiled down into talismanic tallow candles (*GB*, 65). The scholarly enterprise of reconstructing the rites of the Arician priesthood is at bottom, we may say, a mechanism for analyzing the mentality that expresses itself in these macabre modern events that Frazer sees as spurts of primitive superstitious magma through the thin crust of civilization. *The Golden Bough* presents itself by the same token as an extended parable of the danger allegedly looming over modern society—or rather, welling up from its subterranean depths. In Frazer's text, Pacific islands, ancient Aricia, and other sites of primitive consciousness are meant to be recognizable in one of their aspects as allegorical images of contemporary Europe.

Frazer does not venture any more distinct thesis as to how "latent superstition" might emerge in such a volcanic form as actually to menace twentieth-century civilization, and if his parable has particular contemporary references, he does not allude to them. He does bring the comparative method to bear ever more intensively upon the theme of human sacrifice that seems to him to lie at the primitive basis of civilization, particularly of Christian civilization. *The Golden Bough* identifies the supposed primitive addiction to human sacrifice not with those commonplaces of Victorian ethnographic literature, the natural cruelty and depravity of "savages," but with a complex of traditional social institutions expressing themselves symbolically in ritual. The suspension of the register of moralistic didacticism in favor of structural analysis marks clearly the emergence of what Ortega would call "dehumanized" post-Victorian sensibility in *The Golden Bough*. But this rhetorical shift does not render the thematics of sadistic official violence any less startling in Frazer's narrative, which risks scandalizing Victorian literary propriety by exploring the forms of ritual killing as exhaustively and in as much graphic bodily imagery as the works of Sade explore those of copulation. In the course of *The Golden Bough*, every imaginable injury is inflicted upon the human body: sacrificial victims are roasted alive in fiery ovens, boiled in oil, dragged over stony ground to their deaths, assailed with stones and clubs, pierced with spears, swords, and pointed sticks, torn with knives and sickles, flayed, disemboweled, dismembered, decapitated, crushed, crucified, eaten. Human blood, the most potent of all magical fluids, is spilt, drunk, wallowed in, splashed on trees and fields to promote fertility.[19]

These themes are constant, but a notable shift in emphasis marks the successive editions of Frazer's work: from emphasis on the primordial sacrifice of the god-king to emphasis on displaced versions of this practice involving the sacrificing of symbolic surrogates and then, evidently in a later evolutionary phase, the sacrificing of victims drawn from the ranks of criminals, aliens, or untouchables. Out of this shifting configuration of ideas evolves the version of sacrificial substitution on which Frazer, its original modern theorist, lays increasing emphasis, finally devoting to it the penultimate part of *The Golden Bough*: the system of the scapegoat, the despised victim who is loaded with the guilt and evil of the nation and then killed or cast out in a purifying sacrifice designed to ensure the resurrection of the blighted community. The concept of the scapegoat is closely associated, Frazer tells us, with the superstitious notion that the community has been infiltrated by "swarming multitudes" of malignant beings passing themselves off as regular citizens. "Thus it comes about," he says, "that the endeavour of primitive people to make a clean sweep of all their troubles generally takes the form of a grand hunting out and expulsion of devils or ghosts" (*GB*, 634).[20]

The same motifs dominate his evolving treatment of fire festivals and the human sacrifices that once, as he theorizes, formed their central element. In the original edition of *The Golden Bough* (1890), he endorses the "solar theory," according to which these ceremonies were sun-charms intended to promote an abundant supply of sunshine and thus to ensure the fertility of the fields. In later editions, he embraces instead the "purificatory" theory of Edward Westermarck, in which the ritual fires are conceived rather as "a fierce destructive power" that acts primarily as, in Frazer's word, a "disinfectant" (*GB*, 744). According to this theory, the function of fire festivals was the magical destruction of witches and the expulsion of vampires and other diabolical aliens from the body politic (744–45). In a notable preface of 1913, Frazer highlights the direct transfer of heathen magic and violence into Christianity at this cultural nexus. He comments here on the "dreadful obsession" of Christianity with exterminating witches (*Balder*, viii) and, touching again the chord of prophecy, warns in the same passage that the belief in witchcraft, deeply associated as it is with Christian religion, only hibernates dangerously beneath "a thin veneer" of civilization, awaiting the opportunity to break out anew the moment rational constraints are relaxed. Frazer plainly warns his readers that the eruption of that volcano of primitive superstitious savagery that will imperil modern Europe will take the form of a revived frenzy of witch-annihilation. Thus *The Golden Bough* resonates ever more distinctly with echoes of Feuerbach's indictment of Christian religion for the current of sadistic violence running through it, causing it to "[delight] in the blood of heretics and unbelievers" and to foster the worship of an impostor divinity in the form of "an unloving monster, a diabolical being" (Feuerbach, *Essence*, 52).

Often the linkage of pagan sacrifice practices with modern Christianity remains remote and unstated in Frazer's narrative. It requires an elaborate evolutionary argument, for example, to identify the god-king Jesus with the bloodthirsty god-king of the Marquesas, or to identify Christian symbolism with that of the Carthaginians' methods of sacrificing children to Moloch, in which, Frazer tells us, "the children were laid on the hands of a calf-headed image of bronze, from which they slid into a fiery oven, while the people danced to the music of flutes and timbrels to drown the shrieks of the burning victims" (*GB*, 327). But Christian parallels come out with shocking vividness in many other places,[21] notably in the section on Aztec sacrifice titled "Killing the God in Mexico," which forms the climax of part 6, *The Scapegoat*, and in some sense, as Hyman says (*Tangled Bank*, 261), of *The Golden Bough* as a whole. Aztec religion amazed the Spaniards who conquered Mexico in the sixteenth century, Frazer maliciously tells us; their curiosity "was naturally excited by the discovery in this distant region of a barbarous and cruel religion which presented many curious points of analogy to the doctrine and ritual of their own church" (*GB*, 680). In one important Aztec rite that Frazer describes in unsparing detail, a young slave girl aged twelve or thirteen was costumed as the Maize Goddess Chicomecohuatl, then, amid much ceremony, beheaded; her gushing blood, supposed to be rich with magic efficacy, was splashed on the offerings of corn, peppers, pumpkins, seeds, and vegetables that the worshippers of the corn-goddess cult had brought to her as offerings. Then her body was flayed, Frazer tells us, and the skin was immediately squeezed into by a priest who appeared in it before the crowds of worshippers, to symbolize, as Frazer supposes, the divine resurrection, itself magically symbolic of the rebirth of vegetation after the death of winter. He describes in detail a series of similar sacrificial festivals in ancient Mexico, often involving this horrendous symbolism of the flayed skin, and often, as he tells us, involving the mass slaughter of vast numbers of captives and slaves, men, women, and children; he quotes a Spanish historian of the sixteenth century who estimates that more people were killed as sacrifices under the auspices of the god-king Montezuma than died of natural deaths (*Scapegoat*, 297).[22] He also makes a point of stressing that this hitherto unsurpassed example of a people whose national religion was devoted to a fanaticism of mass bloodletting involved no primitive tribe, but a people of a proto-European cultural level (*Scapegoat*, 305). And throughout this section, he keeps reminding us of the parallelisms of ancient Mexican and Christian religious ceremonies. For example, he reports that one ghoulish high Aztec festival occurred on Easter day (*GB*, 680, 684).

The Aztecs thus appear in *The Golden Bough* as Frazer's fullest allegorical vision of that catastrophic upheaval in European life he repeatedly prophesies, in which the unleashing of a wave of purifying religious violence and the revival of a fanatical cult of divine national leaders, men masquerading as gods, will evidently go hand in hand. I do not mean to credit Frazer with clairvoyant foreknowledge of the rise of National Socialism. But any percipient contemporary reader must have known perfectly well who the primary victims would be if the apocalyptic events of Frazer's premonitions ever came to pass—if a twentieth-century version of the Aztec empire of blood were ever to be established in Europe. In the context of European Christendom, the only possible candidate for the role of sacrificial witch-devil and scapegoat was of course the Jew. Never could this principle have been clearer than in the last quarter of the nineteenth century and the first decade of the twentieth. Starting with the founding of the League of Anti-Semites in Germany in 1879 and the wave of hysterical, destabilizing anti-Jewish agitation that engulfed that country in the 1880s and 1890s, the craze of persecution spread rapidly across the Continent with the complicity of political parties and governments, and of course with the avid participation of Christian churches. The result was a chain reaction of outbreaks of terrifying large-scale violence against Jews, sometimes rhetorical, sometimes murderously physical, notably in Austria, Hungary, Romania,

Russia, and France (including, very significantly, Algeria).[23] This prolonged psychotic upheaval was precisely contemporaneous with the composition of *The Golden Bough*. That it forms the great unstated contemporary correlative of Frazer's anthropology seems so likely that the apparent failure of Frazer scholarship to make the point is surprising. It is noteworthy in this connection that Frazer, in common with his much-admired Renan (author of the multivolume *History of the People of Israel* [1887–93]), had a particularly profound affection and admiration for Jewish culture, that his most intimate friend in the 1890s, amid the chronic anti-Semitism of Cambridge academic society (see Ackerman, *Frazer*, 183), was the great Jewish scholar Solomon Schechter, and that he was acutely sensitized to the malign workings of modern anti-Semitism. In 1913, for example, he emerged from his usual scholarly isolation to issue an impassioned public denunciation of the persecution of Jews by Tsarist authorities (Ackerman, 183), and then incorporated this material into the third edition of *The Golden Bough*. Here he declares that rumors of ritual murders committed by Jews in modern times, a subject of supposedly serious scholarly investigation and controversy at the turn of the century, were undoubtedly "mere idle calumnies, the baneful fruit of bigotry, ignorance, and malice" (Frazer, *Scapegoat*, 396).

It is difficult, then, not to guess that the striking shift toward themes of purificatory religious violence that occurred in *The Golden Bough* between the editions of 1890 and 1900 reflected contemporary events such as the outbreak of a conflagration of anti-Semitism in the Dreyfus affair, that watershed of European politics and of the European soul, in the 1890s. As Frazer drafted the second edition, the French press was issuing torrents of crazed anti-Semitic invective and mobs rampaged throughout France shouting the slogan of their movement, "Death to the Jews" (Arendt, *Origins*, 111). At precisely this moment, in 1896, Frazer married a French widow of obscure origins of whom it has been asserted by Angus Downie, a longtime confidant of the Frazers and a biographer of Sir James, that she was herself Jewish (Ackerman, *Frazer*, 124). Whatever the truth of this claim, Frazer certainly could have found in the French dress rehearsal for the Holocaust an abundance of materials strikingly relevant to the themes of *The Golden Bough*. When, for example, the rabble-rouser Max Régis called on a Paris mob to "water the tree of freedom with the blood of the Jews" (Arendt, *Origins*, 111–12), he seemed to call for a revival of precisely the fertility sacrifices documented at vast length by Frazer. Or when subscribers to the notorious 1899 anti-Dreyfusard manifesto known as the "Henry Monument" called (among many other infamous fantasies) for the skins of Jews to be made into bedside rugs, saddles, and other objects, they seemed to cause Frazer's nightmare of a modern revival of Aztec religious pathology to come astonishingly true (Quillard, *Monument Henry*, 77, 85, 95).[24]

Frazer's critique of Christian religion concentrates, in any case, upon its potential for violence and for authoritarian tyranny—the potential underlined soon afterward by Adolf Hitler in his declarations that the Nazi persecution of the Jews was an expression of "true Christianity" (*Mein Kampf*, 422). Amid all its shiftings of course, *The Golden Bough* argues not just that Christian belief is intellectually primitive (this is the least of its failings), but that it bears permanently lodged within it that instinctual-seeming "dreadful obsession" with sacrificial killing it inherits from its evolutionary past. In elaborating as audaciously as he does the precarious logic of the comparative method, Frazer is not merely proposing, as he claims, a technique of scientific analysis able to function as productively as possible "in default of direct evidence" (as though there could ever be such a thing as "direct evidence" within the orbit of "a theory of symbols"). Implicitly he is seeking to construct and to exemplify in practice a mode of thought that breaks decisively from the ancestral heritage of superstition and from the ideology of purificatory violence that necessarily accompanies it in social as in scholarly life. To free ourselves from that noxious mentality, he suggests, nothing less than a disavowal of the categorical imperatives of mechanistic logic and of absolute objectivity will suffice.

Frazer nowhere spells out this nexus of his argument, however; nor can we take it for granted, fundamental as it may be to *The Golden Bough*. Relativity has no inherent ideological structure. We are reminded of this principle once again by the monitory case of Nietzsche, who brings modern-style relativistic epistemology into the service of a cult of "the conquering and *master* race—the Aryan race"—and, inevitably, of the malignant anti-Semitism Nietzsche imbibed from German demagogues of the 1880s (Nietzsche, *The Genealogy of Morals*, 26).[25] "In default of direct evidence" and in default of any assumption of ideological constants, then, one's only recourse at this juncture is to appeal to some version of the comparative method itself. An illuminating point of reference for analyzing the Frazerian argument can be found, for example, in the writings of W. K. Clifford.

As we have seen, Frazer presents comparatist analysis as a science of "inference" that never claims to attain more than "a more or less probable" account of its subject (*GB*, 2–3); certainty is beyond its grasp. He thus implicitly identifies comparatist anthropology with broad trends of probabilistic science that appeared in the latter half of the nineteenth century, particularly in connection with the molecular-kinetic theory of gases and with Darwin's theory of evolution. Both of these epoch-making theories, as Clifford was one of the first to perceive, based themselves on principles of statistical reasoning that are fundamentally incompatible with mechanistic concepts of causality. Corroborating Spencer's doctrine "that human intelligence is incapable of absolute knowledge" (Spencer, *First Principles*, 50), Clifford extended the jurisdiction of probabilistic principles to all of science, arguing in 1873 that all knowledge is "of the nature of inference, and not of absolute certainty" (Clifford, *Lectures*, 1:350–51; see also 2:316). When Frazer repeatedly proclaims the sovereignty of a Cliffordian uncertainty principle over his own researches, he thus allies himself to a key theme of the radical philosophy of science of the time.[26]

Clifford explicitly equates this theme with a moral and ideological code, expounding the same linkage proclaimed, again, by Spencer and then reduplicated in Dewey, Schiller, and other turn-of-the-century writers. The unattainability of "necessary and universal truths" makes imperative, Clifford declares (*Common Sense*, 226), a philosophical attitude of nearly unlimited receptiveness and fluidity; all obedience to received ideas, "conventional habits of thought," and conventional propriety is corrupting (*Lectures* 1:116–17). In the field of moral values, similarly, evolutionary relativity prevails, says Clifford. There is no such thing as a body of general moral rules that one can know with certainty; "Kantian universality is no longer possible." We must therefore rid ourselves of faith in the existence of "immutable and eternal verities" other than the paramount one that "no maxim can be valid at all times and places for all rational beings" (*Lectures*, 2:279, 283). The same radical uncertainty principle that rules the cognitive world rules the moral—and by extension the political—worlds as well.

From this doctrine follows directly Clifford's Feuerbachian polemic against theistic religion, both for its authoritarian dogmatism and for the predilection for violence and cruelty that is, he suggests, ingrained indelibly within it. If one allows room for preserving any trace of religious superstition, however seemingly innocuous, then, he says, "room may also be found for the goddess Kali, with her obscene rites and human sacrifices." Philosophically sophisticated science must be committed to the rigorous, uncompromising disavowal of certainty, and for this reason it is necessarily antithetical to all systems of despotic authority and to their habit of pronouncing "sleepless vengeance of fire upon them that have not seen and have not believed." Christianity in its attenuated modern guise may seem harmless, says Clifford, but it represents "the slender remnant of a system which has made its red mark on history, and still lives to threaten mankind"; it is "the seed of that awful plague which has destroyed two civilisations" and only awaits an opportune moment to break out in a new epidemic (*Lectures*, 1:298–300). In a fierce essay of 1877 titled "The

Ethics of Religion," he specifically identifies the dangerous evils of Christian religiosity with its compulsion "to declare with infallible authority what is right and what is wrong" and to set up "an infallible authority" over the moral life of the community. This principle of totalitarian control based on the ideal of infallibility—the point of juncture of religion and orthodox science—leads naturally and inevitably, Clifford says, to the widespread worship of "criminal deities," to theologies of vicarious sacrifice, and to the sadistic vindictiveness exhibited in one representative devotional work called *A Glimpse of Hell*, from which he cites imagery of a child writhing for eternity upon burning coals in punishment for the crime of not being born Christian (Clifford, *Lectures*, 2:218–20). In the same essay, shortly after a passage in praise of the Jews, "this heroic people," though without any reference to the outbreak at precisely this moment of rabid anti-Semitic agitation in Germany, Clifford issues a warning of the possibility of a revival of superstitious despotism in modern times. "It seems to me quite possible that the moral and intellectual culture of Europe, the light and the right . . . may be clean swept away by a revival of superstition," he declares; the moment seems rife, he says, with danger of a huge catastrophe produced by "the strength of a past civilisation perverted to the service of evil" (*Lectures*, 2:229, 233–34). In such an eventuality, he says in a passage I have quoted already from another piece, "the wreck of civilised Europe would be darker than the darkest of past ages" (*Lectures*, 2:256).

Frazer's own apocalyptic prophecies echo Clifford's unmistakably, as does the passage at the end of *The Golden Bough* where he evokes the Middle Ages, the age of triumphant Christian belief, in Cliffordian terms as "a dark crimson stain," the color of blood, in the fabric of history (*GB*, 827). These powerful concordances help us to read between the lines of Frazer's text Clifford's clearly stated connection between, on the one hand, denying on principle the possibility of certainty in scientific analysis and, on the other, resistance to ideologies of authoritarian violence. To affirm the possibility of discovering conclusive truths and "of reducing the vast, nay inconceivable complexity of nature and history to a delusive appearance of theoretical simplicity" (Frazer, *Magic Art*, x) is willy-nilly to reinforce the ideology of purificatory violence, whether directed at unorthodox ideas or at supposed witches and demons in our midst: some such profoundly tendentious Cliffordian/Spencerian theme underlies Frazer's allegorical tableau of himself as a scientific investigator struggling hopelessly, like Sisyphus or the daughters of Danaus, to decipher cultural symbolism by means of the comparative method. This ever-inconclusive project may seem a woeful fate for a scholar, and Frazer's tone as he evokes it is decidedly pessimistic. Yet compared, say, to doing obeisance to the Napoleon cult that

Spencer identifies as the defining political aberration of modernity and that Nietzsche ardently endorses, glorifying Napoleon as a "combination of the monster and the superman" (*Will to Power*, 2:405; see 2:376, 397, etc.), or compared to participating in the kinds of organized bloodletting that inevitably accompany, according to *The Golden Bough*, the worship of "men who have masqueraded as gods," the image of the disenchanted modern-style scientific investigator as Sisyphus takes on an unexpectedly idealistic and heroic aspect.

The freeform, inconclusive anthropological text with its self-exposed inconsistencies and theoretical vacillations in plain view, taking as its organizational principle that of "a series of separate dissertations loosely linked together by a slender thread of connexion with [the] original subject" (*Magic Art*, vii), becomes by the same token the embodiment of an ethic of uninhibited intellectual freedom, improvisation, and creativity, one emancipated from the false scientific ideal of rigid logical order. It expresses the overriding principle of libertarian philosophy from Mill to, say, Paul Feyerabend: in Schiller's words, that "it is vain to prohibit the play of mental activity with the given, and unwise to restrict its freedom," and that an enlightened science will make itself known by its readiness to acknowledge "its *tentativeness* and *lack of certainty*, and readiness to *run the risk* of its fictions, postulates, and hypotheses" (Schiller, *Formal Logic*, 341, 399).

In a 1911 preface to *Taboo and the Perils of the Soul* (part 2 of the third edition of *The Golden Bough*), Frazer echoes Clifford nearly verbatim in explicitly invoking, for once, the moral implications of the "comparative method," presenting it as the sovereign corrective to the prejudice that our own values are permanently true and applicable in a categorical way to all human beings. "The old view that the principles of right and wrong are immutable and eternal is no longer tenable," he boldly declares. "The moral world is as little exempt as the physical world [and, he might have added, the intellectual world] from the law of ceaseless change, of perpetual flux" (*Taboo*, vi). Premising the scientific study of cultural values on the principles that no absolute or objective standard exists and that modes of human behavior can only be assessed relatively to one another and to historical circumstances does not come free of cost: it may lead to unresolvable indeterminacy, to "theoretical confusion," and even to something like textual anarchy. But following this difficult path is the only remedy, says Frazer, for the blighting vices of "narrowness and illiberality" (*Taboo*, viii) and for the sanguinary consequences of seeing the world through a lens "dimmed by thick mists of passion and prejudice" (Frazer, *Balder*, vi).

Einstein: Relativity and Moral Renovation

The comparatism of *The Golden Bough* and the relativity physics of "On the Electrodynamics of Moving Objects" [hereafter EMB] (1905) are both driven by the intention not just of clarifying a set of technical problems but of reconstructing basic principles of scientific rationality, and along similar lines. The reconstruction is not total after all, for it retains as its central trope the pledge of allegiance to the objectivity of factual data (though as we have seen, this pledge is only hypothetical in the case of Frazerian anthropology, which is obliged to make up its data by means of analogy and conjecture as it goes along) and to rigorously value-free logical analysis. But neither of these new sciences will be fully intelligible in historical terms if they are divorced from their powerful and convergent motives of moral renovation.

I have cited before, as a point of reference for the contrary argument developed in this book, Arnold Sommerfeld's assertion that special relativity "has, of course, absolutely nothing whatsoever to do with ethical relativism" (Sommerfeld, "Einstein's Birthday," 99). The translatability of scientific language out of and into other languages must be rigidly contained at this point, according to Sommerfeld and other commentators; otherwise, the mandate of modern science would evidently be placed in jeopardy, not to mention the danger that would be incurred by the fabric of society were ethical relativism, that perverse thing, to be allowed to shelter beneath the prestigious mantle of science. A less dogmatic and less tendentious approach to the ethics of relativity is suggested by Bertrand Russell's statement (in specific reference to the abolition of the concept of "force" in general relativity) that "if people were to learn to conceive the world in the new way . . . it would alter not only their physical imagination, but probably also their morals and politics"—a "quite illogical" effect but one none the less likely for all that (*ABC of Relativity*, 195–96). *Contra* Sommerfeld, physics, morality, and politics are all bound up together in the relativity hypothesis, according to Russell.

Indications of the illicit linkage between Einstein's physical theorizing and (using the vexed term perhaps a little loosely) "ethical relativism" are in fact quite clear. They stand out with special clarity in his 1933 volume *The World as I See It (Mein Weltbild)*, a mixed collection of scientific and extrascientific essays that strongly implies correlations among the different departments of his thinking. In this collection, explication and interpretation of relativity theory and commentary on the history of physics are intertwined with writing about Judaism, about German politics in the terrible year 1933, and about the relations of fascism and science, as though to demonstrate at length that no definite lines of demarcation can properly be drawn among these different aspects of a single complexly integrated and historically situated *Weltbild*. In particular, *The World as I See It* invokes a crucial topos that we have encountered in a series of different iterations: it sets the development of relativity theory against the backdrop of an imagined impending European calamity, just as the deployment of the comparative method in *The Golden Bough* is set by Frazer, or as modern probability theorizing was earlier set by Clifford.

Thanks to the ascendancy of nationalism and militarism, Europe in 1933 is in imminent danger of sinking, Einstein declares, into "a state of universal anarchy and terror"; "slavery for the individual and the annihilation of our civilisation threaten us," he proclaims (*World*, 207, 204). The source of the terrible danger is of course the brutal authoritarianism and the sinister racial ideology that have appeared in Europe with the accession of "intolerant, narrow-minded and violent people" (*World*, 167) to power in Germany. It is the love of justice and of personal independence natural to Jews that renders this people, Einstein says, necessarily anathema to Hitlerism. "Those who are raging today against the ideals of reason and individual liberty and are trying to establish a spiritless state-slavery by brute force rightly see in us their irreconcilable foes," he declares (*World*, 143). There is perhaps a sense in which the Nazi abhorrence of relativity as "Jewish physics" was for Einstein himself not entirely devoid of meaning.[27] Of course, one may question the justification of reading such statements about the rise of National Socialism back into the turn-of-the-century period when Einstein was first cogitating the problem of the ether and all that it implied for physics, and of associating special relativity in this way with the imagery of European annihilation that marks the development of relativistic themes in different fields of late-Victorian scholarship and identifies this movement with what its authors thought of as a desperate historical struggle against, in Schiller's phrase, "the intolerance of Absolutism" (*Studies*, xiii). This degree of anachronism need not, I think, trouble anyone who assumes that the themes of an author's imagination are largely continuous and can be read back and forth in his or her work fairly freely. For a young man of Einstein's disposition, almost fanatical, by his own account, in his hatred of abusive authority, who came of age in the day of Bismarck and of the Dreyfus affair, the sense that the world of civilized order was menaced by authoritarian violence could hardly have been a new development in 1933. From an interpretive point of view, in any case, not much would be at stake in deciding whether Einstein was conscious of certain moral and political implications of his theory from the start or only became so later on—as *The World as I See It* implies, at a minimum, he had by 1933.[28]

In the essays on moral, political, and historical topics in this volume, Einstein's *idée fixe* is the principle of the imperative need in human society for the acceptance of intractably different points of view, sets of values, ways of life. The key term in the moral lexicon of *The World as I See It* is thus the somewhat rueful term *tolerance*, which, like Keats's term "negative capability," implies both the impossibility of dispelling contradictions and the necessity of learning not to need to dispel them. "Nationalities want to pursue their own path, not to blend," Einstein writes, for example. "A satisfactory state of affairs can only be brought about by mutual toleration and respect" (*World,* 156). According to the paradigm of "toleration," one holds to one's own frame of reference while at the same time—paradoxically, from the point of view of what Schiller calls "Formal Logic"—recognizing the validity of frames incompatible with it, and refusing to condone attempts to make one's own frame prevail over others. One embraces this form of ethical relativism (a term Einstein does not use), first, from an awareness that nothing is one thing just by itself, and that human groups in a profound sense cannot exist without other groups. One also takes this logically hazardous position, he makes clear, from fear of physical violence. No influence is more dangerous in politics than the ingrained "intolerance of national majorities," writes Einstein. "Against that intolerance we shall never be safe" (*World,* 168).

This fundamental principle of the tolerance of incompatible points of view is for Einstein essentially identical to the code of freedom he professes in *The World as I See It* in the same radical terms invoked by Spencer, Mill, Stallo, Pearson, Schiller, and other writers in the Victorian relativistic tradition. "Political liberty," he says, for example, "implies liberty to express one's political opinions orally and in writing, and a tolerant respect for any and every individual opinion" (*World,* 173). He consistently professes this creed in which "tolerance and freedom of thought" form the central requirements and the irreplaceable means of "liberating the individual from degrading oppression" (*World,* 244, 159). From such principles, he derives his cardinal doctrine in the field of international relations, that of the evil consequences flowing from the doctrine of "the unlimited sovereignty of the individual country" (206)—that is, from the habit of imagining a country as an absolute, self-defining entity, a "thing in itself," rather than as constituted by (and as responsible to) its reciprocal relations with other entities. This rule applies in particular to the Jewish state in Palestine, which as its overriding task, says Einstein, must "solve the problem of living side by side with our brother the Arab in an open, generous and worthy manner" (150; see also 147–48, 162). The goal of Israel must not be one of a state in which one group exercises potentially unlimited dominance over others, but rather, that of "a stable community . . . built up out of groups of different nationality" (148). The code of the supreme importance of ethical relativity and of egalitarian coexistence is further identical for Einstein with his intense antagonism toward military establishments and compulsory military service, itself based on his abhorrence of "brute force, so unworthy of our civilisation"; "force always attracts men of low morality," he declares (203, 207–8, 204, 240).

The establishment of the regime of toleration may not be Einstein's ultimate ideal, after all. Just as he identifies scientific theory-making with striving to gain a view of the miraculous "harmony" immanent in nature, so he describes politics as ideally attaining a state beyond the mere mutual toleration of irreconcilable adversaries. This state he describes as a "higher community of feeling, now thrust into the background by national egotism" and as a social order representative of "the ideal solidarity of all living things" (*World*, 228, 146). This is the conception of all-embracing reciprocity, of all beings as "mutually dependent, and in that sense integrated" (Spencer, *First Principles*, 253), that has underwritten modern relativity speculation from the outset. In invoking it as the antithesis of the Hitlerian regime of absolutism and purificatory violence, Einstein conceives it not just as the basic *donnée* of philosophical science, but as a supreme moral and political ideal. The paradox, if it is one, is that this condition of inescapable interdependency is defined by Einstein as equivalent to freedom.

The ideal of freedom is expressed in even more vividly paradoxical terms in Einstein's philosophy of scientific inquiry, which, in conjunction with special relativity theory itself, gives culminating expression to the two long-connected, in fact identical, postulates of nineteenth-century relativity literature (and, as we saw, of Frazerian anthropology in particular): that of the symbolic character of reality and that of the unattainability of direct access to the external world.

Diverging from his youthful infatuation with the radical empiricism of Ernst Mach, Einstein ever more forcefully proclaimed his belief in what he spoke of as "the gulf—logically unbridgeable—which separates the world of sensory experiences from the world of concepts and propositions" (Einstein, "Remarks on Bertrand Russell," 287).[29] Einstein draws from this principle of estrangement more drastic implications for scientific practice than Spencer (who finally just finessed it, as we saw) ever did. If the world of sense experience and that of concepts are in fact sundered from each other by an unbridgeable gulf, then *theories cannot possibly be derived rationally from the analysis of empirical data*. This was exactly Einstein's conclusion. "There is no logical path" leading to the laws of physics, he states unequivocally. "Concepts have reference to sensible experience, but they are never . . . deducible from [it]" (*World*, 22, 84). The connection of concepts and propositions with observed fact "is purely intuitive, not itself of a logical nature" ("Autobiographical Notes," 13). Obviously such declarations fly in the face of the orthodox ideology of science, and present scientific rationality under the aspect of an almost unfathomable riddle. If logic is wholly incapacitated as a means of inquiry into the world of nature, how is any systematic knowledge of this world possible?

Instead of offering a solution to this riddle, Einstein offers a historical analysis according to which the achievements of recent physical theory derive precisely from a breakdown of the regime of logical explanation in science—that is, of causal explanation based on classical mechanical models.[30] He particularly identifies this profound change in scientific thinking with Faraday's and Maxwell's creation of the concept of the electromagnetic field as the "ultimate entity" of physics. The concept of the ether, elaborated in the nineteenth century in order to salvage the principle of mechanical explanation in optics, never proved viable, as Einstein explains. Thus from a science of material points there emerged a wholly reconstructed modern-style science of "continuous fields, not mechanically explicable" (*World*, 64, 65). In the emergence of Frazerian evolutionary anthropology, a similar development could be traced: the principle of racial determinism and the vogue of craniometry were displaced in the study of "primitive man" by the conception of transhistorical cultural and cognitive fields enabling one to posit as continuous social phenomena from societies remote from one another in both space and time, among which no material linkage and no system of "explanation" could possibly be constructed.

What distinguished Einstein's physics from Lorentz's and FitzGerald's theories of the contraction of moving objects, which used the same equations, was precisely Einstein's incomprehensible-seeming disavowal of mechanical explanation (that is, of explanation based on the supposed effects on the physical structure of objects by movement through the ether). For Einstein, the relativity effect was purely a kinematic one having solely to do with the relations of space and time; these relations were representable mathematically but they were "not mechanically explicable." What the appearance of this new paradigm signified for contemporaries was precisely that sacrifice of the natural thing to the symbol that Spencer proclaimed in the mid-nineteenth century and that powerfully determined the methodology of Frazerian comparatist anthropology. As Niels Bohr put it, relativity and quantum theory both proceed by means of "the renunciation of the absolute significance of conventional physical attributes of objects" and by substituting for these attributes "not-directly-visualizable symbolism" (Bohr, "Discussion," 238). In the wake of Maxwell's innovations, says Einstein similarly, "the equations alone appeared as the essential thing" in physics (*World*, 64), which increasingly moved in a world of abstract symbolism containing "no material actors" (Einstein and Infeld, *Evolution*, 152). The distinctive property of Einsteinian physics, thus says Cassirer, is its imagining of reality "as a system, not of existing things or properties, but of abstract intellectual symbols, which serve to express certain relations of magnitude and measure" (*Einstein's Theory*, 357): henceforth, even the most exhaustive scientific knowledge "never offers us the objects themselves, but only *signs* of them and their reciprocal relations" (Cassirer, *Substance and Function*, 303). To the extent that the new world of scientific symbolism can be brought into relation with empirical data, it is only by means of an arduous and extended process of mathematical transformation. In the physics deriving from Einstein, says Philipp G. Frank, "the general principles have been formulated by using words or symbols which are connected with observational concepts by long chains of mathematical and logical argument" (Frank, "Einstein, Mach, and Logical Positivism," 274). In modern theoretical science, says Einstein similarly, "the hypotheses with which it starts become steadily more abstract and remote from experience. . . . Meanwhile the train of thought leading from the axioms to the empirical facts or verifiable consequences gets steadily longer and more subtle" (Einstein, *World*, 91). Such statements are hardly more suitable to relativity physics, we may note, than they are to the analytical method of that other great document of late-Victorian symbolist science, *The Golden Bough*, with its long, impossibly complicated, "labyrinthine" chains of cultural analogies linking theories (the theory of imitative magic, say) to the phenomena (the forbidden tree at Nemi, say) they are supposed to explain. Such are the inevitable stylistics of scientific literature once the dissolution of things into symbolically expressible relations has occurred.

For Frazer, the inability of scientific reason operating under such a regime to attain anything other than "more or less plausible" theories was ultimately a cause of anguish, as we saw. For Einstein, too, the development of a science of symbolism evidently originated in his conviction of the insurmountable estrangement of things themselves from logical analysis. Special relativity comes into being, we may say, in what Susan Stewart calls "the gap between signifier and signified that is the place of generation for the symbolic" (*On Longing*, ix). But for Einstein, the gulf between conception and empirical reality defines, above all, the space in which human intellect, imagination, and intuition exercise their amazing creative power. Scientific theories, he thus repeatedly declared, are in fact "free inventions of the human mind" and have fundamentally what he called a "purely fictitious character" (*World*, 35, 34). They involve, according to Einstein, "a free play with symbols according to . . . arbitrarily given rules of the game" ("Remarks on Bertrand Russell," 289). "All our thinking," he asserts, "is of this nature of a free play with concepts"; "all concepts, even those which are closest to experience, are . . . freely chosen conventions" ("Autobiographical Notes," 7, 13). "The axiomatic basis of theoretical physics cannot be extracted from experience but must be freely invented"; the scientific theorist thus "should be encouraged to give free rein to his fancy, for there is no other way to the goal" (*World*, 36, 92). Scientific theories, in other words, are "free creations of the human mind" (Einstein and Infeld, *Evolution of Physics*, 33), imaginary productions pure and simple. We are bedeviled by a "prejudice," says Einstein, summing up this line of his mature thinking,

> that facts by themselves can and should yield scientific knowledge without free conceptual construction. Such a misconception is possible only because one does not easily become aware of the free choice of such concepts, which, through verification and long usage, appear to be immediately connected with the empirical material. ("Autobiographical Notes," 49)

Lying in the unconscious background of this statement, no doubt, is Nietzsche's declaration that "what passes for truth in every age" is nothing in reality but a "mobile army of metaphors, metonyms, and anthropomorphisms: in short a sum of human relations which . . . after long usage seem to a nation fixed, canonic and binding" (Nietzsche, "On Truth and Falsity," 180). Scientific genius for Albert Einstein consists, at all events, in precisely the mode of playful free imaginativeness that can liberate us from the mystified regime of "deep-rooted prejudices" parading as "truth" (Einstein and Infeld, *Evolution*, 196; see also 187).[31]

That this philosophy of science is permeated with moral and political implications is too clear to need much comment. In its stress on "the essentially constructive and speculative nature of . . . scientific thought" (Einstein, "Autobiographical Notes," 21), it sharply divorces itself from "purificatory" conceptions of scientific reason; but most of all, it evinces by the same token the same exorbitant, almost subversive commitment to uninhibited individual freedom that marks the relativity tradition from Spencer and Mill onward. It is continuous with Einstein's own affirmation of "the free development of the individual" as the highest social value, with the politics of liberty and toleration he advocates in *The World as I*

See It, and with his devotion in his private life to "an almost fanatical . . . desire for personal independence" (*World*, 249, 143). Like F. C. S. Schiller (*Formal Logic*, 399–400), he conceives scientific research as the realm of human activity in which the ideal of unimpaired freedom and imaginative creativity is most fully realized. Yet his "epistemological credo" ("Autobiographical Notes," 11) is obviously a precarious one, and even suggests that intellectual freedom can flourish only in circumstances where knowledge is fraught with uncertainty. Resting as it does on the tenet of the radical inaccessibility of the objects of study (the workings of physical nature, separated from logical thought by an unbridgeable gulf) in much the same way that Frazer's affirmation of the free speculativeness of the comparative method rests on the presumption of the permanent inaccessibility of his own objects (the Arician priesthood and the primitive mind), Einstein's idea of science leads to theoretical considerations as vertiginous as Frazer's.

Einstein was keenly aware of the logical incoherence of his attempt to account for scientific knowledge, though he expressly did not draw from it Frazer's view of his own research as heroically futile Sisyphean labor, a hopeless striving to gain a view of "what can never be known." "I still believe," he states with evident defensiveness, "in the possibility of a model of reality—that is to say, of a theory which represents things themselves" (*World*, 39). But he does stress the paradoxical and precarious nature of any such conception of science. For one thing, since we can obtain knowledge of nature only obliquely, "by speculative means," he says, sounding very much like Frazer, "it follows . . . that our notions of physical reality can never be final" (*World*, 60). Nor, to put the same point more strongly, is it easy to see how scientific thinking could ever accord with experience if it is true "that the universe of ideas cannot be deduced from experience by logical means, but is, in a sense, a creation of the human mind" (Einstein, *Meaning of Relativity*, 2). One could only anticipate in such a case an utter disconnect between scientific ideas and outward realities (as Spencer taught) or an uncontrollable proliferation of different theoretical constructs, each as valid as the next (as Poincaré taught). "Theoretically," this latter condition should indeed obtain, says Einstein. "But evolution has shown that at any given moment, out of all conceivable constructions, a single one has always proved itself absolutely superior to all the rest. Nobody who has really gone deeply into the matter will deny that in practice the world of phenomena uniquely determines the theoretical system, in spite of the fact that there is no logical bridge between phenomena and their theoretical principles" (*World*, 22–23). This head-spinning statement makes clear, at least, that scientific theories are not "absolutely" or "uniquely" governed by the world of phenomena after all, since it pointedly stresses that the superiority of the chosen theory is relative to the particular moment of the evolution of science—not to mention its further stress on the intractable paradox of phenomena somehow governing theories in the absence of logical bridges connecting them. In other places Einstein expresses much less confidence that a single theoretical system is bound in practice to seem at a given moment "absolutely superior." Conflicting theoretical narratives may in fact, he says, each correspond to all available data with sufficient coherence as to render the choice between them undecidable. He illustrates this point by citing the supposed impossibility of deciding experimentally between the theories of evolution by natural selection and by the inheritance of acquired characteristics (*Meaning of Relativity*, 124).[32]

That the world could be scientifically comprehensible is finally for Einstein an unfathomable mystery before which analytical language falls mute. He thus describes his "conviction . . . of the rationality or intelligibility of the world" as a state "akin to religious feeling," which is to say, a belief unjustifiable by the norms of reason (*World*, 29; see also 139). Hence his assertion that "the fact that the totality of our sense experiences is such that by means of thinking . . . it can be put in order . . . is one which leaves us in awe, but which we shall never understand. One may say 'the eternal mystery of the world is its comprehensibility'. . . . The fact that [experience] is comprehensible is a miracle" ("Physics and Reality," 351). Like Newman, Einstein can explain genuine knowledge in the last analysis only by supernatural intervention. His account of the intellectual work of science turns out, after all, to be almost identical to Frazer's vision of himself as Sisyphus, striving at a hopeless task—but one now rendered successful by "miracle."

Of course, in Einstein's writing, for all the portrayal of physical science as a "free play with symbols," homage is often paid to the necessity of grounding scientific theory in correspondence with empirical fact. But this correspondence is bound to be an ambiguous ideal in a relativistic intellectual environment, where facts are never fully separable from interpretation. Given Einstein's doctrine that concepts in science "can not be deduced by means of a logical process from the empirically given" ("Reply," 678), it is not surprising that his attitude toward the sanctity of the principle of experimental verification was often strikingly cavalier compared to the decisive weight he always attached to such stylistic criteria as the "logical simplicity" and "inner perfection" ("Autobiographical Notes," 23)—the intellectual beauty—of scientific theories. Elie Zahar questions how seriously he in fact took the strong falsificationist statements he sometimes uttered (Zahar, *Einstein's Revolution*, 88). The apparently conclusive experimental disproof of special relativity by Walter Kaufmann in 1906 was simply dismissed as uninteresting by the young scientific aesthete Einstein, and the results of D. C. Miller in the 1920s, similarly incompatible with relativity theory, were likewise a matter of indifference to him (Clark, *Einstein*, 143–44, 473; see also Feyerabend, *Against Method*, 40–41). Had empirical observation failed to confirm the general theory, he declared to his student Ilse Rosenthal-Schneider, "Then I would have been sorry for the dear Lord—the theory *is* correct" (Clark, *Einstein*, 287). In this as in other more overt respects, the advent of relativity entailed notable slippage in the code of rigidly determinate rationality supposed to prevail in scientific research.

How might Einstein's ethic of tolerance and of freedom be said to express itself in the equations of special relativity? The question, to which, in a sense, the entire inquiry of the present book leads, may be a fruitless one. That special relativity "can not be deduced by means of a logical process" from a set of ethical principles, or vice versa, can be taken for granted, and for some this is all that need be said on the subject. Certainly any further pursuit of it will come to naught if one accepts unqualifiedly the determination of many scholars (some of whom are discussed in the introduction) to banish relativity, in effect, from Einsteinian relativity theory and to define it categorically as a theory of invariant natural "absolutes." We get a very different impression of special relativity, however, if we consider "The Electrodynamics of Moving Objects" independently of Hermann Minkowski's subsequent mathematical elaborations of it and of Einstein's subsequent endorsement of Minkowski's interpretation.[33] The "real world" of post-Minkowskian special relativity, the world of "rigid and absolute" physical values (Einstein, *World*, 91) where the relativity effect is abolished, lies in the four-dimensional continuum of space-time, a place with no observable location. Minkowski showed that highly sophisticated mathematical operations incorporating both temporal and spatial values yield for any event a physical "interval" which is uniform and in this sense "absolute" for all observers under the special conditions covered by the theory. But in the empirical world, the world of human experience, of observable time and space, all is relativity, according to Einstein's theory: every physical occurrence is "infinitely variable and infinitely ambiguous," allowing "no possibility of an exact objective determination of the state of physical reality" (Cassirer, *Einstein's Theory*, 363). Rigid rods stretch and contract, clocks race or slow down, events occur in different sequential order, depending on the relative motion of observers. In 1905, no Minkowski continuum is invoked to cushion the tremendous philosophical impact of the discovery of such effects or of what they implied, the proposition that the characteristics of physical entities and events—time, space, motion, simultaneity—could no longer be considered fixed intrinsic values but, rather, could be accurately determined only in relation to an observer's point of view.[34]

It is perfectly true that special relativity does not accord any role whatever to the subjectivity or free agency of the observer. Relativistic variations of time and space are strictly calculable from one reference frame to another under the specified conditions of the theory (which is valid only for bodies in unaccelerated translational motion in a straight line). There is therefore nothing indeterminate, capricious, or incalculable about the relativity effect, according to the special theory. There seems *à plus forte raison* to be no room at all in such a theory for the intervention of moral values. But observed magnitudes are wholly relativized by Einstein nonetheless in a way that transforms the intellectual universe—which is to say, the universe itself. To speak of such magnitudes independently of how they are measured from a particular point of view is henceforth to speak unscientific nonsense. The consequences Einstein draws from this principle are astounding. "Electric and magnetic forces," he declares, for example, "do not have an existence independent of the state of motion of the coordinate system" (EMB, 159). *Physical objects and events have no inherent properties capable of being perceived:* that special relativity produces only undetectably small observational differences for ordinary phenomena does not render this proposition less revolutionary—even though it had often been enunciated in the half century preceding Einstein's statement of it.

Einstein's accomplishment was therefore to give definite mathematical form, within rigorously specified physical conditions, to the prevalent themes of several generations of nineteenth-century theorists who, declaring the nonexistence of absolute time, space, and motion, had labored to abolish, in Karl Pearson's words, "all preconceptions of the absolute" (Pearson, *Ethic*, 428) from the field of scientific knowledge. Einstein's postulation of the absolute value of the speed of light—the principle, as I have stressed, "on which the special theory of relativity rests" (Einstein, *World*, 77)—served in effect as the device enabling him to construct a mathematical physics that would accord with this increasingly dominant trend of advanced nineteenth-century thought. He did not, that is, postulate for some good empirical reason the (wildly implausible, almost incomprehensible) principle of the uniform speed of light in all inertial reference frames and then discover as a logical consequence that, lo and behold, all other physical properties were thus relativized, just as Sir William Hamilton, Mansel, Spencer, Stallo, Helmholtz, Mach, Pearson, Poincaré, and other avant-garde Victorian writers had declared. It was just the other way around—of course. "The explanation given to any relation can survive and develop within a given society only if this explanation is stylized in conformity with the prevailing thought style," said Ludwik Fleck, for "in science, just as in art and in life, only that which is true to culture is true to nature" (Fleck, *Genesis and Development*, 2, 35). The extended episode of intellectual history we have studied exemplifies Fleck's postulate with unmistakable clarity and casts grave doubt on any portrayal of Einstein's theory as "a Minerva-like creation." For physics to appear valid in the light of contemporary speculation, *relativity had to be a given*. Einstein's challenge over the ten years leading up to 1905 was to find the logical mechanism, or to imagine the principle of nature, that would entail relativity as a result (as it was Darwin's challenge to imagine the mechanism that would entail the foregone conclusion of evolutionary change). This philosopher's stone, hidden in plain sight in Maxwell's equations of the electromagnetic field, was the invariant speed of light, which in special relativity is emancipated from the concept of the ether and becomes the fundamental law of the physical universe.

"The theory of relativity is often criticized for giving, without justification, a central theoretical rôle to the propagation of light, in that it founds the concept of time upon the law of the propagation of light," Einstein says (*Meaning of Relativity*, 28). No response to this criticism will be complete that fails to recognize the way the deification of light in special relativity gives to Einstein's theory the character of an invocation of a world of moral and religious significance—not that this significance could in any way be directly inferable from "Electrodynamics of Moving Bodies." In *Fire within the Eye*, David Park recounts the discursive tradition, flowing powerfully from the Bible and from Plato, in which "light" functions as the paramount metaphor of the ultimate values of Western civilization, standing for truth, for the Good, for justice, for the presence of God. In imagining the propagation of light as the process that transcends and sets aside the mechanistic interpretation of reality and as the seat of the sole "absolute" value of the physical universe, Einstein fuses modernistic physical theory with this ancient discursive tradition. If modern physics is indeed "a theory of symbols," and if all scientific theories are the result of "free conceptual construction," as we have Einstein's word for it that they are ("Autobiographical Notes," 49), we can in any case hardly avoid taking this possibility seriously into account.

If moral and religious themes can indeed be derived from Einsteinian theory, they are at the antipodes of those that Feuerbach's, Clifford's, and Frazer's analyses identify with the Christian tradition. As I have emphasized, special relativity in its pre-Minkowskian form of 1905 did not conceptualize a space-time manifold that could serve as a basis for "absolute" physical measurements. Rather, it claimed to demonstrate that differently situated scientific observers would necessarily interpret a given physical event differently; and, with that revolutionary thesis once set forth, it sought to provide an instrument for translating the mathematical symbolism proper to one physical frame of reference into that proper to another (this instrument being, as it transpired, the set of equations known already as the "Lorentz transformations"). In other words, Einstein's physics claims to determine the interpretation that will be made of a given event in a given foreign context, just as Frazer's anthropology, where the frames in question are human cultures rather than "Galilean coordinate systems," claims with its own transformation equations (the one equating ceremonial blood symbolism with original practices of human sacrifice, for example) to recover the interpretation given to particular cultural phenomena from "the native's point of view." In both sciences, the defining operation is the systematic translation of symbolic materials from one frame of reference into others. We may note in passing that Frazerian analysis is built, just as much as Einsteinian analysis is, on a doctrine of time as not uniform and absolute but as radically differential. Time as it is manifested in Frazer's pre-Einsteinian scheme passes much more quickly in some geographical areas (Europe, for example) than in others, where it may seem, culturally speaking, virtually to stand still (Tierra del Fuego, for example), just as Einstein speculates that a clock at the Earth's equator will run slower than an identical one at one of the poles (EMB, 153). In other words, references to time in *The Golden Bough* are always radically spatialized, and spatial locations are radically temporalized, for the purposes of the comparative method: in Frazerian analysis, as in special relativity, the all-important proposition is that "time is robbed of its independence" (Einstein, *Relativity*, 56).

Implicit in Einstein's wonderful demonstration of the relativity of space and time, and repeatedly made explicit by him and by other commentators in later texts, was the principle that a rigorously scientific physics, one emancipated from the influence of "deep-rooted prejudices," needed to take as its guiding rule the suppression of all conscious and unconscious preference in favor of one's own frame of reference—the demanding principle put forward earlier by Spencer, then echoed by Frazer and Malinowski, as the precondition of the authentically scientific study of human societies. Special relativity thus effected its revolution by proclaiming the gospel of what Einstein (*World*, 69) strikingly calls "the equal legitimacy of all inertial systems." According to special relativity, "there is no such thing," he declares elsewhere, "as a 'specially favoured' (unique) co-ordinate system" that affords a view of things scientifically truer than any other (Einstein, *Relativity,* 53). We are almost irresistibly prone to imagine our own frame of reference to be the sole genuine one, says Eddington similarly, "but this egocentric outlook should now be abandoned, and all frames treated as on the same footing" (*Nature of the Physical World,* 61; see also 15, 16, 21, 113). As far as the scientific measurement of space is concerned, neither the dweller on the sun nor a flea on earth can be said to have the more valid view, says Bertrand Russell, in the endlessly repeated phraseology of relativity literature: "each is equally justified" (Russell, *ABC of Relativity,* 53). This axiom of interpretation forms the fundamental motto of relativity physics; and with its insistent language of legitimacy, justification, and anti-egocentrism, it stamps a motive of radical ethical reform—one closely aligned with the ethic of "tolerance" preached in *The World as I See It*—very plainly and unmistakably upon the Einsteinian reconstruction of physical science. Ethical and intellectual reform exactly coincide.

Various statements by Einstein confirm the guess that the development of special (and later, general) relativity was governed from the start by the philosophical imperative to expose the prejudice in favor of privileged, uniquely authoritative points of view, to demonstrate the non-innocuous character of this prejudice, and to teach scientists how to live in an intellectual world cleansed of it for the first time.

The pressing issue for theoretical physicists in the 1890s concerned the status of the "luminiferous ether," that mysterious medium conceived in order to make possible a mechanical model of the undulatory propagation of light and then instituted in Victorian scientific thought as the supreme reference frame of physical nature, the Body Alpha (see Whittaker, *History,* 28)—despite the vexatious impossibility of detecting it experimentally. "To start with," Einstein reminisces in *The World as I See It,* "it disturbed me that electrodynamics should pick out *one* state of motion [i.e., that of the ether] in preference to others, without any experimental justification for this preferential treatment. Thus arose the special theory of relativity" (*World,* 138; see also 127–28). Relativity in this lapidary statement is simply the antithesis of a policy of "preferential treatment"—simply the intellectual outcome, we might say, of mapping an overwhelming moral and political imperative directly onto the natural world. It is not that giving unjustified preference is logically inadmissible (there is nothing illogical about the ether or about Lorentz's theory of the contraction of objects moving through it) but that it is repugnant to a certain moral sensibility in the same way that awarding wealth and privilege to favored individuals on the basis of a system of social caste is repugnant. On what justification does the "ether frame" claim to be scientifically better than other frames and deserving of special deference? A frame's a frame for a' that. Such statements of Einstein's as the one quoted above lend credence to Elie Zahar's claim that Einsteinian relativity originates nowhere else but in "the metaphysical thesis that there exist no privileged frames of reference" (*Einstein's Revolution*, 269; see also 285)—no frames, to put it in Frazerian terms, masquerading as gods. In special relativity, the privileged frame whose position of authority was summarily abolished was that of the ether; general relativity was motivated by the further need to undo the privileged status that in special relativity is granted to "inertial" frames.[35]

To make this argument is not to claim that there exists an automatic equivalence between, on the one hand, conceiving the world in symbolic and relativistic terms and, on the other, anti-authoritarian politics. Nor is it to claim that Frazer and Einstein relate to the militant nineteenth-century Protagorean tradition in just the same way, or that their relationships to it are any more unequivocal than are those of the other figures discussed in this book. Neither Frazer nor Einstein would portray himself as a doctrinaire philosophical relativist as authors like Stallo, Mach, or Schiller would. Each constructs his scientific universe around "absolutes": Frazer at least gives lip service to the presumption of cultural evolution toward the discovery of scientific "truth" and has even been misread as offering support for the kinds of imperialistic usurpations that Spencer's Anti-Aggression League sought to outlaw; Einstein, for his part, deifies the speed of light and increasingly aligned himself intellectually with deterministic scientific absolutism even as his pacifism and his anti-authoritarianism became more militant in the interwar period. Both Frazer and Einstein betray reactionary strands of thinking that set them at odds with subsequent trends of theory in their respective disciplines, in other words. None of this overturns the proposition that it is in their interlocking bodies of work that the nineteenth-century relativity movement, its potential of ideological critique undissipated, comes to some of its most impressive attainments.

AFTERWORD

Protagoras and History-Writing

> The writer's sense of fact, in history especially, and in all those complex subjects which do but lie on the borders of science, will still take the place of fact, in various degrees. Your historian . . . with absolutely truthful intention, amid the multitude of facts presented to him must needs select, and in selecting assert something of his own humour, something that comes not of the world without but of a vision within. So Gibbon moulds his unwieldy material to a preconceived view.
>
> —Walter Pater, *Appreciations*

The object of this study has been to help in recovering and annotating a notable movement of thought that for various reasons (I have guessed at what they might be) has largely been consigned to oblivion. It would be disingenuous to pretend that my aim in undertaking this project has merely been to fill in the gaps of a lacunary historical record; nor do I imagine that most readers would so describe my book. It will seem to some to be a manifesto for relativism garbed in the sheep's clothing of intellectual history. Certainly it has refused to take for granted that criminalization of Protagoras that has represented a foundational institution of Western thinking from the outset. Such a refusal necessarily brings about a reconfiguring of the late-Victorian intellectual scene. It makes the line of thought connecting Spencer to Einstein appear, for example, to form a key intellectual axis. Is not this appearance merely the effect of a tendentious historical revisionism? Readers will gauge for themselves the persuasiveness of the evidence that has been cited in these chapters. But in order to suggest the implications of a Protagorean outlook for my own field of cultural and intellectual history, I want briefly to invoke one final Victorian author, a scholar worthy to figure in the roster of illustrious thinkers that makes up the *dramatis personae* of this book.

In the first phase of his public career, George Grote (1794–1871) was a militant activist on behalf of Benthamite and democratic reformist principles, and closely associated himself with the milieu of subversive freethought.[1] One of his early works, for instance, was a redaction of a set of Bentham's papers on natural religion that was pseudonymously published by Richard Carlile, one of the leading nineteenth-century martyrs to the cause of free speech and freedom of the press; Carlile already at that date (1822) was imprisoned in Dorchester Gaol, where, relentlessly persecuted by the authorities for such crimes as republishing Tom Paine's "blasphemous" *Age of Reason*, he was to spend more than nine years of his life.[2] Grote campaigned on behalf of the crusade for parliamentary reform and then for the principle of voting by secret ballot, as later he struggled for equality of education for women at the University of London.[3] Following the passage of the Reform Bill in 1832, he was elected to Parliament, where for three sessions he led, not very effectively, the Benthamite party known as the "Philosophical Radicals."[4] Having abandoned his political career in 1841 to devote himself to classical studies, he published between 1846 and 1856 his great twelve-volume *History of Greece* [hereafter *HG*], a work celebrated in its day and ever after for bringing the study of ancient civilization under the aegis of a new standard of rigorous scientific research. (It insisted, for example, on the need to draw a sharp distinction between legendary materials and historical information vouched for by reliable testimony.) Grote's advocacy of exactingly scientific historiography did not, however, imply adopting toward his subject a pose of disinterested value-free objectivity any more than his devotion to the study of antiquity meant giving up his previous career of polemical political commitment. Rather, he frankly conceived the *History* as a parable of radical politics, a glorification and defense of Greek democracy—much maligned by previous historians such as Mitford—and a sustained polemic against the joint evils of authoritarianism and religious superstition. "He avowed himself . . . as the historian of Grecian freedom," says his fellow Benthamite Alexander Bain; "never weary of the theme of human liberty, he re-touches it on each occasion with fresh and glowing colours" (Bain, "Intellectual Character," 69, 84). None of his original readers could have failed to see that Grote's sharply revisionist portrayal of ancient Greece amounted to an extended fable of the political struggle being waged in mid-Victorian Britain.

Possibly the most provocative section of the *History* was its eulogy of the Sophists in volume 7 (1850), which it would be an error to read without finding there an implicit statement of a creed of scientific historiography. With much illustrative reference to modern scholarly authorities, Grote shows the Sophists to have been uniformly presented by historians as "the moral pestilence of their age," condemned for their "corrupt and immoral" practices and for having "poisoned and demoralised, by corrupt teaching, the Athenian moral character" (*HG*, 7:37, 46, 52)—for having by their insidious influence essentially caused the downfall of classical Greek civilization. (We have of course noted many latter-day reprises of this same persistent myth, the role of evil subversives commonly being filled by relativists or Jews—or both at once.) Grote denies the justice of these charges; he denies even that the portrayal of Protagoras by Plato justifies the hostile view of him taken by his many modern detractors (*HG*, 7:59–63). He in fact depicts the heretic philosopher and his school as among the leading Greek champions of the struggle for political freedom and for the right to express unpopular ideas that he himself waged in his own historical sphere. "It was the blessing and glory of Athens," he says, making explicit the contemporary reference frame of his story, "that every man could speak out his sentiments and his criticisms with a freedom unparalleled in the ancient world, and hardly paralleled even in the modern, in which a vast body of dissent both is, and always has been, condemned to absolute silence" (*HG*, 7:30). The "negative," critical method of teaching aimed primarily at revealing the flaws in received ideas, for their devotion to which the Sophists have been much despised, is not only legitimate but essential to the preservation of freedom in an age tending toward intellectual despotism, says Grote. "It is not simply to arrive at a conclusion . . . and then to proclaim it as an authoritative dogma, silencing or disparaging all objectors—that Grecian speculation aspires" (*HG*, 7:27). Reviling the Sophists for their supposed immorality in cultivating the techniques of argument and persuasion for their own sake (independently of the justice of a particular cause) is no less misguided, says Grote, for if the art of swaying opinion by means of free public debate be discredited, "you leave open no other ascendancy over men's minds, except the crushing engine of extraneous coercion with assumed infallibility" (*HG*, 7:41–42 n.). Nine years later, Mill echoes this language in *On Liberty*, where, in a phrase cited already, he points with alarm to the widespread action of "the engines of moral repression" in modern society. But no glossing by Mill is necessary to see how urgently Grote's defense of the Sophists is driven by the motive of seeking to reform the repressive political and ideological culture of nineteenth-century Britain, where, as the case of Richard Carlile attests, the expression of unorthodox opinions, particularly with regard to the sanctities of the Christian religion, was likely to provoke violent persecution.

F. M. Turner thus aptly describes the commentary on ancient philosophy in Grote's "self-consciously radical history" as "one of the major statements of mid-Victorian philosophic and political radicalism" (Turner, *Contesting Cultural Authority*, 328, 334). But Turner fails in his discussion of Grote's depiction of Protagoras to be specific about the philosophical principles the great Sophist devoted his career to expounding. Perhaps he judged that the reader's sympathy for Grote would not survive if it were stated too plainly that Grote's hero Protagoras stood not only for the principle of the free discussion of heterodox ideas but for something more definite, the wicked philosophy of relativism. Possibly from the same calculation of rhetorical expediency, Grote himself partly veils this point in the *History of Greece*. He makes it emphatically, however, when he returns to the subject of the Sophists in his four-volume study of Greek philosophy, *Plato, and the Other Companions of Sokrates* [hereafter *P*] (1865).

As before, Grote sharply rejects in this work the conventional scholarly portrait of the Sophists as "cheats, who defrauded pupils of their money while teaching them nothing at all, or what they themselves knew to be false," and even cautions the reader against too readily assuming that Plato intended to portray every aspect of Protagoras's teachings as "vile perversion of truth" (*P*, 2:266 n., 303). Focusing in his discussion of the *Theaetetus* upon the notorious doctrine that man is the measure of all things, and stressing once again the contemporary reference underlying all his study of the ancients, Grote explicitly identifies "the principle of relativity laid down by Protagoras" (3:118) with the versions of it set forth by modern writers such as Bain, Mansel, Mill, and Sir William Hamilton. In its account of Protagorean philosophy, this episode in Grote's study thus "brings out forcibly," as he himself says of a passage in Plato, "the negation of the absolute, and the affirmation of universal relativity in all conceptions, judgments, and predications" (3:127). "Truth Absolute there is none, according to Protagoras," says Grote definitively (3:138). Nor does he pretend to any posture of detached neutrality in his presentation of this seditious theme. Rather, he identifies himself unequivocally with the avant-garde intellectual politics of his day and implicitly sets all his activity as a historian under the sign of radical relativism. Far from being a cynical fraud, "the Protagorean doctrine of Relativity" is "true and instructive," he declares, and is in fact "the only basis upon which philosophy or 'reasoned truth' can stand" (3:148, 131). "Existence absolute, perpetual, and unchangeable is nowhere to be found: and all phrases which imply it are incorrect, though we are driven to use them by habit and for want of knowing better," he says (3:130). He is quick to deny, at the same time, the ascription to Protagoras by both Plato and Sextus Empiricus of the doctrine that "all opinions are true," since this would contradict, he says, "the general thesis of Relativity" in its genuine Protagorean form (3:138 n.). "The formula of Relativity does not imply that every man believes himself to be infallible" (3:145), Grote insists; it simply means that a true thing may only be said to be true *"to me"* (3:140). What relativity specifically prohibits, indeed, is granting to one's own views a coercive authority over others. It prohibits by the same token any ideology of appealing difficult questions to higher authorities such as a class of scientific experts imagined to possess an "infallible measure" of truth: "these very . . . Experts," says Grote, forecasting Newman's detailed case histories in *Grammar of Assent* five years later, "are perpetually differing among themselves" (3:143). "All men do not agree in . . . what they speak of as Truth and Falsehood," he memorably concludes. "No infallible objective mark, no common measure, no canon of evidence, recognised by all, has yet been found. What is Truth to one man, is not truth, and is often Falsehood, to another" (3:150).

Like other Victorian relativity theorists surveyed in this book, Grote argues further that *the natural outcome of any philosophy of Truth Absolute is a system of coercion and authoritarian violence*—though he does not ask, as he might have, whether absolutist philosophy is merely the invented pretext for the violence, and thus a sham and ruse from the outset, or whether violence in some fashion derives from and acts as a compensation for the infuriating failure of such misconceived philosophy. "I might indeed clothe my own judgments in oracular and vehement language," Grote wryly observes by way of defining the discursive habits characteristic of absolutism: "I might proclaim them as authoritative dicta . . . I might denounce opponents as worthless men, deficient in all the sentiments which distinguish men from brutes, and meriting punishment as well as disgrace" (*P*, 3:140). The usual treatment of the Sophists themselves in such terms forms of course the case immediately at hand; Grote might have cited here, for instance, the violent denunciation of Protagoras (and implicitly of his defender and alter ego, Grote himself) as a "hater of truth" by the rival classicist Benjamin Jowett (quoted in Turner, *Contesting Cultural Authority*, 354). Grote comes at the end of his recklessly courageous apologia for Sophism, in any case, to an early instance of what was to become a defining topos of Victorian relativity literature, a vision of the establishment of a totalitarian dictatorship devoted to the brutal repression of all deviations from official thinking. This is the sort of system exemplified by Plato himself in his late treatise the *Laws*, where the erstwhile champion of free disputation undergoes, says Grote, a horrible metamorphosis into an inquisitorial tyrant, "monopolising all teaching and culture of his citizens from infancy upwards, barring out all freedom of speech or writing by a strict censorship, and severely punishing dissent from the prescribed orthodoxy" (*P*, 3:148). Implicit in the very concept of Truth Absolute is the logic of this calamitous social outcome, Grote declares; it is the great oppositional mandate of Protagorean relativity, by contrast, to protest whenever "the dogmatist enacts his canon of belief as imperative, peremptory, binding upon all" (3:153). For Grote, the doctrine that a man is the measure of all things forms the basis of any effective defense of freedom. "If you pronounce a man unfit to be the measure of truth for himself, you constitute yourself the measure, in his place," declares the unappeasable Benthamite radical. "[You establish] the King, the Pope, the Priest, the Judges or Censors, the author of some book, or the promulgator of such and such doctrine" as an unchallengeable dictator. The political creed of this authority is inevitably the following: "I, the lawgiver, am the judge for all my citizens: you must take my word for what is true or false: you shall hear nothing except what my censors approve—and if, nevertheless, any dissenters arise, there are stringent penalties in store for them" (3:149).

In his chosen idiom of classical scholarship, Grote thus rehearses unmistakably, one final time in these pages, the intimate linkage of the Victorian relativity movement with the cause of emancipatory moral and political reform. His writing on these themes underlines anew—the point almost lies beneath the dignity of serious scholarship to make—how preposterous it always has been to portray the spokesmen of this movement as bearing any resemblance to purveyors of nihilism and "vile perversion of truth" or to "corrupters of . . . morality." More to the point at hand, Grote also exemplifies again the need to dispel from scientific ideology the idea that radical intellectual relativism is in any way incompatible with the pursuit and strong affirmation of scientific knowledge. Quite the contrary, says Grote: scientific investigation predicates itself on a metaphysic of "Truth Absolute," of certain knowledge and infallible proof cleansed of relativity, only at the sacrifice of its own vital principles of logical rigor and of innovative, self-critical inquiry. Of what is true, "no infallible objective mark . . . has yet been found. What is Truth to one man, is not truth, and is often Falsehood, to another": this, for Grote, is the only sound motto for the practice of science. Objectivity remains a defensible scientific ideal, he proclaims, only if divorced unequivocally from the pernicious concept of infallibility.

Grote does not expressly incorporate the relativity postulate—nothing exists but relations—into his own practice of research, even though he uses ancient history as the forum for such a full and militant exposition of Protagorean philosophy. Yet in fact his iconoclastic history of Greece offers a parable of that philosophy after all, in the boldness with which it dramatizes the fact that a shift in authorial point of view—here, from political conservatism to Benthamite radicalism—causes every significant element of Greek history to be wholly transformed. Illustrious statesmen in orthodox histories like Mitford's thus become odious tyrants in Grote's; well-known instances of "the indelible barbarism of democratical government" become evidence of "cautious and long-sighted view of the future—qualities the exact reverse of barbarism" (*HG*, 5: 61, 62)[5]; infamous philosophers of relativity become paragons of wisdom. The systematic reenactment of this relativity effect forms the key feature of Grote's historical writing, and it is precisely this effect that his polemical exposition of Protagorean relativity theory is meant to conceptualize. The form detectable in past events, or rather in those textual residues that ambiguously symbolize them, depends in a radical fashion upon our point of view. No such thing as an absolute historical truth exists, and "all phrases which imply it are incorrect, though we are driven to use them by habit and for want of knowing better." To base historical research on this premise is not to justify (as intellectual vulgarization would have it) perverting facts to suit preconceptions, Grote asserts; it is simply to lessen our susceptibility to the thralldom of ideology. It is to recognize that no serious historical inquirer can evade the responsibility for constructing an imagined reality of his or her own that is both intellectually and morally as satisfactory as possible, and that differing interpretive constructions of any body of evidence necessarily will come into being, as a consequence, in any space faithfully pledged to scientific reason. This is the lesson that Pater distills, in 1888, in the passage I have used as the epigraph of this chapter.

If Grote's work as a classicist thus formed for the educated Victorian public a striking and precocious object lesson in the transformative potential of relativity, it forms a no less striking lesson for those concerned in the scholarly debates of today that are called the "science wars," particularly in their applicability to the writing of historical narrative. Grote would have read with sympathy but also with keen uneasiness, for example, the essay "Writing History, Facts Optional" by the respected political historian Tony Judt that appears on the op-ed page of today's *New York Times* (13 April 2000). Writing in the first instance against Holocaust revisionism as practiced by David Irving and others, Judt declares that certain facts of history such as the Nazi genocide are to be regarded by responsible scholars as simply beyond questioning: they are established by such a mass of unimpeachable testimony that no legitimate doubt could possibly be raised, now or ever, about their historical truth. They have achieved a condition, we may say, of infallible authenticity. Judt does not employ the undiplomatic phrase "absolute truth" to categorize this class of sacrosanct historical fact, but formally defining certain propositions about the Holocaust (and presumably about a host of other, similarly attested historical realities) as precisely this, "imperative, peremptory, binding upon all" (to use Grote's phrase), forms the express motive of the essay. Indeed, "Writing History, Facts Optional" turns out not really to be written to defend the historical factuality of the Holocaust, of which we are assumed to be as unshakably convinced as Judt is, but in order to hold up to "punishment as well as disgrace" (Grote again) those who would subversively question the creed of Truth Absolute. "The fashionable academic cant," Judt says, "holds that since all facts are 'facts,' all history-writing a subjective 'representation,' all pasts 'constructed,' it makes no sense to seek a consensus about what really happened in history. There is no objective 'truth.' I have my goals and you have yours, and we choose our past accordingly." Insidiously, the essay thus suggests that it is the spread of such thinking among relativistic postmodernist theoreticians that fosters or condones disreputable intellectual projects like Holocaust denial.

My book provides no ground for a rebuttal of such a suggestion, but it surely has shown that the fashionable academic cant in question can claim, at the least, a distinguished ancestry closely linked to all the achievements in different intellectual fields from physics to history-writing that have occurred under the aegis of "relativity" in the course of the last century and a half. The concerted critique of "fact," "objectivity," and "truth" has been central to progressive scientific thought, as we have seen, for at least that long. This critique was expressly made by the great historian George Grote, and it was crystallized in Niels Bohr's dictum, cited at the outset of this book, that "a complete elucidation of one and the same object may require diverse points of view which defy a unique description." On the ground of venerability alone, we may perhaps say that this philosophical outlook has long been immune to being dismissed with contemptuous phrases—all the more so in this instance since the conspiracy against history-writing described by Judt is more or less imaginary to begin with and certainly is not implicated in the repellent activities of David Irving.

In fact, Judt's dismissal of contemporary Sophism is much more equivocal than he seems prepared to recognize, and not only for the obvious reason that he defines the supposedly perverse category of historical "revisionism" exclusively with reference to the likes of Irving, leaving no apparent space for such a radically revisionist project of history-writing as that, say, of George Grote himself. So sophisticated a writer as Judt must needs be made uneasy by the antiquated-seeming line of polemics he has taken up on behalf of that elusive category, "objective 'truth,'" so he adds, as though in passing, a significant qualification. "History is always being revised," he concedes, "as new data come to light and new generations ask new questions." Grote would ask where the difference lies between, on the one hand, the portrayal of history as immersed necessarily in endless relativistic indeterminacy (as Judt recognizes it to be) and, on the other, the fashionable cant—except that Judt declares some facts of history to be "imperative, peremptory, binding upon all" nonetheless? At the end of "Writing History," he answers this vexing question, which threatens to deflate his whole polemic, with a remarkable pronouncement: in evaluating competing interpretations of historical data, he says, we must rely at last on our "intuitions" that "some statements are true, some are false." This is undoubtedly a justified proposition (or would be, if only we knew what an intuition is), but it is one that gives only the most dubious support, obviously, to this author's stated code of definite historical fact grounded in "objective 'truth'" and shielded from the variability of "subjective" interpretations. He installs what amounts to Newman's "illative sense" at the center of his brief for scientific objectivity, in other words, but seems oblivious to Newman's long, profound disquisition on the unattainability of certain knowledge once the scientific method has given up infallibility and has acknowledged intuition to be its final court of appeal.

Judt's brief essay exemplifies, in any case, the tenacious habit among publicists for science of sheltering beneath one form or other of "the moralization of objectivity," in which the dogma of belief in unquestionable truth is overtly or tacitly, and with whatever degree of lucidity, equated with scientific reason. Grote argues contrariwise that the rhetorical apparatus of absoluteness and infallibility can never be invoked in a wholly innocuous way—never in such a way, that is, as not to embroil scientific rationality in unreason and, at the same time, as not to imply a potential recourse to purificatory violence. This twofold proposition forms the presiding moral of the modern relativity movement.

What exactly was the process by which this strongly moralized and politicized doctrine was instituted in effect as a law of nature and as the basis of modernistic science in the various domains that in the late nineteenth century were transformed by the theory of "universal relativity"? A full answer to this question, to the extent that one is conceivable, belongs to the detailed technical histories of the disciplines and lies outside the reach of this book. The evidence submitted here is meant to indicate, however, that no inquiry into the history of scientific discovery can fully succeed if it is severed from broad histories of discourse and held hostage to any implausible ideology of the objectivity and autonomy of science. Authors like Spencer, Veblen, Ludwik Fleck, and Michel Foucault, in their insistence on the mutual interdeterminacy of cultural and natural structures, offer the only sound basis I can see for research into that zone where intellectual, ideological, and natural phenomena seem to fuse inextricably. Such writers suggest that facts of nature, once they have undergone transformation into scientific data, are enmeshed in systems of symbolic exchange and therefore, among other things, in the manifold sociocultural relations that these systems exemplify, concretize, perpetuate. Poincaré underlines in a memorable passage in *The Value of Science* (1905) the necessity of setting some form of this principle at the center of any sophisticated conception of science. "What is objective must be common to many minds and consequently transmissible from one to the other," he says, "and as this transmission can only come about by . . . discourse . . . we are even forced to conclude: no discourse, no objectivity" (*Value*, 347–48).[6] The objective world of nature that science contemplates is a world created by and for the purpose of symbolic transmission, declares the great mathematician; it is subject to whatever the laws and the local contingencies of communication may be. A science in a wholly demystified state would incorporate in its fundamental axioms, Poincaré suggests, an awareness of its own structuring by this principle. This is what he means by "no discourse, no objectivity," what Mach means in observing that "in great measure it is really the intelligence of other people that confronts us in science" (*Popular Scientific Lectures*, 196), and what Hannah Arendt means, I think, by remarking that "truth can exist only where it is humanized by discourse" (Arendt, "On Humanity," 30). This is a doctrine to which George Grote, the modern-day disciple of Protagoras, would wholly subscribe.

Notes

PREFACE

1. Jacques Derrida thus speaks of the mode of relativity he calls *différance* as "the juncture . . . of what has been most decisively inscribed in the thought of what is conveniently called our 'epoch,'" and traces its intellectual ancestry to the work of five authors: Nietzsche, Saussure, Freud, Levinas, and Heidegger ("Differance," 130, and see also 145–60).

2. In 1871, W. S. Jevons, an important figure in the modern relativity movement, identifies the cult of revered authorities as the bane of contemporary speculation: "In matters of philosophy and science authority has ever been the great opponent of truth" *(Theory of Political Economy*, 273–74).

3. This scenario, typical of any number of similar episodes in cultural history, need not involve any supposition of cynical or dissembling intent. In a study of René Girard's alleged failure to recognize his debt to the same romantic thinking he indicts for its besetting naiveté, Paul de Man suggests that the relationship of any critical theorist to his or her precursors is likely to be fraught with amnesia. "The particular mistake" in which Girard is trapped, says de Man, "that of locating a cause of error his own judgment seems able to dispel in the source from which he is, in fact, receiving the very light that allows him to judge correctly, is characteristic of all thought that occurs in a state of crisis, that is reestablishing contact with its own origin" *(Romanticism*, 26). We may not want to subscribe in a doctrinaire way to de Man's pessimistic, crypto-Freudian formula, but it does illuminate the relation of much current theory to Victorian thought. My thanks to Michal Ginsburg for calling my attention to this text of de Man's.

4. In *Ideology and Rationality in the History of the Life Sciences*, Georges Canguilhem defines ideology as "any system of ideas resulting from a situation in which men were prevented from understanding their true relation to reality," and approvingly rehearses Marx's thesis that "ideologies are reassuring fables, unconsciously complicit in a judgment determined by self-interest" *(Ideology and Rationality*, 30, 31). He warns against confounding scientific ideology simplistically with political class ideology or with "false science" (32), but his discussion of it remains hostage to his fundamental contrast of this category—as exemplifed, say, by Spencer's theory of evolution (36–37)—with that of true science, defined by its use of rigorous methodologies of verification. How

this invidious contrasting of true with false under the aegis of "ideology" may impair lucidity is suggested in Canguilhem's comments on Spencer, which slide quickly into undocumentable fables of interpretation such as the assertion that Spencer's use of the evolutionary concept of the "primitive" served as a balm for "the conscience of colonialists" (37). Such a declaration in the guise of scientific analysis has its own unmistakable aura of the ideological: this is merely an expression of dislike of colonialism masquerading as close analysis of the structure of Spencer's theory. One notes for the record that Spencer was a fiercely polemical opponent of colonialist usurpations, as witnessed by his attempt to found an "Anti-Aggression League" in his later years and by his participation on the Jamaica Committee. His vilification of "the actions of European soldiers and colonists who out-do the law of blood-revenge among savages, and massacre a village in retaliation for a single death" (Spencer, *Autobiography*, 2:545) gives an idea of the degree of his conscious sympathy with colonialism.

INTRODUCTION

1. *The Chronicle of Higher Education*, 16 May 1997, A13.

2. The closest to a comprehensive history of relativity may be Ernst Cassirer's 1910 *Substance and Function*, which demonstrates the trend in philosophy, mathematics, and natural science away from the idea of substance to the new one of relations. In *The Victorian Frame of Mind*, Walter Houghton offers a welcome emphasis on strains of "relativism" in the Victorian intellectual milieu; however, he treats "relativism" as roughly synonymous with "doubt," not as a positive and productive intellectual method. Classic works of intellectual history such as *Modern European Thought: Continuity and Change in Ideas, 1600–1950*, by Franklin L. Baumer, and two books by Frank M. Turner, *Between Science and Religion: The Reaction to Scientific Naturalism in Late Victorian England* and *Contesting Cultural Authority: Essays in Victorian Intellectual Life*, are full of relevant material, but they never admit the existence of a relativity movement as such and indeed rarely make even passing mention of this trend of thinking.

3. For example, under feudalism, "a settled scheme of predaceous life, involving mastery and servitude, gradations of privilege and honor, coercion and personal dependence" (Veblen, "Place of Science," 10), the categories of natural things are cast in terms of graded hierarchies of prestige and influence as they are in the sciences of alchemy and astrology, the deity is conceived as a despotic king, and so forth. Veblen elaborates a scheme of thought fully stated already by Spencer. See, for instance, *The Principles of Sociology*, where Spencer defines the regime of scientific imagination that necessarily characterizes "militant," regimented societies, where everything is seen in relation to the influence of powerful leaders. The essential scientific ideas of "impersonal causation" and of evolution can hardly arise in such a social milieu, Spencer argues. "The natural genesis of social structures and functions is an utterly alien conception, and appears absurd when

alleged. The notion of a self-regulating social process is unintelligible" (*Principles*, 2:600).

4. Mill similarly declares that his relativistic science of economic values "has nothing to do" with moral evaluations of the uses to which different goods may be put (*Principles*, 1:521).

5. For Grote's discussion of ethical reciprocity, see *Plato*, 4:99–132. Opposing the relativistic definition of justice favored by Protagoras and the Sophists, who base their arguments on the good consequences of just actions, Plato, says Grote, "desires, above every thing, to stand forward as the champion and panegyrist of justice . . . in itself," to praise "justice *per se* . . . absolutely and unconditionally." Grote argues instead for a non-absolutist Protagorean ethics keyed to the "reciprocity of service and need" that forms "the basis of social theory" (*Plato*, 4:101, 102, 132).

6. See, for example, Spencer, *First Principles*, 206, 496, and see chapter 5 below.

7. It might be said that the Newtonian principle of gravitation, in which each body attracts and is attracted by each other body, offers the original version of a modern system of nature based on a (nonrelativistic) law of reciprocity.

8. Contradictorily enough, they also meant to stress its supposedly crypto-Marxist "materialism" (Frank, *Einstein*, 251–256).

9. For a useful account of German physics in the Hitler years, see Paul R. Josephson, *Totalitarian Science and Technology*, 56–66, 70–74. Josephson does not suggest that relativity physics held any philosophical significance for National Socialist ideologues apart from its association with Einstein's Jewishness.

10. Lewes gave strong if sometimes inconsistent support to relativistic theory in his *Problems of Life and Mind* (1873–79). "The principle of the Relativity of Knowledge . . . is sometimes resisted on the ground of its leading to universal scepticism," he says. "The fact, however, is otherwise." "The world is to each man as it affects him," says Lewes; "to each a different world" (*Problems*, 1:184, 185). Yet despite the "indisputable" principle that "all knowledge must be relative" and that "absolute knowledge, or absolute truth, is a contradiction in terms," Lewes staunchly asserts that "irreversible certainty" is attainable: "that is absolutely true which cannot be otherwise," he declares (*Problems*, 2:71). My thanks to George Levine for pointing out these passages to me. For an elucidation of Lewes's thinking about "Objective Truth," see Peter Allan Dale, *In Pursuit of a Scientific Culture*, 108.

11. Mivart was an anti-Darwinian whose objections to the theory of evolution are discussed at length in chapter 7 of late editions of *The Origin of Species*.

12. The Spenserian equivalent of this imagery is the scene of the abode of Despair, "an hollow cave, / . . . / Darke, dolefull, dreary, like a greedy grave," and ultimately the "yawning gulfe of deepe Avernus hole" (*Faerie Queene*, I.ix.33, I.v.31).

13. The translations from Tonnelat are my own.

14. Alexander Bain's comments on the rhetorical category of the "Infinite," closely akin as it is to that of the "Absolute," are exactly to the point. "The Infinite is a phrase most various in its purport: it is for the most part an emotional word, expressing human desire and aspiration; a word of poetry, imagination, and preaching, not a word to be discussed under science" (*Practical Essays*, 63).

15. See Christopher Ray, *The Evolution of Relativity*, 86–87. Ray notes that J. D. North identifies eight or nine different meanings of "absolute" in space-time literature, that J. Earman cites a dozen, and that Michael Friedman specifies three primary meanings, in which "absolute" is opposed to "relational," to "relative," and to "dynamic," respectively.

16. According to the Oxford English Dictionary, "absolutism" in the sense of "the practice of absolute government; despotism; an absolute state" was coined in 1830 by Gen. Perronet Thompson.

17. See, e.g., W. K. Clifford, *Lectures*, 1:175.

18. The identification of the rise of twentieth-century relativism with the subversive influence of Jews is renewed in a long anti-Semitic pamphlet titled "Anti-Semitism, —Found," anonymously distributed to university campuses in the Chicago area in October 1995. "Moral relativism" is defined in this publication as standing for the principle "if it feels good, do it!" and "cultural relativism" for the doctrine that "a skull sucking aborigine is exactly equivalent to Thomas Jefferson." The racial malignancy of the pamphlet is aberrant in the discursive environment of contemporary academia, but the representations of relativism are perfectly consistent with those offered in many publications to which respected academics are willing to sign their names.

19. The same imagery surges into Rodney Needham's exhaustive unfolding of the implications of linguistic relativity for the human sciences in *Belief, Language, and Experience* (1972). After such an inquiry, says Needham, the intellectual landscape looks "like that in a dream." "With each step in the analysis . . . infirmities and obscurities have proliferated, and in nothing have we found any certainty. . . . Every apparent ground of absolute knowledge or judgement has crumbled or dissipated, and only shifting relativities remain" (235, 243).

20. Bloom announces in his opening sentence that American university students have today only a single belief: "that truth is relative." From this flawed concept flows all the degradation of mind, sensibility, and moral purpose Bloom sees in contemporary America. "Relativism has extinguished the real motive of education, the search for a good life," and it has generated in today's youth a pervasive religious collapse, "apathy about the state of their souls." It could hardly be otherwise, since relativism "means accepting everything and denying reason's power"; its position, says Bloom, is "to deny the possibility of knowing good and bad" and to turn away from "the quest for knowledge and certitude" (*American Mind*, 1, 34–35, 40, 41). Relativism and nihilism are synonymous. The same matrix of themes is traced in Gertrude Himmelfarb's *De-Moralization of Society* (1995), which praises the moral virtues cultivated by Victorian society,

comparing them favorably with "the relativized and subjectified . . . 'values'" of twentieth-century America; in contrast to our own, the Victorian ethos was one "that does not denigrate or so thoroughly relativize values as to make them ineffectual and meaningless" (9, 250).

21. See, for example, the recent volume of essays on philosophical hermeneutics edited by Lawrence K. Schmidt titled *The Specter of Relativism*.

22. Authentic "*freedom* of thought," declares Nietzsche in *The Genealogy of Morals*, requires "*taking leave* of the very belief in truth" and replacing the doctrine of truth with the relativity principle: that "there is only a seeing from a perspective, only a 'knowing' from a perspective," only "interpretation" (195, 153, 196). By stressing the universality and affirming the intrinsic moral value of the Will for Power, however, Nietzsche introduces a set of dogmatic, crypto-theological absolutes into this relativistic field, and identifies his cult of freedom with a cult of tyrannical domination. A crucial element in the latter argument is his identification of the Jewish people as a conspiracy of unclean subversives who must be suppressed in order to unleash the power of the aristocrat, "the magnificent *blonde brute*, avidly rampant for spoil and victory" (40). No teachings could be more remote from those of Protagorean relativism. See below, chapter 5, note 25.

23. "Truth, says the cultural relativist, is culture-bound. But if it were, then he, within his own culture, ought to see his own culture-bound truth as absolute. He cannot proclaim cultural relativism without rising above it, and he cannot rise above it without giving it up": Quine, quoted in Siegel, *Relativism Refuted*, 43. See also Burnyeat, "Protagoras and Self-Refutation in Plato's *Theaetetus*."

24. To claim, in the face of the well-documented loathing of relativism by twentieth-century tyrannies of the Right and Left alike, that moral relativism bears the responsibility for the iniquities of Hitlerism or resembles it in any way seems nearly to fall outside the bounds of legitimate opinion. Such a close-grained study of the fascist mentality as Daniel Jonah Goldhagen's *Hitler's Willing Executioners* lends little support to the assertion of a link between "moral relativism" and fascist cruelty. Rather, it presents a mountain of evidence tending to associate the nihilistic moral callousness of Hitlerism with long traditions of racial hatred and persecution, activated by the establishment of a totalitarian state—pathologies with which moral relativism has, to say the least, nothing in common.

25. In *Science in a Free Society*, Feyerabend declares his sympathy for relativism and pronounces it "the path of growth and freedom" (145). He does disavow that form of "philosophical relativism" that affirms "the doctrine that all traditions, theories, ideas are equally true or equally false" (83)—essentially a nonexistent philosophical school to begin with.

26. Barbara Herrnstein Smith's excellent *Contingencies of Value*, in which she argues "that literary value is radically relative" (11) and that "the search for essential or objective value" is always misguided (15), forms an instance of this amnesia. None of the Victorian propounders of such doctrines earn citation in this book.

27. Abraham Pais, in "*Subtle is the Lord,*" reviews Einstein's contradictory testimony about his knowledge of the Michelson-Morley experiment prior to 1905 (116–117), but concludes that he "unquestionably" did know of it and that his knowledge influenced the creation of special relativity (115). However, Pais judges that in 1905 Einstein did not know of the Lorentz transformations or of Poincaré's technical work on relativity issues (121). See below, chapter 1, note 25.

28. G. K. Chesterton made this point perfectly: "No one can understand tradition, or even history, who has not some tenderness for anachronism" (*The Victorian Age*, 38; see also 7–8, 37).

CHAPTER ONE

1. Relativistic arguments were distinctly formulated by Enlightenment thinkers like the Encyclopedists Diderot and d'Alembert, if only as preliminary maneuvers or as adjuncts to an overriding universalist project of knowledge. See, for example, d'Alembert's "discours préliminaire" to the *Encyclopédie*, where he states that one could hypothetically imagine "as many different systems of human knowledge" as one could imagine world maps based on different projections—which is to say, an unlimited number of them (xv).

2. For an informative treatment of Hamilton and, especially, Mansel, see Bernard Lightman, *The Origins of Agnosticism*. Lightman portrays the archconservative Anglican Mansel as the "missing link" (31) in the history of Victorian agnosticism.

3. See Mill, *Sir William Hamilton*, 1:66.

4. Cf. Fredric Jameson's declaration that postmodernism "marks the end of philosophy" ("Postmodernism and Consumer Society," 112).

5. The Benthamite George Grote underlines this point distinctly in his comparative analysis of Protagorean and Platonic concepts of justice. See Introduction above, n. 5.

6. The great fault of "Hedonistic Utilitarianism," says Green, lies in "taking the good to be relative to something external to itself [i.e., pleasure]; to have its value only as a means to an end wholly alien to, and different from, goodness itself." Such a system contradicts the true moral ideal of a categorical imperative able to "command something to be done universally and unconditionally" (Green, *Prolegomena*, 205, 206).

7. Franklin L. Baumer labels the nineteenth-century rationalistic movement "The New Enlightenment," stressing, for example, the direct continuity of the Benthamites of the 1820s with the *philosophes* of the eighteenth century (*Modern European Thought*, 302–36).

8. "The name," says Bain's intellectual heir Derrida, "is always caught in a chain or a system of differences" (Derrida, *Of Grammatology*, 89).

9. In *A Pluralistic Universe* (1909), James again assails the doctrine of the inherent self-contradictoriness of ideas, here ascribing it to Hegel (95).

10. The psychologist James Ward attacked Bain and associationist (and physiological) psychology generally for upholding a view of the human mind as merely a passive creation of its environment. Ward stressed the volitional, world-shaping capacity of human subjectivity. "The world limits me in manifold ways, but it is also dependent upon me," wrote Ward in 1911. "For I am not wholly passive and inert: I am able to react upon it and do in fact in some measure modify it" (quoted in Turner, *Between Science and Religion*, 237; see 201–45).

11. Marx's comments on Feuerbach or Derrida's on Saussure (not to mention, say, Barbara Herrnstein Smith's on Richard Rorty) form exemplary instances of the pattern. So, too, does James Ward's critique of Bain (see note 9).

12. See Alan Lightman, *Great Ideas in Physics*, 182–242. The dumbfounding results of this experiment are "impossible to fathom," even by "the best physicists in the world," says Lightman (200).

13. Mill refers in a letter to Mansel's "detestable to me absolutely loathsome book" (quoted in Packe, *Life of Mill*, 443).

14. Robert M. Young, in *Darwin's Metaphor: Nature's Place in Victorian Culture*, has insisted on the need to resituate Spencer at the center of the history of evolutionist thought in the nineteenth century. "The failure of historians of science seriously to consider Spencer" impairs our understanding of the history of evolutionist thinking, says Young, since "his theory, along with the generalizations and extrapolations based on it, was probably more influential in the general debates of the late nineteenth century than those of any of the other evolutionists" (185, 184).

15. See, for example, Spencer's survey of his pre-1859 statements of evolutionary theory in the preface to the fourth edition of *First Principles* (vii–viii), and especially in his *Autobiography* (1:448–52, 538, 587; 2:5–14, 57, 194–99). Darwin's decisive contribution to the theory was of course the discovery of its causal mechanism, the natural selection of favorable variations. Spencer had in fact come close to formulating a theory of natural selection as early as 1852: see *Autobiography* 1:450–51.

16. The "orthodox dogmatism" of the day was characterized, says Ely, "by an absolutism of theory" that Karl Knies subdivided into two forms, "perpetualism which implies that a policy holds good for all times, and cosmopolitanism which holds that any one policy can be applied to all lands." For Knies, the former was the more obnoxious (Ely, *Ground under Our Feet*, 58).

17. In *The Cosmic Web: Scientific Field Models and Literary Strategies in the Twentieth Century*, N. Katherine Hayles usefully identifies this theme with the "field concept" of contemporary physics. "Perhaps most essential to the field concept is the notion that things are *interconnected*," she notes. "In marked contrast to the atomistic Newtonian idea of reality, in which physical objects are discrete and events are capable of occurring independently of one another

and the observer, a field view of reality pictures objects, events, and observer as belonging inextricably to the same field; the disposition of each, in this view, is influenced . . . by the disposition of the others" (9–10).

18. Cassirer quotes Helmholtz as one early formulator of this principle. "Each property or quality of a thing is in reality nothing but its capacity to produce certain effects on other things," Helmholtz stated (Cassirer, *Substance and Function*, 305). J. B. Stallo quotes Leibniz as declaring "whatever does not act does not exist" (Stallo, *Concepts and Theories*, 170). C. S. Peirce about the same time developed the conception that the *behavior* of an object is the essence of its meaning. For similar statements, see also Nietzsche, *The Will to Power*, 2:66, and Ortega y Gasset, *The Dehumanization of Art*, 67.

19. Spencer declares in his *Autobiography* (1:172) that his profound "consciousness of physical causation" formed his basic principle of thought.

20. Bernard Lightman surveys the tradition among Spencer's commentators of doubting his "sincerity" in professing respect for religion (*Origins of Agnosticism*, 203 n. 48), but argues that Spencer's theism was in fact genuine (88–90). I prefer to set aside such questions of authorial psychology (as to which, no very meaningful evidence does or ever can exist) and to seek instead to describe the ideological structures of the texts themselves. Seen from this angle, Spencer emerges very clearly, I think, as a member of the devil's party.

21. The theme of interdependency or "entangledness" in Darwin has been well emphasized by Gillian Beer and by George Levine, both of whom note the figurings of the theme in Dickens's *Bleak House* and other works of Victorian fiction. See Beer, *Darwin's Plots*, 22–24, 45–47, 170; Levine, *Darwin and the Novelists*, 17–18, 119–20, 147–49.

22. "To a living being then," wrote Samuel Butler, "there can be no absolute 'it is' without the skeleton of an 'It is not' in some one or other of its cupboards" (*Collected Essays*, 1:125).

23. The paradox is played out vividly in E. E. Evans-Pritchard's *The Nuer* (1940), the most systematically relativistic of all works of anthropology. There is "always contradiction in the definition of a political group, for it is a group only in relation to other groups," says Evans-Pritchard (147). Thus the quasigenocidal warfare between the Nuer and their ancestral enemies, the Dinka, is interpreted in this work as a kind of harmonious commonality between the two tribes, a joint institution expressing "one of the most fundamental characteristics of . . . all social groups: their structural relativity" (125–32, 135).

24. Spencer's insistence on the slow pace of evolutionary change leads him to disavow schemes for revolutionary social change, however. See, for example, *The Study of Sociology*, 399–403.

25. Sir Edmund Whittaker notoriously analyzed what he called "the relativity theory of Poincaré and Lorentz" in his *History of the Theories of Aether and Electricity:* "Einstein published a paper [in 1905] which set forth the relativity theory of Poincaré and Lorentz with some amplifications, and which attracted

much attention" (40). Widely denounced by historians of physics, Whittaker's interpretation with reference to Poincaré "contains much more than a simple grain of truth," says Elie Zahar (*Einstein's Revolution*, 149). "In 1905, Poincaré had gone far beyond the results obtained by Einstein," according to Zahar (150). Einstein's biographer Abraham Pais describes the remarkable speculative work done by Poincaré between 1898 and 1905, in which he enunciates basic elements of relativity theory such as the impossibility of demonstrating absolute motion, the impossibility of giving an objective meaning to simultaneity, and the prospect of "a new mechanics . . . in which the velocity of light would become an impassable limit"; but Poincaré never grasped the need to abandon the ether and ultimately "never understood the basis of special relativity," says Pais (*Subtle is the Lord*, 126–29, 128, 21). In *Science and Hypothesis*, however, Poincaré makes clear his guarded but unequivocal rejection of the ether. "I do not believe, in spite of Lorentz," he says, "that more exact observations will ever make evident anything else but the relative displacements of material bodies"—that is to say, no medium filling empty space will ever be detected scientifically. "Some day, no doubt, the ether will be thrown aside as useless," he concludes (*SH*, 172, 212). See below, chapter 5.

26. As Samuel Butler phrased it fifteen years previously, we "solve the riddle, 'What is Truth?' by giving the answer, 'Convenience' " (*Collected Essays*, 1:160).

27. In this account I am glossing over the inconsistencies in Poincaré's own rhetoric in respect to the category of "truth." He certainly never suggests that scientific inquiry is unable to produce genuine knowledge merely because it relies on conventions and definitions, and he cannot divest himself after all of the language he has mercilessly deconstructed. "Experiment is the sole source of truth," he says at one point, for example. "It alone can teach us something new; it alone can give us certainty" (*SH*, 140).

28. This argument of Feuerbach's is paralleled in Regina Schwartz's *The Curse of Cain: The Violent Legacy of Monotheism*. See also Nietzsche, *Genealogy of Morals*, 109.

CHAPTER TWO

1. Abraham Pais insists, however, that Einstein was not temperamentally or intellectually a revolutionary, and never took issue with authority simply for its own sake (*Subtle is the Lord*, 17, 38–39).

2. The terrorist known as the Professor had been the son of "an itinerant and rousing preacher of some obscure but rigid Christian sect" whose religious fanaticism had passed to his son in the form of "a frenzied puritanism of ambition" and of "the subconscious conviction that the framework of an established social order cannot be effectually shattered except by some form of collective or individual violence" (*Secret Agent*, 76–77).

3. By 1930, the date of his *Logic for Use*, he not only has heard of Einstein but has incorporated Einsteinian relativity into his own "voluntarist theory of

knowledge." See *Logic for Use*, 149, and the numerous references to Einstein cited in the index.

4. See page 8 above in reference to Veblen's analysis of the culture of "predaceous" medieval society.

5. Spencer, along with all the leading evolutionists of the day, including Darwin, Wallace, Huxley, and "others less known," joined Mill on the Jamaica Committee to demand official condemnation of Eyre's actions (Spencer, *Autobiography*, 2:168)—an incident bearing clear witness to the thesis that the scientific revolution in Victorian Britain was inseparable from a movement of political emancipation. Another of Spencer's noble quixotic enterprises was his attempt to found an Anti-Aggression League to oppose the rise of militarism and imperialism (see Spencer, *Autobiography*, 2:446).

6. In this extended indictment, Spencer oddly does not mention (as Mill does in *On Liberty*) the government assault upon religious unorthodoxy by means of the blasphemy laws.

7. "What passes for truth in every age," says Nietzsche, is nothing but a "mobile army of metaphors, metonyms, and anthropomorphisms: in short a sum of human relations which become poetically and rhetorically intensified, metamorphosed, adorned, and after long usage seem to a nation fixed, canonic and binding" ("Truth and Falsity," 180).

8. Feuerbach's monster comes to life in a series of later avatars. One of them is Nietzsche's morbidly diseased ascetic priest, whose "awful historic mission" it is to exercise *"lordship over sufferers"* and who must be "impregnable . . . in his will for power, so as to acquire the trust and the awe of the weak so that he can be their hold, bulwark, prop, compulsion, overseer, tyrant, god." He comes in this role, says Nietzsche, "to represent practically a new type of the beast of prey," one who is "venerable, wise, cold, full of treacherous superiority" (*Genealogy*, 162–63). Another Feuerbachian refraction is the goddess Conversion, who stands for the principle of evangelical religiosity in Virginia Woolf's *Mrs. Dalloway.* This mode of religion professes altruistic ideals but is in fact a cult of sheer domination. "Conversion," which "feasts on the wills of the weakly," says Woolf's narrator, "offers help, but desires power." She is active in building churches and hospitals, "but conversion, fastidious Goddess, loves blood better than brick, and feasts most subtly on the human will" (*Mrs. Dalloway*, 151–52).

9. Thus Derrida writes that "the signified concept is never present in itself, in an adequate presence that would refer only to itself. Every concept is necessarily and essentially inscribed in a chain or system, within which it refers to another and to other concepts, by the systematic play of differences" ("Differance," 140).

10. See Stillman Drake's discussion of this passage, comparing Stallo's refutation of Neumann with Mach's less radical one (Drake, "J. B. Stallo and the Critique of Classical Physics," 27–29).

11. "We first have to clarify what is to be understood here by 'time,'" says Einstein. "We have to bear in mind that all our propositions involving time are

always propositions about *simultaneous events*," such as the simultaneity of a watch hand pointing to seven and the arrival of a train; and a clock "located at the Earth's equator must be very slightly slower than an absolutely identical clock . . . at one of the Earth's poles" ("Electrodynamics of Moving Bodies," 141, 153).

12. Thus the "tilt toward idealism" that Franklin L. Baumer describes in twentieth-century scientific philosophy and associates with figures such as Planck, Jeans, and especially Eddington (*Modern European Thought*, 470–72) has significant nineteenth-century antecedents.

CHAPTER THREE

1. Einstein's special-relativity paper was rejected as his thesis for the position of Privatdozent at Bern University in 1907 in part on the grounds that it was incomprehensible (Hoffmann, *Albert Einstein*, 86).

2. F. H. Bradley in his *Principles of Logic* (1883) insists in similar terms on the self-evidence of the law of contradiction. "Truth is unchangeable, and, as discrepant assertions alter one another, they can not be true" (Bradley, *Principles*, 1:147).

3. On many points of theory, T. H. Huxley was at odds with the anti-evolutionist Mivart, but he no less fervently invoked the creed of the rigorously logical character of scientific knowledge. The goal of physical research, he wrote in 1887, is "the discovery of the rational order which pervades the universe" (Huxley, *Method and Results*, 60).

4. He may have been quoting the psychologist James Ward. Bain's "differential theory," says Ward in 1886, "however plausible at first sight, must be wrong somewhere since it commits us to absurdities" ("Psychology," 49). The same point is made in 1910 by Henry Sturt in an essay tracing the career of the doctrine of "the relativity of knowledge." The notion that "a piece of experience is entirely constituted by its relation to other experiences," he says, is clearly illogical. "Such an extreme relativity, as advocated by T. H. Green in the first chapter of his *Prolegomena to Ethics*, involves the absurdity that our whole experience is a tissue of relations with no points of attachment on which the relations depend" (Sturt, "Relativity of Knowledge," 59). He fails to note that Green himself makes the same point. "We cannot reduce the world of experience to a web of relations in which nothing is related," says Green (*Prolegomena*, 45).

5. In 1906, the respected physicist Walter Kaufmann announced experimental results concerning the constitution of the electron that were, he declared, "incompatible with the Lorentz-Einstein postulate," and were not conclusively disproved until 1914–16. See Pais, *Subtle is the Lord*, 159.

6. For a discussion of this passage, see Cassirer, "Structuralism in Modern Linguistics," 103.

7. W. K. Clifford echoes Mill in 1873: "How am I to know that the angles of a triangle are exactly equal to two right angles under all possible circumstances; not only in those regions of space where the solar system has been, but everywhere else?" (*Lectures*, 1:335).

8. The date of *The Will to Power* is difficult to establish. Composed between 1884 and 1888, Nietzsche's fragmentary text was posthumously published in various forms, including materials of disputed authenticity, between 1895 and 1906. Schiller does not refer to him in *Formal Logic*.

9. In *Philosophical Explanations*, Robert Nozick disavows the violent ideal of philosophical reason in terms that seem almost to quote Schiller. "The terminology of philosophical art is coercive," says Nozick; "arguments are *powerful* and best when they are *knockdown*, arguments *force* you to a conclusion. . . . A philosophical argument is an attempt to get someone to believe something, whether he wants to believe it or not. A successful philosophical argument, a strong argument, *forces* someone to a belief." "Is that a nice way to behave toward someone?" he asks (Nozick, *Philosophical Explanations*, 4, 5).

10. Postmodernism, says Stanley Aronowitz, for example, bases itself upon a "rejection of reason as a foundation for human affairs"; Chantal Mouffe declares her politics to be exempt from the "specific form of rationality" pursued by the Enlightenment. See the interesting discussion of these passages in Neil Larsen, *Reading North by South*, 165–66. In the same vein, Paul Feyerabend denounces "the idea of a [scientific] method that contains firm, unchanging, and absolutely binding principles" (*Against Method*, 14).

11. See, for example, William James's evocation of "the enormously rapid multiplication of theories" in the modern age, a factor that "has well-nigh upset the notion of any one of them being a more literally objective kind of thing than any other" (James, *The Meaning of Truth*, 206).

12. Schiller elaborates this point in *Studies in Humanism:* "What is fact for one science, and from one point of view, is not so for and from another, and may be irrelevant or a fiction. If, therefore, rival theorists are determined to occupy different points of view, and to stay there without seeking common ground, they can controvert each other's 'facts' for ever" (*Studies*, 371).

13. See, for example, Clifford Geertz, *The Interpretation of Cultures*, 49: "There is no such thing as a human nature independent of culture. . . . We are . . . incomplete or unfinished animals who complete or finish ourselves through culture."

14. The conception of the human mind as not a passive receptor of stimuli operating by means of involuntary associations, or as necessarily obedient to logical rules of thought, but as an active personal agency with voluntary powers of selection and even creation, was reiterated in various places in the years following *Grammar of Assent*. For example, C. Lloyd Morgan's *Animal Life and Intelligence* (1890–91) makes a strongly Spencerian argument that the phenomenal world is a mental "construct" symbolizing an unknowable external reality, and makes a point of refusing "to reduce the human mind . . . to the condition of a mere passive recipient instead of a vital and active agent in the construction of man's world" (Morgan, *Animal Life*, 312–13, 332). In *The Genealogy of Morals* (1887) and other works, Nietzsche gives this doctrine an especially polemical expression, insisting

on "the active and interpreting functions" of the mind as against the mythology of "pure reason"; to eliminate the role of will and the emotions in human cognition, he declares, would be "intellectual *castration*" (*Genealogy*, 153).

15. Bain makes a point of stressing the possibility of alternate representations of the motion of the earth (*Logic*, 2:391), as does Karl Pearson in his 1892 book *The Grammar of Science* (236).

16. Spencerian agnosticism in the philosophy of science is hard to overcome "by mere argumentation," concedes Engels. "But before there was argumentation, there was action," and the success of scientific knowledge when put to the purposes of practical action is "an infallible test" of its validity, he declares (Engels, *Socialism*, 13–14). All these writings feed at last into Sartre's existentialism. "There is no reality except in action," says Sartre (*Existentialism and Humanism*, 41).

17. For a rehearsal of some of the same arguments against the rationality of science, now called "naturalism" and identified with "Agnosticism, Positivism, Empiricism," but set in the context of an extended affirmation of religious faith, see Balfour's later and better-known work, *The Foundations of Belief: Being Notes Introductory to the Study of Theology* (1895). Balfour focuses in this book upon "the inner antagonism which exists between the Naturalistic system and the feelings which the best among mankind . . . have hitherto considered as the most valuable possessions of our race," and debunks the supposed primacy of reason in human life. "It is Authority which supplies us with essential elements in the premises of science . . . it is Authority rather than Reason which lays deep the foundations of social life . . . it is Authority rather than Reason which cements its superstructure" (Balfour, *Foundations*, 6, 77, 238). For an excellent discussion of *The Foundations of Belief* and of Huxley's hostile review of it, left incomplete at his death, see Bernard Lightman, "Fighting Even with Death."

18. A full account of these developments would include not only a study of their continuity with Benthamite utilitarianism (in part through the intermediation of Mill) but also a study of C. S. Peirce's elaboration in the 1870s of the modern logic of relations, in which an object is defined not in terms of indwelling qualities but in terms of evolutionary relations among its different states and between itself and other objects, and in which logical propositions, rather than expressing timeless verities according to the model of Kantian philosophy, are conceived as devices for overcoming the unpleasantness of doubt and attaining the desired condition of "belief."

19. In 1907, the somewhat conservative commentator on physics Lucien Poincaré (not to be confused with Henri), in a book expressing sympathy, in the face of admittedly compelling relativistic theory, for the common-sense "notion of an absolute length" and for the ether hypothesis, testifies to the scorn of professional scientists of the day for the "rather puerile subtlety" of philosophical discussions of science (*The New Physics*, 22, 174, 8).

20. That is, as Jacques Derrida phrases it long afterward, "the so-called 'thing itself' is always already a *representamen* shielded from the simplicity of intuitive

evidence" and from "the plenitude of . . . an absolute presence" (*Of Grammatology*, 49, 69).

21. "Is there a knowledge . . . scientific or not, that one can call alien . . . to violence?" asks Derrida, answering the question in the negative, as Butler did (*Of Grammatology*, 127).

22. For a thoughtful study of Butler's philosophical thinking, see Frank Miller Turner, *Between Science and Religion: The Reaction to Scientific Naturalism in Late Victorian England*, 164–200.

23. On the variability of human nature, see Spencer, *The Study of Sociology*, 118, 347, 390; *The Principles of Sociology*, 1:683–84, etc. See also Clifford, *Lectures*, 2:278–79, 282.

24. "While the bodily natures of citizens are being fitted to the physical influences and industrial activities of their locality, their mental natures are being fitted to the structure of the society they live in," said Spencer in 1872–73. "For every society, and for each stage in its evolution, there is an appropriate mode of feeling and thinking" (*Study of Sociology*, 347, 390).

25. Bronislaw Malinowski, in the founding text of modern ethnographic anthropology, invokes in 1922 exactly Spencer's relativistic postulate (without mentioning him) but ignores his and Lévy-Bruhl's warnings of methodological self-contradiction. The goal of ethnographic study, he says, is "to grasp the native's point of view, his relation to life, to realise *his* vision of *his* world" (*Argonauts of the Western Pacific*, 25). For Rodney Needham, a latter-day admirer of Lévy-Bruhl, the conundrum of anthropological science in reference to the belief-systems of foreign cultures remains as daunting as ever. Statements of belief have a deceptively logical form, says Needham, practically quoting Newman or Balfour, but "their assertion or acceptance does not depend on logical validity but rather upon what we might call a circumstantial cogency. Logic and belief are independent" (Needham, *Belief, Language, and Experience*, 75). Therefore the belief systems of other peoples are closed to scientifically reliable logical analysis. Given the absence of an objective standard of human experience, there must always be, he concludes, in language with a long Victorian pedigree, "an intrinsic uncertainty in all our premises concerning the nature of man" (210).

26. This insight, and indeed the whole of *Science and Hypothesis*, places the great scientist utterly at odds with the polemics of the likes of Paul R. Gross and Norman Levitt, who declare science to be a system of "straightforward . . . reasoning" and of unambiguous objectivity, and who dismiss relativistic critiques of science as nothing but the "resentment" of frustrated leftists (*Higher Superstition*, 21, 27).

27. That the disavowal of logical reason may also serve to shield authoritarian institutions from critical scrutiny has been seen in the exemplary cases of Mansel and Newman, and is restated by Jürgen Habermas in regard to Heidegger's assertion in his 1953 "Letter on Humanism" that "there is a truth more rigorous than the conceptual" (Heidegger, "Letter," 235). Such a principle, says Habermas, serves to

bolster "the claim that a few people have a privileged access to truth, may dispose of an infallible knowledge, and may withdraw from open argument" ("Work and Weltanschauung," 456).

CHAPTER FOUR

1. Karl Pearson invokes it, for example, in an 1887 essay. "Darwin has destroyed the old Ptolemaic system of the spiritual universe," he says. "We can no longer regard all creation as revolving about man as its central sun" (Pearson, *Ethic*, 324). Sir James Jeans rehearses the standard narrative at the outset of his 1943 *Physics and Philosophy* (1–2), naming as the great scientific enemies of human narcissism Copernicus, Darwin, and Newton (a holy Trinity like Freud's own), and connecting this historical progress a little ambiguously to the contemporary revolution in physics. See also, in addition to other instances cited in this chapter, Foucault, *The Order of Things*, 348.

2. Lucien Lévy-Bruhl, in *Ethics and Moral Science* (1903), gives a similarly guarded assessment of the prospects for soon eliminating the contaminant of human narcissism from modern thinking. Ideally, he says, phenomena under scientific study should be "desubjectivized," cleansed of "their specifically human character" and thus given "objectivity" (*Ethics*, 5). Yet so deeply ingrained is the anthropocentric principle in human thought, says Lévy-Bruhl, that even the intellectual revolutions of Copernicus, Kepler, Galileo, and Darwin have failed to extinguish it, most particularly in the field of ethical philosophy. "Anthropocentrism has . . . been able to subsist . . . by no longer taking the earth, but human reason, for the centre of the universe," he declares. "The struggle against anthropocentrism is far from ended," concludes Lévy-Bruhl, "its strongest positions are not yet touched. It has only lost, so to speak, its outworks" (*Ethics*, 164–65). "Anthropomorphism plays a considerable historic role" in the genesis of physical thinking, allows Henri Poincaré in 1902, "but it can be the foundation of nothing of a really scientific or philosophical character" (*Science and Hypothesis*, 106–7). Ernst Cassirer analyzed Einsteinian relativity theory in just these terms in 1921, as an advance of "true objectivity" against the resistance of primitive human-centered modes of thinking. "The anthropomorphism of the natural sensuous picture of the world, the overcoming of which is the real task of physical knowledge, is . . . [in relativity theory] forced a step further back" (*Einstein's Theory*, 381–82).

3. "The aspect European existence is taking on in all orders of life points to a time of masculinity and youthfulness," says Ortega, with apparent approval. "Life is a petty thing unless it is moved by the indomitable urge to extend its boundaries." To those who might recoil from the political implications of such a code of virile expansionism, Ortega issues a stern injunction. "What ought to be done does not depend on our personal judgment," he says; "we have to accept the imperative imposed by the time. Obedience to the order of the day is the most hopeful choice open to the individual" ("Dehumanization," 24, 52, 13). That an

ardent anti-authoritarian like Ortega could find such language appealing is the clearest possible sign of its centrality to the modern Western mentality.

4. Another indication of the complexly cooperative relations existing between religion and science in the nineteenth century is found in the way evolutionary science initially framed itself as a new form of natural theology. See Young, *Darwin's Metaphor*, 10–11, 16. "The evolutionary debate produced an adjustment within a basically theistic view of nature rather than a rejection of theism," says Young (16). This is the general theme of Bernard Lightman's enlightening work *The Origins of Agnosticism*, which portrays such spokesmen of scientific naturalism as Huxley, Clifford, Stephen, and Tyndall as "the new natural theologians" (Lightman, *Origins*, 156). For a similar argument, see Frank M. Turner, *Contesting Cultural Authority*, 18–20. My own discussion bears not so much on the transmission of theological doctrine as on that of a certain structure of puritanical moral sensibility, and it describes not an avowed union of interests like that of early evolutionism and the natural-theology tradition, but a covert and illicit one.

5. Similarly, F. C. S. Schiller seeks in 1907 to probe the "emotional origin" of scientific ideology, the key principle of which he declares to be "the sacrifice of 'personal preference' to 'objective truth.'" This operation has not a rational, but a moral basis, he argues: a drive to "heroic self-sacrifice" that expresses itself in the scientist or in the devotee of science as "an emotional desire to mortify himself (or, more often, others), the satisfaction of which appears to him as a good." From this point of view, it seems "wrong" and "wicked" to allow the possibility that such disreputable factors as human desires could ever condition a scientist's analysis of objective "facts" (Schiller, *Studies*, 92).

6. See, for example, the 1895 *Fortnightly* review of Pearson's *Grammar of Science* by St. George Mivart, in which Pearson is charged with upholding a philosophical idealism in which he only affects to believe (Mivart, "Denominational Science," 434), or Elie Zahar's analysis of the ambiguity of Poincaré's wavering between conventionalism and an essentialist concept of truth (*Einstein's Revolution*, 151), or William James's oscillations between denying and affirming "absolute" properties of objects—or even Einstein's denial and subsequent affirmation of the existence of the ether. Self-emancipation from established structures of thought is easier to proclaim than to achieve.

7. *The Brothers Karamazov* (1880), book 5, chapter 5.

8. In an 1887 essay, Pearson declares that for the nineteenth century, Christian symbols are not viable: "they denote in the present serfdom of thought, and serfdom of labour, and serfdom of sex." Socialism, he says, proposes as its ideals "freedom of thought, and freedom of labour, and freedom of sex" (Pearson, *Ethic*, 446). His strongly collectivist political program sets him at odds with libertarian predecessors like Spencer and Mill, however. He argues in "The Moral Basis of Socialism" that the great flaw of modern society lies in its "loss of veneration for the State," and he stresses the need for the socialist state to be strongly led by an aristocratic elite, "the selected few" (*Ethic*, 322).

9. See, for example, N. Katherine Hayles, "Constrained Constructivism," a sophisticated inquiry into "the anthropomorphic grounding that underlies the idea of observables" in science (Hayles, 27–28). Without mentioning him, Hayles follows Pearson closely in her basic distinction between unmediated reality, "inherently unknowable and unreachable by any sentient being," and "the constructed concepts that for us comprise the world" (30). What Pearson invokes as "the chaos beyond sensation . . . the sphere outside knowledge" (*GS*, 108) is what Hayles calls "the flux" (30). She even stages an allegorical walk with her dog, illustrating how different the world appears from human and canine perspectives (30–31)—and mirroring Pearson's similar episode of a walk with another dog, illustrating that "objective reality does not consist of the same sense-impressions for man and dog" (*GS*, 102–3).

10. Pearson confirms this assessment, noting in the preface to the 1911 third edition of *The Grammar of Science* that the book demonstrated to him "how the heterodoxy of the 'eighties had become the commonplace and accepted doctrine of to-day" (*GS*, v).

11. Frazer uses another version of Feuerbach's paradigm to account for the creation of myth, which occurs, he argues, once the original rationale of social practices has been forgotten and the need is felt to manufacture a new one (*Golden Bough*, 553). The alienation of people from their own customs is the origin of the fantastic narrative of myths, as the alienation of labor is the origin of the "fantastic form" of commodities imagined "as independent beings endowed with life" (Marx, *Capital*, 72).

12. In an 1884 essay on socialism, Pearson derides arguments by Conservative politicians to the effect that "it was . . . a law of nature (if not of God) that society should have a basis of misery." "It is the fault of our present social system, and not a law of history, that the toilers should be condemned to extreme misery and poverty," he declares (Pearson, *Ethic*, 350, 352).

13. "The limit to the velocity of signals is our bulwark against that topsy-turvydom of past and future, of which Einstein's theory is sometimes wrongfully accused" (Eddington, *The Nature of the Physical World*, 57–58).

14. The sort of change Pearson sees coming in late-Victorian England is suggested in another essay of 1887 where he reflects on an impending revolution in the system of relations between the sexes: "in order that a woman . . . may save her own soul," he says, "she must have economic independence" (*Ethic*, 433).

15. He denies the scientific validity of the idea of causality, for example (Clifford, *Lectures*, 1:122, 170–72).

16. Eddington argues precisely this, asserting that the fundamental laws and the great numerical constants of nature (the velocity of light, the gravitational constant, the cosmical number, and so forth) are wholly derivable on *a priori* grounds, being in fact "*wholly subjective*"—a consequence, he says, of the proposition that modern or "epistemological" physics "directly investigates *knowledge*, whereas classical physics investigated or endeavoured to investigate an

entity (the external world) which the knowledge is said to describe" (*Philosophy of Physical Science*, 57, 49).

17. Pearson was consistent in asserting the Spencerian doctrine that "the things-in-themselves which . . . sense-impressions symbolise, the 'reality,' as the metaphysicians wish to call it, at the other end of the nerve, remains unknown and is unknowable" (Pearson, *GS*, 63; see also 61–62, 67, 73, 108).

18. When he declares all time measurements to be "propositions about *simultaneous events*" (EMB, 141), for example, he echoes Mach's statement in 1893 that "determinations of time are merely abbreviated statements of the dependence of one event upon another, and nothing more" (Mach, *Popular Scientific Lectures*, 204).

19. Derrida's concept of the linguistic "trace," argues Christine Froula in a brilliant essay, is historically inseparable from scientific relativity. "Morphologically no less than chronologically, it comes 'after the physics' of Einstein and Heisenberg, and it attempts to formulate a language for the cosmos revealed by that physics" (Froula, "Quantum Physics," 288).

20. "Everywhere physical thought must determine for itself its own standards of measurement before it proceeds to observation," says Cassirer in 1921, showing how deeply he has absorbed the lessons of writers like Newman and Schiller. "Each measurement contains a purely ideal element; it is not so much with the sensuous instruments of measurement that we measure natural processes as with our own thoughts" (*Einstein's Theory*, 364, 365).

21. For an especially distinct statement of the dependency of special relativity on the postulate of the constant speed of light, see Einstein and Infeld, *The Evolution of Physics*, 195–96: "If the velocity of light is the same in all [coordinate systems], then moving rods must change their length, moving clocks must change their rhythm, and the laws governing these changes are rigorously determined." All these amazing phenomena begin to occur, in effect, as consequences of Einstein's postulate.

CHAPTER FIVE

1. Modern thinking in anthropology effectively defines itself as a repudiation of Frazer. In Mary Douglas's account, Frazer is eclipsed chiefly by the developing sociologically-oriented heritage of Durkheim; I. C. Jarvie, James A. Boon, Marilyn Strathern, and others give versions of this narrative as Oedipal drama in which Frazer's student Malinowski, as the exponent of scientific fieldwork and functionalism, plays the role of inspiring sponsor of genuinely modernist anthropology—anthropology, that is, that defines itself in a concerted antagonism to Victorian received ideas. "Malinowski plotted and directed the revolution in social anthropology—aiming to overthrow the establishment of Frazer and Tylor and their ideas; but mainly it was against Frazer" (Jarvie, *Revolution in Anthropology*, 173).

2. Stallo quotes this passage with great admiration (*Concepts and Theories*,

200–201). Einstein, of course, shockingly overturns Helmholtz's principle as the founding move of special-relativity theory.

3. Marginalist economics is elaborated philosophically by Georg Simmel in essays such as "A Chapter in the Philosophy of Value" (1900) and "Exchange" (1907). Simmel proclaims relativistic economics at this rather late date with the dramatic rhetoric of one announcing a revolutionary new principle. "*Economic value as such does not inhere in an object in its isolated self-existence*," he says, for example, "*but comes to an object only through the expenditure of another object which is given for it*"; exchange is thus "the economic realization of the relativity of things" ("Exchange," 54, 69). These ideas obviously retained their challenging radical character despite having been old hat for decades. For an interesting discussion of "the relativization of value inherent in marginalism," see Lawrence Birken, "From Macroeconomics to Microeconomics," 260.

4. See, for example, Herbert Dingle, "The Scientific Outlook in 1851." Dingle remarks that the philosophy of Newtonian mechanics "was fundamental and it was universal, but it was so deep-seated that only on the rarest occasions did it appear in explicit statements"; its basis was "the search for the universal, inviolable, causal laws that governed the course of events in the real external world" (Dingle, "Scientific Outlook," 86, 90). See to the same effect P. M. Harman, *Energy, Force, and Matter*, 8–10, though Harman qualifies this picture by remarking that "the conceptual innovations of nineteenth-century physics—energy conservation, the theory of the physical field, the theory of light as the vibrations of an electromagnetic ether, and the concept of entropy—cannot meaningfully be described as 'Newtonian'" (11).

5. He thus stresses repeatedly in his *Autobiography* and elsewhere his consciousness of physical causation as the supreme principle of science (*Autobiography*, 1:172, 2:7, etc.).

6. References to *The Golden Bough* in this chapter are, for convenience, chiefly to the 1922 one-volume abridged edition, abbreviated *GB*. References to the two-volume first edition of 1890 are abbreviated *GB1*. References to the twelve-volume third edition of 1911–15 are to titles of individual volumes.

7. "No legend, no allegory, no nursery rhyme, is safe from the hermeneutics of a thorough-going mythologic theorist," Tylor complains, for example (*Primitive Culture*, 1:319).

8. Frazer collected Spencer's books in great number and was "massively" influenced by him, according to Robert Ackerman (*Frazer*, 22, 34).

9. For discussions of the comparative method that stress its eighteenth-century origins, see Harris, *Rise of Anthropological Theory*, 150–57, and Ackerman, *Frazer*, 46–47.

10. Frazer's biographer Robert Ackerman echoes this criticism, commenting disapprovingly on "the piling of conjecture upon conjecture that is the essence of Frazer's argumentative strategy" (Ackerman, *Frazer*, 170).

11. Marc Manganaro has described this gesture on Frazer's part as a duplicitous strategy of domination, allowing him to posture as a scientific authority but also to withdraw, if challenged, to the position of the literary artist whose writing is to be evaluated not on the basis of "scientific validation," of "scientific truth or ethnographic accuracy," but on that of artistic merit (Manganaro, "'Tangled Bank' Revisited," 116, 121). This dichotomy becomes more complicated if we see Frazer as engaged in the project of destabilizing the orthodox norms of scientific validation itself.

12. I discuss the development of the anthropological concept of culture, though without reference to Frazer, in *Culture and Anomie: Ethnographic Imagination in the Nineteenth Century*.

13. See Girard, *Deceit, Desire, and the Novel*.

14. "It would not be difficult to show that Frazer often comes near to developing a psychoanalytic insight into the unconscious and subconscious motives of human behavior" (Malinowski, *A Scientific Theory of Culture*, 189). The same concept plays a central role in Saussure's linguistics. "Speakers are largely unconscious of the laws of language," he declares (*Course in General Linguistics*, 72).

15. This principle is invoked distinctly by Malinowski in his 1922 study of the Kula system in the Trobriand Islands. "The natives obey the forces and commands of the tribal code," he says, following Frazer though lapsing into mechanistic language that Frazer avoids, "but they do not comprehend them," and thus are unable to describe their own practices in the form of "abstract, general rules" (*Argonauts*, 11, 12; see also 83–84, 397). Ludwik Fleck similarly declares in 1935 that "the individual within the collective is never, or hardly ever, conscious of the prevailing thought style, which almost always exerts an absolutely compulsive force upon his thinking" (*Genesis and Development*, 41). Lévi-Strauss invokes the same principle—neglecting, of course, to recognize its derivation from Frazer—in reference to his own study of phenomena like myth. "It is doubtful, to say the least," he says, to suppose that primitive peoples, beyond their fascination with mythological stories, "have any understanding of the systems of interrelations to which we reduce them" (*Raw and Cooked*, 12).

16. "I cannot but feel that in some places I may have pushed [my theory] too far," says Frazer in the preface to the original edition of 1890. "If this should prove to have been the case, I will readily acknowledge and retract my error as soon as it is brought home to me" (*GB1*, ix).

17. One contemporary attitude resembling the one I here ascribe to Frazer would be the "radical empiricism" of William James, which bases its antagonism to philosophical absolutism on the principle "that a disseminated, distributed, or incompletely unified appearance is the only form that reality may yet have achieved" (*Pluralistic Universe*, 44).

18. His wife stated, for instance, that at the outbreak of the great war in 1914, he did not bother to read the newspapers (Ackerman, *Frazer*, 263).

19. As Edward Westermarck says, primitive religions are characteristically driven by "the ascription of mysterious propensities to blood, and especially to sacrificial blood, which . . . made it a most efficient conductor of curses" (Westermarck, *Origin*, 2:619).

20. Among a host of illustrative cases, Frazer reports an African instance of this kind of purification in which the sacrificial victim was a young woman. "They dragged her alive along the ground, face downwards, from the king's house to the river, a distance of two miles, the crowds who accompanied her crying, 'Wickedness! wickedness!' " The intent, declared Frazer's informant, was "to take away the iniquities of the land" (*GB*, 660).

21. Take for example the later section devoted to the cult and mythology of the Asiatic deity Attis, who like Adonis and other such crypto-fertility figures dies annually in order to enact the fundamental drama of resurrection. Again, by tracing the succession of different editions, we can observe here what appears to be a significant shift in Frazer's thinking. In the editions of 1890 and 1900, the account of Attis stresses his character as a benign fertility deity whose rites were performed to promote the fruitfulness of crops. In the 1914 third edition, the rite of Attis has become a scene of orgiastic masochistic violence symbolizing the violent death of the god. On the next day of the festival, Attis rises from the dead, says Frazer, and "the resurrection of the god was hailed by his disciples as a promise that they too would issue triumphant from the corruption of the grave." The sacrilegious intent of Frazer's commentary grows ever more flagrant as he recounts, for instance, the baptism of blood in which a novice in the mysteries of Attis descended into a pit covered with a wooden grate, where a bull was then slaughtered with a sacred spear. "Its hot reeking blood poured in torrents through the apertures," reports Frazer, "and was received with devout eagerness by the worshipper on every part of his person and garments, till he emerged from the pit, drenched, dripping, and scarlet from head to foot, to receive the homage, nay the adoration, of his fellows as one who had been born again to eternal life and had washed away his sins in the blood of the bull" (*GB*, 408).

22. There is no evidence, Frazer comments in the 1913 preface of *The Scapegoat*, that the Aztecs regarded their sacrificial victims as scapegoats. "The intention of slaying them seems rather to have been to reinforce by a river of human blood the tide of life which might else grow stagnant and stale in the veins of the deities" (*Scapegoat*, vi).

23. For an eloquent compressed account of these events, see Lucien Wolf's article "Anti-Semitism" in the *Encyclopedia Britannica*, 11th ed. (1910–11). For a more complete history of anti-Semitism in nineteenth-century Germany, see Massing, *Rehearsal for Destruction*.

24. That superstitious quack theology can continue to serve as the pretext for persecution is evidenced startlingly in today's *New York Times* (16 January 1999), which reports the Rev. Jerry Falwell's announcement that the Antichrist is now

alive somewhere in the world, planning his subversion of true religion, and that he is "of course . . . Jewish."

25. In *The Aryan Myth: A History of Racist and Nationalist Ideas in Europe*, Léon Poliakov comments on "the way [Nietzsche's] thought was exploited in Fascist and racist propaganda," and concedes that "a careful reading" of Nietzsche does disclose a couple of vague remarks about racial stocks in decline, but apparently nothing worse. Poliakov states, without citation, that Nietzsche seemed to rank Jews as the leading category of mankind, describes his evocations of the "blond beast" and the "master race" as examples of "Nietzschean slogans which, taken out of context, can be fitted into racist catechisms," and cites an anecdote of Nietzsche, already prey to madness, expressing a wish to have all anti-Semites shot (Poliakov, *Aryan Myth*, 299–300). He does not refer to such a passage as the one from the first essay of *The Genealogy of Morals* (1887) on the subversion of "aristocratic" values by the historic Jewish conspiracy. "All the world's efforts against the 'aristocrats,' the 'mighty,' the 'masters,' the 'holders of power,' are negligible by comparison with what has been accomplished against those classes by *the Jews*—the Jews, that priestly nation which eventually realised that the one method of effecting satisfaction on its enemies and tyrants was by means of . . . an act of the *cleverest revenge*." The Jews undermine the noble classes of society, says Nietzsche, by their avid "thirst for revenge" and by their devotion to "hate, Jewish hate." Christ is in fact simply a supremely clever invention of the Jews to foist their malevolent, cunning scheme to impose the values of the poor and base, the slaves, upon European aristocratic leadership; Christianity is unmasked as an instrument of "those very *Jewish* values and those very Jewish ideals." Through this stratagem, the Jews have transmitted to lethal effect their "seductive, intoxicating, defiling, and corrupting influence." The master race has suffered "a blood-poisoning" in which "everything is obviously becoming Judaised, or . . . vulgarised." The Jews thus are the very incarnation of the spirit of "vindictive hatred and revengefulness" and are at the origin of "a certain *diseased taint*" in society; "their weakness causes their hate to expand into a monstrous and sinister shape, a shape which is most crafty and most poisonous." Such passages (*Genealogy of Morals*, 26–35) could indeed serve, even without much wrenching out of context, as an inspiration for racist demagoguery. All this language, entirely characteristic of the anti-Semitic movement that swept Germany in the 1880s, recurs almost verbatim and *ad nauseam*, to mention just one point of reference, in Hitler's oratory against "the Jewish poison" (Hitler, *Speeches*, 25) and in *Mein Kampf*. One cannot in fairness accuse the National Socialists of putting words in Nietzsche's mouth. For his affiliation with the racist politics of contemporary agitators like Wilhelm Marr, Adolf Stöcker, and Herrmann Ahlwardt, see Wolf, "Anti-Semitism," 137.

For another exculpatory reading of Nietzsche that fails to refer to texts like those I have cited, see Hans Sluga, *Heidegger's Crisis*, 41–49. Citing an admiring passage about the Jews from *Human, All Too Human* (1876–80), Sluga praises Nietzsche for his supposedly unwavering "rejection of antisemitism" and insists that "[a] gulf . . . separates him from his later Nazi admirers" (Sluga, *Heidegger's*

Crisis, 46). The anti-Semitic ranting of *The Genealogy of Morals* (1887) is in fact directly continuous with that of *Mein Kampf* or of the Henry Monument.

26. For an illuminating discussion of Clifford as an early apostle of probabilism, see Bernard Lightman, *Origins of Agnosticism*, 168–72.

27. Philipp Frank has stated that it was in the period following his appointment in Prague in 1910 that Einstein, "perhaps for the first time since his childhood," became aware of the problems of the Jewish community (Frank, *Einstein*, 83). This statement merits intense skepticism. In view of the wave of ferocious anti-Semitic persecution that swept the length and breadth of Europe in the last quarter of the nineteenth century and the first decade of the twentieth, and in view of the radicalized intellectual circles frequented by Einstein in Zurich starting in 1896 and of his own "almost fanatical" hatred of injustice and authoritarianism, it seems inconceivable that he could have been oblivious to the dire portents hanging over European Jewry during the decade of his development of special relativity.

28. My discussion of the ideological subtext of Einstein's physics parallels in some respects Lewis S. Feuer's argument in *Einstein and the Generations of Science*. This book is often stimulating, but beyond generally demonstrating that the young Einstein inhabited a milieu full of Marxist revolutionary currents, and beyond characterizing relativity theory as motivated by a spirit of generational rebellion "against the science of the establishment" (Feuer, *Einstein*, 47), it never specifies connections between Einstein's social and scientific ideas; nor, for that matter, does it offer convincing evidence to demonstrate that he was in fact driven by the revolutionary zeal that animated some of his radicalized companions in Zurich.

29. Einstein's formulation echoes Kant in *The Critique of Judgment*: "Between the realm of the natural concept, as the sensible, and the realm of the concept of freedom, as the supersensible, there is a gulf fixed, so that it is not possible to pass from the former to the latter (by means of the theoretical employment of reason), just as if they were so many separate worlds, the first of which is powerless to exercise influence on the second" (Kant, *Critique of Judgment*, 14). As we shall see below, Einstein specifically identifies the realm of scientific theory as "the realm of the concept of freedom."

30. For a thorough account of this process that may well have influenced Einstein's thinking, see Duhem, *The Aim and Structure of Physical Theory* (1906), 55–104.

31. Paul Feyerabend, with his credo of "theoretical anarchism" and his call for absolute "freedom of artistic creation" in science, is of current philosophers of science probably the one who most closely replicates Einstein's position (Feyerabend, *Against Method*, 9, 38). Feyerabend himself systematically echoes the radically libertarian philosopher of science F. C. S. Schiller, who declares, for example, that his goal "is to import democracy into Science, or rather anarchism" (*Formal Logic*, 400).

32. He does not mention perhaps the most pertinent instance of all in this context, the experimental undecidability between the relativistic theory of length

contraction and the mechanical theory of Lorentz and FitzGerald, which explains the same phenomenon in terms of molecular compression.

33. In *Einstein and the Generations of Science*, Feuer emphasizes the distinction between the original 1905 formulation of special relativity and Minkowski's in 1908, which argues for the invariance of physical "intervals" within the space-time continuum, which Minkowski called "the absolute world." See Feuer, *Einstein*, 76. Arnold Sommerfeld quotes Einstein as saying at the time that he himself did not understand Minkowskian mathematics, and that only later did he come to regard the four-dimensional schema as indispensable (Sommerfeld, " Einstein's Seventieth Birthday," 102).

34. Paul R. Gross and Norman Levitt, ardent antirelativists though they are, make this point distinctly (*Higher Superstition*, 102).

35. In "A Relativistic Account of Einstein's Relativity," an essay that shows almost a hyperacute sensitivity to the moral and political implications of special relativity, Bruno Latour argues in effect that Einstein in his theory reintroduces a privileged and indeed an autocratic frame, the one occupied by the scientist whose prerogative it is to coordinate within a single invariant system of calculation the differing reports of physical events sent in to headquarters by observers in different inertial frames of reference. From Latour's perspective, the theory looks like an allegory of a system of subjugation of a working class of information gatherers for the sake of the power and prestige of a physicist-king—or rather, of "an Einstein-God" (Latour, "Relativistic Account," 36). "Independent and active observers" are banished from this system, which has openings only for docile, stupefied, "dominated," rigidly disciplined "delegates" of the authoritative "enunciator"; these lowly workers are transformed "into dependent pieces of apparatus that do nothing but watch the coincidences of hands and notches" (36, 21, 22). The degradation of the observers in this regime is "the price to pay for the freedom and credibility of the enunciator" (19). Thus the abolition of privileged frames of observation is merely a device to ensure the overwhelming status of the scientific intellect that presides over "the chain of command"; Einstein the supposed revolutionary is unmasked as an arch-reactionary; "relativity and absolutism merge" (23, 24–25, 36).

Latour in this perversely brilliant essay offers one of the most extreme statements of the dogma that "relativity . . . is the exact opposite of relativism, as many commentators of Einstein have pointed out" ("Relativistic Account," 14). In common with all such statements, it effaces the astounding proposition of special relativity that the world of experience is intractably different when viewed from different frames of reference. It fictionalizes the class of Einsteinian "observers" in a far-fetched and plainly invidious fashion, supposing as it professes to do that Einstein's account presents knowledge of relativistic science as the exclusive, jealously protected possession of "the enunciator." And it attributes to Einstein a politics sharply at odds with all of his explicit political discourse, as many citations in this chapter indicate.

AFTERWORD

1. For an excellent account of Grote's career, see Turner, *Contesting Cultural Authority*, 326–61.

2. Joss Marsh gives an admiring account of Carlile's career in *Word Crimes*, 60–77.

3. "The Ballot," he declared in 1837, "will be an act of emancipation for all dependent voters," protecting each one "against the tyranny of the great man in his neighbourhood." "I contend," he said in 1862, "that every female . . . has a right to choose for herself among the various types of education, which of them will best suit her own aptitudes, tastes, or plans in life" (quoted in Bain, "Intellectual Character," 30, 32, 166).

4. Grote's parliamentary career was a disappointment to Mill, who found him lacking in "courage and energy" when faced with controversy, and who blamed his ineffective leadership of the radical cause for "ten years of relapse into Toryism" in the wake of Reform. See Mill, *Autobiography*, 177 n.

5. Grote here is contradicting Mitford's analysis of the Athenian decision, at the outset of the Peloponnesian War, to establish a special treasury only to be drawn upon in case of ultimate military emergency, and to punish with death any citizen who would propose diverting this money to other purposes.

6. Einstein later echoes this mighty aphorism of Poincaré's. "By the aid of language different individuals can, to a certain extent, compare their experiences. Then it turns out that certain sense perceptions of different individuals correspond to each other. . . . We are accustomed to regard as real those sense perceptions which are common to different individuals, and which therefore are, in a measure, impersonal" (*Meaning of Relativity*, 1–2). Objective reality is therefore a function of linguistic communication, says Einstein.

Works Cited

Ackerman, Robert. *J. G. Frazer: His Life and Work*. Cambridge: Cambridge University Press, 1987.

Adams, Henry. *The Education of Henry Adams*. Ed. Ernest Samuels. Boston: Houghton Mifflin, 1973 [1907].

d'Alembert, Jean Le Rond. "Discours préliminaire des editeurs." In *Encyclopédie, ou dictionnaire raisonné des sciences, des arts et des métiers*, ed. Jean Le Rond d'Alembert and Denis Diderot, 1:i–xlv. Elmsford, N.Y.: Pergamon Press, 1969 [1751].

Appell, Paul. *Henri Poincaré*. Paris: Plon, 1925.

Arendt, Hannah. "On Humanity in Dark Times: Thoughts about Lessing." In *Men in Dark Times*, trans. Clara and Richard Winston, 3–31. New York: Harcourt, n.d.

———. *The Origins of Totalitarianism*. New York: Harcourt, n.d.

Arnold, Matthew. *Culture and Anarchy*. Ed. J. Dover Wilson. Cambridge: Cambridge University Press, 1966 [1869].

———. "The Literary Influence of Academies." In *Poetry and Criticism of Matthew Arnold*, ed. A. Dwight Culler, 259–79. Boston: Houghton Mifflin, 1961.

Bachelard, Gaston. "The Philosophical Dialectic of the Concepts of Relativity," trans. Forrest W. Williams. In *Albert Einstein: Philosopher-Scientist*, ed. Paul Arthur Schilpp, 565–80. Evanston, Ill.: Library of Living Philosophers, 1949.

Bain, Alexander. "The Intellectual Character and Writings of George Grote." In *The Minor Works of George Grote*, ed. Alexander Bain, [1]–[170]. London: Murray, 1873.

———. *Logic*. 2 vols. 2d ed. London: Longmans, 1873 [1870].

———. *Practical Essays*. New York: Appleton, 1884.

———. *The Senses and the Intellect*. 3d ed. New York: Appleton, 1874 [1855].

Bakhtin, M. M. *The Dialogic Imagination*. Ed. Michael Holquist. Trans. Caryl Emerson and Michael Holquist. Austin: University of Texas Press, 1981.

Balfour, Arthur James. *A Defence of Philosophic Doubt: Being an Essay on the Foundations of Belief*. 2d ed. London: Hodder and Stoughton, 1920 [1879].

———. *The Foundations of Belief: Being Notes Introductory to the Study of Theology.* New York: Longmans, 1895.

Baumer, Franklin L. *Modern European Thought: Continuity and Change in Ideas, 1600–1950.* New York: Macmillan, 1977.

Beer, Gillian. *Darwin's Plots: Evolutionary Narrative in Darwin, George Eliot and Nineteenth-Century Fiction.* London: Routledge, 1983.

———. *Open Fields: Science in Cultural Encounter.* Oxford: Clarendon, 1996.

Bentham, Jeremy. *An Introduction to the Principles of Morals and Legislation.* Oxford: Clarendon, 1879 [1789].

Berlin, Isaiah. *The Crooked Timber of Humanity: Chapters in the History of Ideas.* Ed. Henry Hardy. New York: Alfred A. Knopf, 1991.

Bernstein, Richard J. *Beyond Objectivism and Relativism: Science, Hermeneutics, and Praxis.* Philadelphia: University of Pennsylvania Press, 1983.

Beyerchen, Alan D. *Scientists under Hitler: Politics and the Physics Community in the Third Reich.* New Haven, Conn.: Yale University Press, 1977.

Birken, Lawrence. "From Macroeconomics to Microeconomics: The Marginalist Revolution in Sociocultural Perspective." *History of Political Economy* 20 (1988): 251–64.

Blake, William. *The Poetry and Prose of William Blake.* Ed. David V. Erdman. Garden City, N.Y.: Doubleday, 1965.

Bloom, Allan. *The Closing of the American Mind.* New York: Simon and Schuster, 1987.

Bohr, Niels. "Discussion with Einstein on Epistemological Problems in Atomic Physics." In *Albert Einstein: Philosopher-Scientist,* ed. Paul Arthur Schilpp, 201–41. Evanston, Ill.: Library of Living Philosophers, 1949.

Bradley, F. H. *The Principles of Logic.* 2 vols. 2d ed. Oxford: Oxford University Press, 1963 [1883].

Bridgman, Percy W. "Einstein's Theories and the Operational Point of View." In *Albert Einstein: Philosopher-Scientist,* ed. Paul Arthur Schilpp, 335–54. Evanston, Ill.: Library of Living Philosophers, 1949.

———. "Introduction." In J. B. Stallo, *The Concepts and Theories of Modern Physics,* vii–xxix. Cambridge: Harvard University Press, 1960.

———. *The Nature of Physical Theory.* New York: Dover, 1936.

Brown, James Robert. *Smoke and Mirrors: How Science Reflects Reality.* London: Routledge, 1994.

Burnyeat, M. F. "Protagoras and Self-Refutation in Plato's *Theaetetus.*" *Philosophical Review* 85 (1976): 172–95.

Butler, Samuel. *Collected Essays.* 2 vols. Ed. Henry Festing Jones and A. T. Bartholomew. London: Cape, 1925.

———. *The Way of All Flesh.* Ed. James Cochrane. Harmondsworth: Penguin, 1971 [1903].

Canguilhem, Georges. *Ideology and Rationality in the History of the Life Sciences*. Trans. Arthur Goldhammer. Cambridge, Mass.: MIT Press, 1988.

Carlyle, Thomas. "On History." In *Victorian Literature: Prose*, ed. G. B. Tennyson and Donald J. Gray, 38–43. New York: Macmillan, 1976.

———. *Past and Present*. London: Dent, 1960 [1843].

Carus, Paul. *The Principle of Relativity in the Light of the Philosophy of Science*. Chicago: Open Court, 1913.

Cassirer, Ernst. *Einstein's Theory of Relativity Considered from the Epistemological Standpoint*. Trans. William Curtis Swabey and Marie Collins Swabey. New York: Dover, 1953 [1921].

———. "Structuralism in Modern Linguistics." *Word* 1 (1945): 99–120.

———. *Substance and Function*. Trans. William Curtis Swabey and Marie Collins Swabey. N.p.: Dover, 1953 [1910].

Chesterton, G. K. *The Victorian Age in Literature*. London: Williams and Norgate, n.d.

Clark, Ronald W. *Einstein: His Life and Times*. New York: Avon, 1971.

Clifford, William Kingdon. *Lectures and Essays*. 2 vols. Ed. Leslie Stephen and Sir Frederick Pollock. London: Macmillan, 1901 [1879].

———. *Seeing and Thinking*. 2d ed. London: Macmillan, 1880.

Conrad, Joseph. *The Secret Agent: A Simple Tale*. Garden City, N.Y.: Doubleday, 1953 [1907].

Culler, Jonathan. *Ferdinand de Saussure*. Harmondsworth: Penguin, 1977.

Dale, Peter Allan. *In Pursuit of a Scientific Culture: Science, Art, and Society in the Victorian Age*. Madison: University of Wisconsin Press, 1989.

Darwin, Charles. *The Origin of Species by Means of Natural Selection, or the Preservation of Favoured Races in the Struggle for Life*. 6th ed. London: Oxford University Press, 1963 [1859].

Daston, Lorraine, and Peter Galison. "The Image of Objectivity." *Representations* 40 (1992): 81–128.

de Man, Paul. *Romanticism and Contemporary Criticism: The Gauss Seminar and Other Papers*. Ed. E. S. Bury, Kevin Newmark, and Andrzej Warminski. Baltimore, Md.: Johns Hopkins University Press, 1993.

Derrida, Jacques. "Differance." In Jacques Derrida, *Speech and Phenomena and Other Essays on Husserl's Theory of Signs*, trans. and ed. David B. Allison, 129–60. Evanston, Ill.: Northwestern University Press, 1973.

———. *Of Grammatology*. Trans. Gayatri Chakravorty Spivak. Baltimore, Md.: Johns Hopkins University Press, 1976.

Dewey, John. *Essays in Experimental Logic*. Chicago: University of Chicago Press, 1916.

———. *Freedom and Culture*. New York: Putnam, 1939.

———. *The Influence of Darwin on Philosophy and Other Essays in Contemporary Thought*. Bloomington: Indiana University Press, 1965 [1910].

Dickens, Charles. *Bleak House*. Ed. George Ford and Sylvère Monod. New York and London: Norton, 1977 [1852–53].

Dingle, Herbert. "The Scientific Outlook in 1851 and in 1951." *British Journal for the Philosophy of Science* 2 (1951): 85–104.

Downie, R. Angus. *Frazer and the Golden Bough*. London: Gollancz, 1970.

Drake, Stillman. "J. B. Stallo and the Critique of Classical Physics." In *Men and Moments in the History of Science*, ed. Herbert M. Evans, 22–37. Seattle: University of Washington Press, 1959.

Duhem, Pierre. *The Aim and Structure of Physical Theory*. Trans. Philip P. Wiener. Princeton: Princeton University Press, 1954 [1906].

Durkheim, Emile. *The Elementary Forms of the Religious Life*. Trans. Joseph Ward Swain. New York: Free Press, 1965 [1912].

———. *The Rules of Sociological Method*. Ed. Steven Lukes. Trans. W. D. Halls. New York: Free Press, 1982 [1895].

Dutton, Denis. "Knowledge Replacement Therapy." *Philosophy and Literature* 21 (1997): 208–21.

Eddington, A. S. *The Nature of the Physical World*. Cambridge: Cambridge University Press, 1932 [1928].

———. *The Philosophy of Physical Science*. New York: Macmillan, 1939.

———. *Space, Time and Gravitation: An Outline of the General Relativity Theory*. Cambridge: Cambridge University Press, 1953 [1920].

Einstein, Albert. "Autobiographical Notes." In *Albert Einstein: Philosopher-Scientist*, ed. Paul Arthur Schilpp, 2–95. Evanston, Ill.: Library of Living Philosophers, 1949.

———. "On the Electrodynamics of Moving Bodies." In *The Collected Papers of Albert Einstein*, 2 vols., trans. Anna Beck, 2:140–71. Princeton: Princeton University Press, 1987.

———. *The Meaning of Relativity*. 5th ed. Princeton: Princeton University Press, 1955.

———. "Physics and Reality," trans. Jean Picard. *Journal of the Franklin Institute* 221 (1936): 349–82.

———. *Relativity: The Special and the General Theory*. Trans. Robert W. Lawson. New York: Crown, 1961 [1916].

———. "Remarks Concerning the Essays Brought Together in This Co-Operative Volume," trans. Paul Arthur Schilpp. In *Albert Einstein: Philosopher-Scientist*, ed. Paul Arthur Schilpp, 665–88. Evanston, Ill.: Library of Living Philosophers, 1949.

———. "Remarks on Bertrand Russell's Theory of Knowledge." In *The Philosophy of Bertrand Russell*, ed. Paul Arthur Schilpp, 279–91. Evanston and Chicago: Northwestern University Press, 1944.

———. *The World as I See It*. New York: Covici-Friede, 1934.

Einstein, Albert, and Leopold Infeld. *The Evolution of Physics: The Growth of Ideas from Early Concepts to Relativity and Quanta*. New York: Simon and Schuster, 1942.

Eliot, George. *Middlemarch*. Baltimore, Md.: Penguin, 1965 [1871–72].

———. *The Mill on the Floss*. Ed. Gordon S. Haight. Boston: Houghton Mifflin, 1961 [1860].

Eliot, T. S. "A Prediction in Regard to Three English Authors." *Vanity Fair* 21 (1924): 29.

Ely, Richard T. *Ground under Our Feet: An Autobiography*. New York: Macmillan, 1938.

Engels, Friedrich. *The Condition of the Working Class in England*. Ed. and trans. W. O. Henderson and W. H. Chaloner. Stanford, Calif.: Stanford University Press, 1968 [1845].

———. *Socialism: Utopian and Scientific*. Trans. Edward Aveling. New York: International Publishers, 1935 [1882].

Ermarth, Elizabeth Deeds. *Realism and Consensus in the English Novel*. Princeton: Princeton University Press, 1983.

Evans-Pritchard, E. E. *The Nuer: A Description of the Modes of Livelihood and Political Institutions of a Nilotic People*. New York: Oxford University Press, 1982 [1940].

Falck, Colin. *Myth, Truth and Literature: Towards a True Post-Modernism*. Cambridge: Cambridge University Press, 1989.

Feuer, Lewis S. *Einstein and the Generations of Science*. New York: Basic Books, 1974.

Feuerbach, Ludwig. *The Essence of Christianity*. Trans. George Eliot. New York: Harper, 1957 [1841].

Feyerabend, Paul. *Against Method*. Rev. ed. London: Verso, 1988.

———. "Nature as a Work of Art." *Common Knowledge* 1 (1992): 3–9.

———. *Science in a Free Society*. London: Verso, 1978.

Fleck, Ludwik. *Genesis and Development of a Scientific Fact*. Ed. Thaddeus J. Trenn and Robert K. Merton. Trans. Fred Bradley and Thaddeus J. Trenn. Chicago: University of Chicago Press, 1979.

Flint, Robert. *Agnosticism*. Edinburgh: Blackwood, 1903.

Foster, Hal, ed. *The Anti-Aesthetic: Essays on Postmodern Culture*. Port Townsend, Wash.: Bay, 1983.

Foucault, Michel. *Discipline and Punish: The Birth of the Prison*. Trans. Alan Sheridan. New York: Vintage Books, 1979.

———. *The Order of Things: An Archaeology of the Human Sciences*. New York: Vintage Books, 1973.

———. "What Is an Author?" In *Textual Strategies: Perspectives in Post-Structuralist Criticism*, ed. Josué V. Harari, 141–60. Ithaca, N.Y.: Cornell University Press, 1979.

Frank, Philipp. *Einstein: His Life and Times*. Trans. George Rosen. Ed. and rev. Shuichi Kusaka. New York: Alfred A. Knopf, 1947.

———. "Einstein, Mach, and Logical Positivism." In *Albert Einstein: Philosopher-Scientist*, ed. Paul Arthur Schilpp, 271–86. Evanston, Ill.: Library of Living Philosophers, 1949.

Fraser, Robert. *The Making of* The Golden Bough: *The Origins and Growth of an Argument*. New York: St. Martin's, 1990.

Frazer, James. *Adonis Attis Osiris: Studies in the History of Oriental Religion*. 2 vols. London: Macmillan, 1914.

———. *Balder the Beautiful: The Fire-Festivals of Europe and the Doctrine of the External Soul*. 2 vols. London: Macmillan, 1914 [1913].

———. *The Golden Bough: The Roots of Religion and Folklore*. 2 vols. New York: Avenel, 1981 [1890].

———. *The Golden Bough: A Study in Magic and Religion*. 2d ed. 3 vols. London: Macmillan, 1900.

———. *The Golden Bough: A Study in Magic and Religion*. Abridged Edition. New York: Collier, 1963 [1922].

———. *The Magic Art and the Evolution of Kings*. 2 vols. London: Macmillan, 1917 [1911].

———. *The Scapegoat*. London: Macmillan, 1914 [1913].

———. *Taboo and the Perils of the Soul*. London: Macmillan, 1919 [1911].

Freud, Sigmund. *Civilization and Its Discontents*. Trans. and ed. James Strachey. New York: Norton, 1961 [1930].

———. "A Difficulty in the Path of Psycho-Analysis." Trans. James Strachey. In *The Standard Edition of the Complete Psychological Works of Sigmund Freud*, ed. James Strachey, 17:137–44. London: Hogarth, 1953–74 [1917].

———. "The Uncanny," trans. Alix Strachey. In *The Standard Edition of the Complete Psychological Works of Sigmund Freud*, ed. James Strachey, 17:218–56. London: Hogarth, 1953–74 [1919].

Friedman, Alan J., and Carol C. Donley. *Einstein as Myth and Muse*. London: Cambridge University Press, 1985.

Froude, J. A. *The Nemesis of Faith*. 2d ed. London: Chapman, 1849.

Froula, Christine. "Quantum Physics/Postmodern Metaphysics: The Nature of Jacques Derrida." *Western Humanities Review* 39 (1985): 287–313.

Geertz, Clifford. "Distinguished Lecture: Anti Anti-Relativism." *American Anthropologist* 86 (1984): 263–78.

———. *The Interpretation of Cultures*. New York: Basic Books, 1973.

Gellner, Ernest. *Relativism and the Social Sciences*. Cambridge: Cambridge University Press, 1985.

Girard, René. *Deceit, Desire, and the Novel: Self and Other in Literary Structure*. Trans. Yvonne Freccero. Baltimore, Md.: Johns Hopkins University Press, 1966.

———. *Violence and the Sacred*. Trans. Patrick Gregory. Baltimore, Md.: Johns Hopkins University Press, 1977.

Goldhagen, Daniel Jonah. *Hitler's Willing Executioners: Ordinary Germans and the Holocaust*. New York: Vintage Books, 1997.

Green, Thomas Hill. *Prolegomena to Ethics*. Ed. A. C. Bradley. 3d ed. Oxford: Clarendon, 1890 [1883].

Gross, Paul R., and Norman Levitt. *Higher Superstition: The Academic Left and Its Quarrels with Science*. Baltimore, Md.: Johns Hopkins University Press, 1994.

Grote, George. *A History of Greece; From the Earliest Period to the Close of the Generation Contemporary with Alexander the Great*. 4th ed. 10 vols. London: Murray, 1872 [1846–56].

———. *The Minor Works of George Grote*. Ed. Alexander Bain. London: Murray, 1873.

———. *Plato, and the Other Companions of Sokrates*. 2d ed. 4 vols. London: Murray, 1888 [1865].

Habermas, Jürgen. "Work and Weltanschauung: The Heidegger Controversy from a German Perspective," Trans. John McCumber. *Critical Inquiry* 15 (1989): 431–56.

Hamilton, William. "Philosophy of the Unconditioned." In Sir William Hamilton, *Discussions on Philosophy and Literature, Education and University Reform*, 9–44. New York: Harper, 1855 [1829].

Harman, P. M. *Energy, Force, and Matter: The Conceptual Development of Nineteenth-Century Physics*. Cambridge: Cambridge University Press, 1982.

Harris, Marvin. *The Rise of Anthropological Theory: A History of Theories of Culture*. New York: Crowell, 1968.

Harrison, Frederic. *The Philosophy of Common Sense*. Freeport, N.Y.: Books for Libraries Press, 1968 [1907].

Hayles, N. Katherine. "Constrained Constructivism: Locating Scientific Inquiry in the Theater of Representation." In *Realism and Representation: Essays on the Problem of Realism in Relation to Science, Literature, and Culture*, ed. George Levine, 27–43. Madison: University of Wisconsin Press, 1993.

———. *The Cosmic Web: Scientific Field Models and Literary Strategies in the Twentieth Century*. Ithaca, N.Y.: Cornell University Press, 1984.

Heidegger, Martin. "Letter on Humanism," trans. Frank A. Capuzzi and J. Glenn

Gray. In *Martin Heidegger: Basic Writings*, ed. David Farrell Krell, 189–242. New York: Harper, 1977.

Heisenberg, Werner. *The Physicist's Conception of Nature*. Trans. Arnold J. Pomerans. London: Hutchinson, 1958.

Helmholtz, Hermann von. "Recent Progress in the Theory of Vision." In *Selected Writings of Hermann von Helmholtz*, ed. Russell Kahl, 144–222. Middletown, Conn.: Wesleyan University Press, 1971 [1868].

Herbert, Christopher. *Culture and Anomie: Ethnographic Imagination in the Nineteenth Century*. Chicago: University of Chicago Press, 1991.

Himmelfarb, Gertrude. *The De-Moralization of Society: From Victorian Virtues to Modern Values*. New York: Alfred A. Knopf, 1995.

Hitler, Adolf. *Mein Kampf*. New York: Reynal & Hitchcock, 1940.

———. *The Speeches of Adolf Hitler: April 1922–August 1939*. Trans. and ed. Norman H. Baynes. London: Oxford University Press, 1942.

Hodgen, Margaret T. *The Doctrine of Survivals: A Chapter in the History of Scientific Method in the Study of Man*. London: Allenson, 1936.

Hoffmann, Banesh, with Helen Dukas. *Albert Einstein: Creator and Rebel*. New York: Viking, 1972.

Holton, Gerald. "Einstein, Michelson, and the 'Crucial' Experiment." *Isis* 60 (1969): 133–97.

———. "Influences on Einstein's Early Work in Relativity Theory." *American Scholar* 37 (1967): 59–79.

———. "Mach, Einstein, and the Search for Reality." *Daedalus* 97 (1968): 636–73.

Houghton, Walter E. *The Victorian Frame of Mind 1830–1870*. New Haven: Yale University Press, 1957.

Huxley, Thomas Henry. *Method and Results*. New York: Appleton, 1897.

Hyman, Stanley Edgar. *The Tangled Bank: Darwin, Marx, Frazer and Freud as Imaginative Writers*. New York: Atheneum, 1962.

James, William. *The Meaning of Truth: A Sequel to "Pragmatism"*. Cambridge: Harvard University Press, 1978 [1909].

———. *A Pluralistic Universe*. New York: Longmans, 1916 [1909].

———. *Pragmatism: A New Name for Some Old Ways of Thinking*. Cambridge: Harvard University Press, 1978 [1907].

———. *The Principles of Psychology*. 2 vols. New York: Dover, 1950 [1890].

Jameson, Fredric. "Postmodernism and Consumer Society." In *The Anti-Aesthetic: Essays on Postmodern Culture*, ed. Hal Foster, 111–25. Seattle: Bay, 1983.

Jarvie, I. C. *The Revolution in Anthropology*. London: Routledge, 1964.

Jevons, W. Stanley. *The Theory of Political Economy*. 3d ed. London: Macmillan, 1888 [1871].

Josephson, Paul R. *Totalitarian Science and Technology*. Atlantic Highlands, N.J.: Humanities Press, 1996.

Judt, Tony. "Writing History, Facts Optional." *New York Times*, 13 April 2000, natl. ed., A27.

Kant, Immanuel. *The Critique of Judgement*. Trans. James Creed Meredith. Oxford: Clarendon, 1952.

Kelvin, Lord (William Thomson). "Lecture XI." In *Kelvin's Baltimore Lectures and Modern Theoretical Physics: Historical and Philosophical Perspectives*, ed. Robert Kargon and Peter Achinstein, 106–14. Cambridge: MIT Press, 1987 [1884].

Kern, Stephen. *The Culture of Time and Space 1880–1918*. Cambridge: Harvard University Press, 1983.

Kuhn, Thomas S. *The Structure of Scientific Revolutions*. 2d ed. Chicago: University of Chicago Press, 1970.

Larsen, Neil. *Reading North by South: On Latin American Literature, Culture, and Politics*. Minneapolis: University of Minnesota Press, 1995.

Latour, Bruno. "A Relativistic Account of Einstein's Relativity." *Social Studies of Science* 18 (1988): 3–44.

Larmor, J. "Introduction." In Henri Poincaré, *Science and Hypothesis*, xii–xxvii. London: Scott, 1905.

Laudan, Larry. *Science and Relativism: Some Key Controversies in the Philosophy of Science*. Chicago: University of Chicago Press, 1990.

Lenin, V. I. *Materialism and Empirio-Criticism: Critical Comments on a Reactionary Philosophy*. In *V. I. Lenin: Selected Works*, 11:87–409. New York: International Publishers, n.d.

Leo, John. "Truth Not Objective on These Campuses." *Grand Haven Tribune*, 2 August 1995, 4.

Lévi-Strauss, Claude. *The Raw and the Cooked: Introduction to a Science of Mythology*, vol. 1. Trans. John and Doreen Weightman. Chicago: University of Chicago Press, 1969.

———. *Structural Anthropology*. Trans. Claire Jacobson and Brooke Grundfest Schoepf. New York: Basic Books, 1963.

Levine, George. *Darwin and the Novelists: Patterns of Science in Victorian Fiction*. Cambridge: Harvard University Press, 1988.

———. "Two Ways to Be a Solipsist." Unpublished manuscript.

Lévy-Bruhl, Lucien. *Ethics and Moral Science*. Trans. Elizabeth Lee. London: Constable, 1905 [1903].

———. *How Natives Think*. Trans. Lilian A. Clare. Princeton: Princeton University Press, 1985 [1910].

Lewes, George Henry. *The History of Philosophy From Thales to Comte*. 4th ed. 2 vols. London: Longmans, 1871.

———. *Problems of Life and Mind: First Series: The Foundations of a Creed*. 2 vols. Boston: Houghton Mifflin, 1891 [1873–79].

Lightman, Alan. *Great Ideas in Physics*. 2d ed. New York: McGraw-Hill, 1997.

Lightman, Bernard. "Fighting Even with Death: Balfour, Scientific Naturalism, and Thomas Henry Huxley's Final Battle." In *Thomas Henry Huxley's Place in Science and Letters*, ed. Alan P. Barr, 323–50. Athens: University of Georgia Press, 1997.

———. *The Origins of Agnosticism: Victorian Unbelief and the Limits of Knowledge*. Baltimore: Johns Hopkins University Press, 1987.

Mach, Ernst. *Knowledge and Error: Sketches on the Psychology of Enquiry*. Trans. Thomas J. McCormack and Paul Foulkes. Dordrecht: Reidel, 1976 [1905].

———. *Popular Scientific Lectures*. Trans. Thomas J. McCormack. La Salle, Ill.: Open Court, 1986 [1893].

Malinowski, Bronislaw. *Argonauts of the Western Pacific: An Account of Native Enterprise and Adventure in the Archipelagoes of Melanesian New Guinea*. London: Routledge, 1932 [1922].

———. *A Scientific Theory of Culture and Other Essays*. Chapel Hill: University of North Carolina Press, 1944.

Manganaro, Marc. " 'The Tangled Bank' Revisited: Anthropological Authority in Frazer's *The Golden Bough*." *The Yale Journal of Criticism* 3 (1989): 107–26.

Mansel, Henry Longueville. *The Limits of Religious Thought Examined*. Boston: Gould and Lincoln, 1859 [1858].

Marsh, Joss. *Word Crimes: Blasphemy, Culture, and Literature in Nineteenth-Century England*. Chicago: University of Chicago Press, 1998.

Marx, Karl. *Capital: A Critique of Political Economy*, vol. 1. Trans. Samuel Moore and Edward Aveling. Ed. Frederick Engels. New York: International Publishers, 1967.

Massing, Paul W. *Rehearsal for Destruction: A Study of Political Anti-Semitism in Imperial Germany*. New York: Harper, 1949.

Maxwell, James Clerk. "A Dynamical Theory of the Electromagnetic Field." In *The Scientific Papers of James Clerk Maxwell*, 2 vols., ed. W. D. Niven, 1:526–97. New York: Dover, 1965 [1864].

McLennan, John F. *Primitive Marriage: An Inquiry into the Origin of the Form of Capture in Marriage Ceremonies*. Ed. Peter Rivière. Chicago: University of Chicago Press, 1970 [1865].

———. "The Worship of Animals and Plants." *Fortnightly Review* 1869–70. (Part 1, "Totems and Totemism": 6, New Series [1869]: 407–27; Part 2, "Totem-Gods Among the Ancients": 6, New Series [1869]: 562–82; Part 2 [concluded]: 7, New Series [1870]: 194–216.)

Merz, John Theodore. *A History of European Thought in the Nineteenth Century*. 4 vols. New York: Dover, 1965 [1896–1914].

Mill, John Stuart. *Autobiography*. Ed. Jack Stillinger. Boston: Houghton Mifflin, 1969 [1873].

———. *An Examination of Sir William Hamilton's Philosophy, and of the Principal Philosophical Questions Discussed in His Writings*. 2 vols. Boston: Spencer, 1865.

———. *On Liberty*. New York: Appleton-Century-Crofts, 1947 [1859].

———. *Principles of Political Economy, with Some of Their Applications to Social Philosophy*. 2 vols. Boston: Little, Brown, 1848.

———. *A System of Logic, Ratiocinative and Inductive: Being a Connected View of the Principles of Evidence and the Methods of Scientific Investigation*. New York: Harper, 1867 [1843].

Milton, John. "Areopagitica." In *Areopagitica and Other Prose Works of John Milton*, 1–41. London: Dent, 1927.

Mivart, St. George. "Denominational Science." *Fortnightly Review* 64 (1895): 423–38.

———. *The Groundwork of Science: A Study of Epistemology*. New York: Putnam, 1898.

———. *On Truth: A Systematic Inquiry*. London: Kegan Paul, 1889.

Morgan, C. Lloyd. *Animal Life and Intelligence*. London: Arnold, 1890–91.

Needham, Rodney. *Belief, Language, and Experience*. Oxford: Blackwell, 1972.

Newman, John Henry. *An Essay in Aid of a Grammar of Assent*. Garden City, N.Y.: Doubleday, 1955 [1870].

———. "The Theory of Developments in Religious Doctrine." In *Fifteen Sermons Preached before the University of Oxford,* by John Henry Newman, 312–51. Westminster, Md.: Christian Classics, 1966 [1843].

Nietzsche, Friedrich. *The Genealogy of Morals: A Polemic*. Trans. Horace B. Samuel. New York: Russell and Russell, 1964.

———. *The Will to Power.* 2 vols. Trans. Anthony M. Ludovici. London: Allen & Unwin, 1924.

———. "On Truth and Falsity in their Ultramoral Sense." In *Early Greek Philosophy and Other Essays*, trans. Maximilian A. Mügge, 171–92. New York: Russell and Russell, 1964 [1873].

Nozick, Robert. *Philosophical Explanations*. Cambridge: Harvard University Press, 1981.

Ortega y Gasset, José. *The Dehumanization of Art and Other Essays on Art, Culture, and Literature*. Princeton: Princeton University Press, 1968.

———. *The Modern Theme*. Trans. James Cleugh. New York: Harper, 1961 [1923].

Packe, Michael St. John. *The Life of John Stuart Mill*. New York: Capricorn, 1970.

Pais, Abraham. *"Subtle is the Lord . . .": The Science and the Life of Albert Einstein*. Oxford: Clarendon, 1982.

Park, David. *The Fire within the Eye: A Historical Essay on the Nature and Meaning of Light*. Princeton: Princeton University Press, 1997.

Pater, Walter. *Appreciations, with An Essay on Style*. Evanston, Ill.: Northwestern University Press, 1987 [1865].

Pearson, Karl. *The Ethic of Freethought: A Selection of Essays and Lectures*. London: Unwin, 1888.

———. *The Grammar of Science: Part I—Physical*. 3d ed. New York: Macmillan, 1911 [1892].

———. *National Life from the Standpoint of Science*. London: Black, 1901.

Pipes, Richard, ed. *The Unknown Lenin: From the Secret Archive*. Trans. Catherine A. Fitzpatrick. New Haven and London: Yale University Press, 1996.

Plato. *Protagoras*. Trans. C. C. W. Taylor. Rev. ed. Oxford: Clarendon, 1991.

———. *Theaetetus*. Trans. John McDowell. Oxford: Clarendon, 1973.

Poincaré, Henri. *Science and Hypothesis*. Trans. W. J. G. London and Newcastle-on-Tyne: Scott, 1905 [1902].

———. *The Value of Science*. Trans. George Bruce Halsted. In Henri Poincaré, *The Foundations of Science*, 201–355. New York: The Science Press, 1921.

Poincaré, Lucien. *The New Physics and Its Evolution*. London: Kegan Paul, 1907.

Poliakov, Léon. *The Aryan Myth: A History of Racist and Nationalist Ideas in Europe*. Trans. Edmund Howard. London: Chatto, 1974.

Pollock, Sir Frederick. "Biographical." In William Kingdon Clifford, *Lectures and Essays*, ed. Leslie Stephen and Sir Frederick Pollock, 1:1–55. London: Macmillan, 1901.

Popper, Karl R. *The Myth of the Framework: In Defence of Science and Rationality*. Ed. M. A. Notturno. London: Routledge, 1994.

———. *The Open Society and Its Enemies*, vol. 2, *The High Tide of Prophecy: Hegel, Marx, and the Aftermath*. 5th ed. Princeton: Princeton University Press, 1966.

Quillard, Pierre. *Le monument Henry*. Paris: Stock, 1899.

Ray, Christopher. *The Evolution of Relativity*. Bristol: Hilger, 1987.

Reichenbach, Hans. "The Philosophical Significance of the Theory of Relativity." In *Albert Einstein: Philosopher-Scientist*, ed. Paul Arthur Schilpp, 289–311. Evanston, Ill.: Library of Living Philosophers, 1949.

Rindler, Wolfgang. *Introduction to Special Relativity*. 2d ed. Oxford: Clarendon, 1991.

Russell, Bertrand. *The ABC of Relativity*. New York: Harper, 1925.

Sartre, Jean-Paul. *Existentialism and Humanism*. Trans. Philip Mairet. London: Methuen, 1948.

Saussure, Ferdinand de. *Course in General Linguistics*. Trans. Wade Baskin. Ed. Charles Bally, Albert Sechehaye, and Albert Riedlinger. New York: McGraw-Hill, 1966 [1916].

Schiller, F. C. S. *Formal Logic: A Scientific and Social Problem*. London: Macmillan, 1912.

———. *Logic for Use: An Introduction to the Voluntarist Theory of Knowledge*. New York: Harcourt Brace, 1930.

———. *Studies in Humanism*. 2d ed. London: Macmillan, 1912 [1907].

Schmidt, Lawrence, ed. *The Specter of Relativism: Truth, Dialogue, and Phronesis in Philosophical Hermeneutics*. Evanston, Ill.: Northwestern University Press, 1995.

Schwartz, Regina M. *The Curse of Cain: The Violent Legacy of Monotheism*. Chicago: University of Chicago Press, 1997.

Shelley, Percy Bysshe. "A Defence of Poetry." In *Critical Theory since Plato*, ed. Hazard Adams, 516–29. Rev. ed. Fort Worth, Texas: Harcourt, 1992.

Shirer, William L. *The Rise and Fall of the Third Reich: A History of Nazi Germany*. New York: Simon and Schuster, 1960.

Siegel, Harvey. *Relativism Refuted: A Critique of Contemporary Epistemological Relativism*. Dordrecht: Reidel, 1987.

Simmel, Georg. "Exchange." In Georg Simmel, *On Individuality and Social Forms*, ed. Donald N. Levine, 43–69. Chicago: University of Chicago Press, 1971 [1907].

Sluga, Hans. *Heidegger's Crisis: Philosophy and Politics in Nazi Germany*. Cambridge: Harvard University Press, 1993.

Small, Ian. *Conditions for Criticism: Authority, Knowledge, and Literature in the Late Nineteenth Century*. Oxford: Clarendon, 1991.

Smith, Adam. "The History of Astronomy." In Adam Smith, *Essays on Philosophical Subjects*, ed. W. P. D. Wightman and J. C. Bryce, 31–105. Oxford: Clarendon, 1980.

Smith, Barbara Herrnstein. *Contingencies of Value: Alternative Perspectives for Critical Theory*. Cambridge: Harvard University Press, 1988.

Sokal, Alan D. "Transgressing the Boundaries: Toward a Transformative Hermeneutics of Quantum Gravity." *Social Text* 14 (1996): 217–52.

Solovine, Maurice. "Introduction," trans. Wade Baskin. In Albert Einstein, *Letters to Solovine*, 5–15. New York: Philosophical Library, 1987.

Sommerfeld, Arnold. "To Albert Einstein's Seventieth Birthday." In *Albert Einstein: Philosopher-Scientist*, ed. Paul Arthur Schilpp, 99–105. Evanston, Ill.: Library of Living Philosophers, 1949.

Spencer, Herbert. *An Autobiography*. 2 vols. New York: Appleton, 1904.

———. *First Principles*. 6th ed. London: Williams and Norgate, 1908 [1862].

———. *The Principles of Psychology*. 2 vols. New York: Appleton, 1896 [1855].

———. *The Principles of Sociology*. 3d ed. 3 vols. London: Williams and Norgate, 1897–1906 [1876–96].

———. *The Study of Sociology*. 22d ed. London: Kegan Paul, n.d. [1872–73].

Spenser, Edmund. *The Faerie Queene*. In *The Complete Poetical Works of Spenser*, ed. R. E. Neil Dodge, 130–677. Boston: Houghton Mifflin, n.d.

Stallo, J. B. *The Concepts and Theories of Modern Physics*. Ed. Percy W. Bridgman. Cambridge: Harvard University Press, 1960 [1881].

———. *State Creeds and Their Modern Apostles: A Lecture Delivered in Rev. Mr. Vickers' Church, Cincinnati, on the Evening of April 3, 1870*. Cincinnati: Robert Clarke, 1872.

Stephen, Leslie. *An Agnostic's Apology and Other Essays*. 2d ed. London: Smith, Elder, 1903.

———. "Art and Morality." *Cornhill* 32 (1875): 91–101.

Stewart, Susan. *On Longing: Narratives of the Miniature, the Gigantic, the Souvenir, the Collection*. Baltimore: Johns Hopkins University Press, 1984.

Strathern, Marilyn. "Out of Context: The Persuasive Fictions of Anthropology." *Current Anthropology* 28 (1987): 251–81.

Sturt, Henry. "Relativity of Knowledge." *Encyclopaedia Britannica*, 11th ed., 1910–11.

Tickner, Lisa. "Men's Work? Masculinity and Modernism." In *Visual Culture: Images and Interpretations*, ed. Norman Bryson, Ann Holly, and Keith Moxey, 42–82. Hanover, N.H.: Wesleyan University Press, 1994.

Tonnelat, Marie-Antoinette. *Histoire du principe de relativité*. Paris: Flammarion, 1971.

Toulmin, Stephen. *Human Understanding*, vol. 1. Princeton: Princeton University Press, 1972.

Turner, Frank Miller. *Between Science and Religion: The Reaction to Scientific Naturalism in Late Victorian England*. New Haven: Yale University Press, 1974.

———. *Contesting Cultural Authority: Essays in Victorian Intellectual Life*. Cambridge: Cambridge University Press, 1993.

Tuttleton, James W. Rev. of *Professing Literature*, by Gerald Graff. *Journal of American History* 74 (1988): 1344.

Tylor, Edward B. *Primitive Culture: Researches into the Development of Mythology, Philosophy, Religion, Language, Art, and Custom*. 2 vols. London: Murray, 1920 [1871].

———. *Researches into the Early History of Mankind and the Development of Civilization*. Ed. and abridged by Paul Bohannan. Chicago: University of Chicago Press, 1964 [1865].

Tyndall, John. *Fragments of Science*. 6th ed. New York: Appleton, 1891 [1871].

Veblen, Thorstein. "The Evolution of the Scientific Point of View." In *The Place of Science in Modern Civilisation and Other Essays*, 32–55. New York: Russell and Russell, 1961 [1908].

———. "The Place of Science in Modern Civilisation." In *The Place of Science in Modern Civilisation and Other Essays*, 1–31. New York: Russell and Russell, 1961 [1906].

Ward, James. "Psychology." *Encyclopaedia Britannica*, 9th ed., 1886.

Watterson, Bill. "Calvin and Hobbes." *The Daily Northwestern*, 3 January 1995.

Wenzel, Aloys. "Einstein's Theory of Relativity, Viewed from the Standpoint of Critical Realism, and Its Significance for Philosophy," trans. Paul Arthur Schilpp. In *Albert Einstein: Philosopher-Scientist*, ed. Paul Arthur Schilpp, 583–606. Evanston, Ill.: Library of Living Philosophers, 1949.

Westermarck, Edward. *Ethical Relativity*. New York: Harcourt, Brace, 1932.

———. *The Origin and Development of the Moral Ideas*. 2d ed. 2 vols. London: Macmillan, 1926 [1906–8].

Whitehead, Alfred North. *Science and the Modern World*. New York: Macmillan, 1925.

Whittaker, Sir Edmund. *A History of the Theories of Aether and Electricity: The Modern Theories 1900–1926*. London: Nelson, 1953.

Wilde, Oscar. "The Decay of Lying." In *The Works of Oscar Wilde*, 597–620. New York: Black, 1927.

Wolf, Lucien. "Anti-Semitism." *Encyclopaedia Britannica*, 11th ed., 1910–11.

Woolf, Virginia. *Mrs. Dalloway*. New York: Harcourt, Brace, n.d. [1925].

Wordsworth, William. *The Prelude, or Growth of a Poet's Mind*. In *The Poetical Works of Wordsworth*, ed. Thomas Hutchinson, rev. Ernest de Selincourt, 494–588. London: Oxford University Press, 1961 [1850].

Young, Robert M. *Darwin's Metaphor: Nature's Place in Victorian Culture*. Cambridge: Cambridge University Press, 1985.

Zahar, Elie. *Einstein's Revolution: A Study in Heuristic*. La Salle, Ill.: Open Court, 1989.

Index